A WALK THROUGH THE YEAR

Books by Edwin Way Teale

GRASSROOT JUNGLES

THE GOLDEN THRONG

NEAR HORIZONS

DUNE BOY

THE LOST WOODS

DAYS WITHOUT TIME

NORTH WITH THE SPRING

JOURNEY INTO SUMMER

AUTUMN ACROSS AMERICA

WANDERING THROUGH WINTER

CIRCLE OF THE SEASONS

SPRINGTIME IN BRITAIN

ADVENTURES IN NATURE

THE JUNIOR BOOK OF INSECTS

THE STRANGE LIVES OF FAMILIAR INSECTS

PHOTOGRAPHS OF AMERICAN NATURE

THE AMERICAN SEASONS (one-volume)

Books Edited by Edwin Way Teale

WALDEN

GREEN TREASURY

AUDUBON'S WILDLIFE

THE THOUGHTS OF THOREAU

THE INSECT WORLD OF J. HENRI FABRE

THE WILDERNESS WORLD OF JOHN MUIR

A WALK THROUGH THE YEAR

by Edwin Way Teale

WITH PHOTOGRAPHS BY THE AUTHOR

We come and go but the land is always here and
the people who love and understand it are the
people to whom it belongs for a little while.

WILLA CATHER

DODD, MEAD & COMPANY NEW YORK

Designed by Sidney Feinberg

1 2 3 4 5 6 7 8 9 10

Library of Congress Cataloging in Publication Data

Teale, Edwin Way, date
    A walk through the year.

    Includes index.
    1. Natural history—New England. 2. Seasons.
I. Title.
QH104.5.N4T43   500.9′74   78-9786
ISBN 0-396-07621-1

Dedicated

To the sun and the moon
and Nellie;

To the pasture rose and the
bluebird and Nellie;

To the starlight and the rainbow
and Nellie;

To all that means the most to
me at Trail Wood—
especially
Nellie.

Contents

IN THE BEGINNING   1

THE WALKS OF SPRING   9

THE WALKS OF SUMMER   109

THE WALKS OF AUTUMN   213

THE WALKS OF WINTER   311

AFTER THE END   399

INDEX   403

Photographs following pages 24, 120, 248, 344

IN
THE
BEGINNING

SEVEN paths lead away from the white cottage with black shutters under the hickory trees. With its pegged beams and great blocks of foundation stone, it stands on its knoll above the flow of Hampton Brook solid and unchanged by blizzard and hurricane. A hundred and seventy years have passed since the first fire was kindled in the great fireplace in the living room. How many drifting summer clouds have passed above it; how many winter winds have pounded it; how many meadow flowers have bloomed and faded around it; how many times has it seen the year of the seasons complete its circle and begin again!

Seven paths winding away, radiating toward different points of the compass. They cross the fields that surround the house. They become trails and thread their way through woods and into swamps, along brooks, across boulder fields, over ridges and into ravines, among ferns and wild flowers, beside a pond and a waterfall, up juniper-clad hillsides, by the stagnant water impounded by a beaver dam, and past my log writing cabin nestled among the aspens. They carry the saunterer to Lost Spring, where deer come to drink, and to a cascade foaming over rocks, where Hampton Brook crosses Old Woods Road. They extend to all the corners of the 130 acres that make up this old New England farm. Winding, branching, and crisscrossing, these trails of Trail Wood run for a total length of more than three miles.

Seven paths that lead us—my wife, Nellie, and me—through the daily changes of the seasons. On them we walk through the year. This is the record of the sequence of those out-of-door days—a book of the open, written largely in the open, set down mainly in the field, the original journal entries recorded on the spot—sometimes sitting on the grass of a hillside, sometimes leaning on an old stone wall or with my back against the bark of a tree, sometimes resting on a mossy log beside a stream deep in the woods, sometimes with cold fingers writing in the lee of a rock or tree trunk.

Henry Thoreau thought "it would be pleasant to write the history of one hillside for one year." And pleasant it has been to record this account of our daily walks, reliving the events of the shifting seasons on this uneven land set amid these tranquil hills. Four years have passed since I recorded our early experiences at Trail Wood in *A Naturalist Buys an Old Farm.* From our house we still see no other house in summer. Our boundary is the skyline fringed with trees.

John Keats, while vacationing in the Surrey countryside, wrote to a friend: "I like this place very much. There is Hill and Dale and a little River." Keats might have been describing the surroundings in which we have settled. Here we, too, have hill and dale and a Little River. Henry Thoreau, in an entry in his journal for July 6, 1852, almost exactly described the lane leading to Trail Wood. "In selecting a site in the country," he wrote, "let a lane near your house . . . cross a sizeable brook." So our lane crosses Hampton Brook before it ascends the slope to the house. Both Keats and Thoreau, apparently, would have approved of our choice. So, no doubt, would have Ralph Waldo Emerson, who expressed his viewpoint in the words: "Thank God I live in the country!"

The pages of this book record that part of our lives when we were free to wander as we desired in the out-of-doors. The days are drawn from the voluminous journals I have kept during what is now nearly twenty years at Trail Wood. Each entry is derived from the same time of year but not, in all cases, from the same year. The only liberty that has been taken is that in some instances entries about the same thing at the same place at the same time of year have been combined.

In earlier ages, when the year began with the month of March, the calendar was nearer in accord with nature. In March winter is coming to an end. In March spring draws nearer and arrives. And it is the coming of spring, the time of nature's beginning again, that in each revolution around the sun forms the natural starting point of the natural year.

On this beginning day of spring, this twenty-first of March, the slant of the land down to the pond below the house is bare, winter worn, a field of yellow and russet. Glimpsed under the arch of a dying apple tree that may well be eight decades old, the pond lies outspread, still largely covered with a lid of ice.

So on this sunny day in March we begin our wanderings through the unfolding season of spring. This, its initial day, forms our point of departure in walking through the year.

Nighthawk Hill · 26
North Boulderfield · 42
Old Cabin Hill · 38
Pussy Willow Corner · 21
Seven Springs · 1
Seven Springs Swamp · 2
Shagbark Hickory Trail · 5
South Boulderfield · 4
Starfield · 25
Stepping Stone Brook · 8
Stone Bridge · 19
Summerhouse · 13
Summerhouse Rock · 12
Twig Hill · 29
Veery Lane · 18
Waterfall · 27
Wet Weather Brook · 20
Whippoorwill Spring · 17
Whippoorwill Cove · 11
Wild Apple Glade · 34
Wild Plum Tangle · 24
Witch Hazel Hill · 41
Writing Cabin · 7

KEY

Azalea Shore · 10
Beaver Rock · 14
Big Grapevine Trail · 3
Broad Beech Crossing · 33
Brook Crossing · 44
Cattail Corner · 16
Chestnut Stub · 37
Deerfield · 45
Driftwood Cove · 9
Fern Brook Trail · 28
Fern Brook Swamp · 30
Glacier Rock · 15
Ground Pine
    Crossing Trail · 31
Ground Pine Crossing · 32
Hellebore Crossover · 43
Hired Man's Monument · 22
Hyla Pond · 39
Hyla Rill · 40
Insect Garden · 23
Lichen Ridge · 6
Lost Spring · 36
Lost Spring Swamp · 35

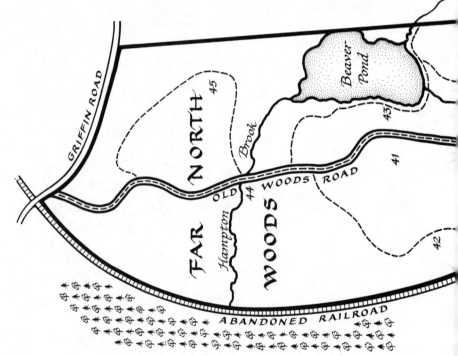

NATCHAUG STATE FOREST

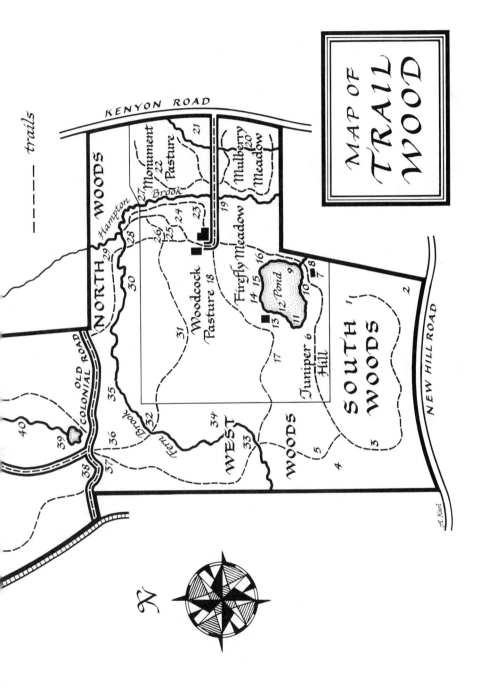

MAP OF TRAIL WOOD

KENYON ROAD

— — — trails

NORTH WOODS

Monument Pasture 22

Mulberry Meadow 20

21

27

Brook

Hampton

23

28

29

26

25

24

30

Firefly Meadow 19

Woodcock Pasture 18

14 15 16

9

8

7

Pond 12

10

11

13

31

17

Juniper Hill 6

SOUTH WOODS

1

2

NEW HILL ROAD

OLD COLONIAL ROAD

35

Brook

Fern

32

34

WEST

33

36

37

40

39

38

WOODS

5

3

4

N

A. Kari

THE
WALKS
OF
SPRING

MARCH 21. In our latitude the first day of spring is only occasionally a spring day. The calendar spring, the astronomical spring of the vernal equinox, may arrive under sullen skies, with cold rain or with a blizzard hurling its snow from the north. But on this first day of the season, as we walk down the slope of Firefly Meadow to the pond, the sun shines, bluebirds call, the fields are bare of drifts, the brooks run high with melted snow. The spring of the calendar and the spring of our woods and fields begin together.

The weather may alter in the night. Snow may come again. But today the sun's rays pour down from an unclouded sky. Warmth is winning over cold. The optimism of spring is in the air.

MARCH 22. We shut the kitchen door and walk past the lilac bushes and along the line of sheds to the barway by the hemlock tree. The path beyond leads across the Starfield, rising to the elevation we call Nighthawk Hill and then descending to the edge of the woods and the beginning of the Fern Brook Trail.

Only a few steps along this trail, before stepping stones carry us across the brook, we stop for a thousandth time to gaze at a double tree rising on our right. Two red maples lift side by side. The diameter of the trunk of one is twice that of the other. Fifteen or

11

twenty feet above its roots, the smaller tree makes a bend like a stovepipe entering a chimney and disappears into the larger tree. What happened to produce so abnormal a circumstance? Nellie and I puzzled over that question many times before a forester friend of ours, Edson Stocking, gave us the explanation. Among quick-growing soft maples, he said, a smaller tree sometimes tilts or is blown over so its top lodges in the crotch of an older maple. The latter grows around it. Then the top dies and falls away. Only the lower part of the trunk is left with its upper end disappearing into the wood of another tree.

Chickadees call around us as we begin our circuit of the Fern Brook Trail, then continue on the traces of the long-abandoned stagecoach route, the Old Colonial Road, that cuts east and west through the North Woods, then up a steep S-shaped curve past a landmark of the woods, the great silvered stub of a chestnut that died decades before we came to Trail Wood, and on to the top of Old Cabin Hill, and so, finally, home by the lowland trail of Ground Pine Crossing. The little birds convoy us as far as the Old Colonial Road—a small troupe of lively blackcaps, darting from tree to tree, stopping to investigate the bark under some limb, calling their names with the bright good cheer we remember from the most blustery days of winter. They keep pace with us. We are old friends of the bitter days, the days when they depended largely on us for food. They recognize us. Or, to be more exact, they recognize our forms and our clothing. For, like most birds, for them it is the same coat or hat that becomes a badge of recognition.

A singular demonstration of this fact occurred one winter at the home of friends, on Bedlam Road, some miles to the west. Periodically, a fish dealer stopped with his refrigerated truck in their yard. As soon as he climbed out and opened the back and got out his scales, he was besieged by chickadees. They fluttered around him. They alighted on his head and shoulders. They appeared to welcome him to the farmyard.

For a while this odd behavior remained a mystery. Then the simple explanation was perceived. His hat and coat and the hat and coat worn by our friend when he put out sunflower seeds on winter mornings were of the same design and color—bright red. In the minds of the little birds, these were the "field marks" of their benefactor.

MARCH 23. The weather has turned cold again. After an evening before the fireplace, I walk down the lane to the bridge in the light of the waning moon. Standing beside one of the low walls, looking down, I am caught in the same net of crisscrossing shadows that has ensnared the bridge, a net cast by the moon shining through the bare branches of the trees.

Behind me, below me, beyond me, the dark water of the brook swirls and slides and tumbles along its rocky bed. Always descending, always renewing itself, always greeting and leaving the beholder behind, the water journeys on. I see flashes of reflected moonlight glinting from its swirls, streaks, and sudden patches of foam whirled away downstream from projecting rocks. The murmuring, the splash and gurgle—all the liquid sounds of the stream rise around me, pushing back the stillness of the silent night.

I wonder about all the other people in other parts of the world who may be, at this same moment, looking down into other streams. The rocks, the swirls, the foam, the sounds are characteristic of watercourses around the world. Gazing down on this small New England brook, flowing through the night, I might be gazing into a stream in Czechoslovakia or Argentina. As little Hampton Brook is so much a part of our lives at Trail Wood, other small streams in other parts of the world are part of the lives of others. Although we will never meet, never know of the existence of these others far away, two things—a flowing stream and the serene light of the same moon—we will share together.

Like the wind, a brook exists only through motion. Down the narrow groove it has worn in the earth, hurrying toward the greater valleys of the rivers that will carry it to the sea, all the dark water foaming and gurgling below me rushes away into the night. The stream flows on and on. So the long life of the ever-renewing brook extends through the years. But it continues without awareness, without sensation, without emotion. Its existence is one of action, of music, of beauty; but it is life without life. The great gift of *our* lives is the gift of awareness.

MARCH 24. The phrase that runs like a refrain through the thirteenth-century French tale of *Aucassin and Nicolette* is "Here I say and tell and relate." So here I say and tell and relate what I have just seen befall an unwary squirrel on this fourth day of spring.

My eye is first caught by its little leaps this way and that as it forages over the ground for food. I watch it for a moment or two before I become aware of a large white cat that has wandered from a neighboring farm. It is crouching low in the grass, its eyes following every movement of the squirrel. For some reason, that usually alert animal seems unaware of its presence. It even turns its back. That is the signal. The cat streaks toward it and makes its leap. With the squirrel in its mouth, it begins to run, making surprising speed with so heavy a burden. Shaking off my surprise, I set out in loud pursuit. The white cat drops the squirrel and scuttles into the protecting maze of the Wild Plum Tangle.

When I come up to it, I find the gray squirrel lying on its side in the grass. Its eyes are wide open. But it is rigid and unmoving. It gives every appearance of being dead. Minute follows minute without any sign of life. I can see no evidence of mortal injury. But ten minutes, fifteen minutes, twenty minutes, half an hour pass while it remains inert in the sunshine.

At last, from where I stand a little distance off, I see a twitch of one hind leg. Then it lifts its head and lets it fall back again. Slowly it revives from its profound state of shock. It struggles to its feet. It flips its banner tail. It makes two or three short, tentative jumps, then flattens out in the grass again. For several minutes it rests. Then another few jumps and a rest. And so, by degrees, it attains a tangle of bushes, then a tree. And so it disappears. It had not sustained any important physical injury. But for half an hour after its narrow escape, the high-strung animal had been in a state bordering on catalepsy.

This is the story of the hunting cat and the lucky squirrel that I say and tell and relate on this twenty-fourth day of the month of March.

MARCH 25. The ice goes out of the pond today. All morning we see it melting, shrinking in the sun. Slowly it retreats to the shaded edge of Azalea Shore. There we watch it growing thin and dark and filled with holes, becoming spongy and turning to slush. By noon only a thin shelving of solid ice remains along the shore. Now, standing on the pond-edge path, in the mid-afternoon, we watch the last remnant of thin ice soften into floating fragments and the fragments melt away, leaving only a scattering of bubbles.

It is three fifteen when the water of the pond is open from end to end. A breeze ruffles its surface—something we have not seen since fall. Looking up, our bass and bluegills now see the sky instead of their winter-long roof of ice and snow. So one of the great turning moments of the Trail Wood year arrives. Leaf-fall in autumn ushers in the months of cold; the going out of the ice in the pond in spring ushers in the months of warmth.

As though waiting in the wings, a kingfisher sweeps in a rattling circuit of the pond, and even before the ice is completely out, a pair of mallards comes quacking in to plow through the cold water in a splashing descent. They are *our* mallards, the pair that stayed with us last year, returned safely from all the adventures of migration. For no sooner have they alighted than they swim to the shore by Summerhouse Rock, where they found cracked corn scattered at the water's edge last summer. Toward sunset, a phoebe calls from the lilac bush. For the second consecutive year, a phoebe has arrived on the exact date when the ice disappeared from the pond. To let it come and go, I open the doors of the center shed, where phoebes nest each spring, doors that will remain open now throughout the season and on into the summer days.

MARCH 26. As though stalking deer, walking in single file, watching where we place our feet, stopping often to listen, Nellie and I cautiously round a curve in the Old Colonial Road. Ahead, nestled below Old Cabin Hill, Hyla Pond, a water-filled depression among the trees twenty feet across and dark with decaying leaves, has shed its winter covering of ice.

Each year, no sooner has the ice melted from this swamp water than the wood frogs, those earliest of our batrachians to mate in spring, congregate here. Warmed by fires within, they sport in the frigid water, filling the air with the low din of their grating, clacking calls. At times the mingled voices resemble the gobbling of turkeys; more often the excited quacking of domestic ducks. However, unlike the clear little voices of the hylas, the raucous chorus of the wood frogs does not carry far. It is produced, like the call of the spring peeper, with the mouth kept shut. So vocal now, the creatures will remain almost entirely silent during the rest of the year. Coming just after their release from winter hibernation, this is their one great gala time, compressed into a few days amid the chill waters of this swampy pool. Then these wood frogs will disperse. They will scatter out through the woods, leaving behind thousands of eggs floating just below the surface of the water.

As it happens, we come upwind, again like the stalker of deer. Our scent means nothing to the calling frogs, but any sound or sight of our approach alarms them. There are times when the breaking of a twig or the scraping of a shoe is enough to silence their chorus. Now, with the wind carrying any small noises we may make to the rear, we are able to work, little by little, nearer the pond. We see the water swirling with constant movement. We hear the confused, harsh commingling of the voices. Through our glasses, we see the frogs floating quietly, darting ahead in sudden rushes, sending rings of ripples spreading outward when they call. We note the brownish-black mask running back along the side of the head above the whitish jaw stripe. Then I, the impatient one, move a little closer for a better view, and the curtain drops on the show before us. Sound and action cease.

But in the sudden silence we hear, off to the north, a larger chorus. The ice is retreating from the edges of the pond where the beavers live, and wood frogs have gathered in the shallows. Carefully we descend the slope through the woods to the eastern side of the pond near the dam. All the frogs seem to be calling on the western side. In a wide detour we circle around and make a stealthy approach from the west. Now all the frogs seem calling from the eastern side.

Where a swamp maple has been uprooted in some windstorm of the past, Nellie and I find a seat on the prostrate trunk. We wait without moving. Five minutes, ten minutes go by. Little ripples on the surface of the shallows spread and overlap. At the center of each ring, like a bull's-eye, is the head of a silent wood frog. Without a movement, without a sound, it keeps us under surveillance. One frog breaks the surface and begins its vigil only a dozen feet away.

We hear the grating of the batrachian voices on the other side of the pond, but here there are only silence and the watchful frogs. For more than twenty minutes we remain seated. But, in the end, the little frogs in the icy water outlast us. We stand up and stretch ourselves, stiff and chilled. All the heads disappear and only overlapping ripples are left. We are halfway over the hill, coming home, when we stop and listen. Behind us the shy, secretive wood frogs that we have watched in silence have begun their calling again.

M A R C H 2 7 . Glowing light is all around me as I walk out into the misty dawn. Minute by minute the silvery sheen grows brighter as the sun climbs among the treetops. Trees and bushes beside the brook are shrouded and indistinct. From somewhere among them, unseen, a small bird gives voice to a strain pure, clear, pensive, touched with sadness, as though with overtones of some age-old lament. It is the song that has been put into the words: "Oh, Canada, Canada, Canada"—the song of the white-throated sparrow, the song that, among all the voices of the birds, affects me most deeply. Such dawns as this, perhaps reminding the little singer of the misty forests of its northern home, always seem to stimulate the singing of the whitethroat.

A friend of mine, George Peters, a climber of mountains, years ago as a young man worked one summer in a lumber camp in the Adirondacks. All through the woods from morning until night, beginning in the light of the earliest dawn and continuing into the crepuscular light of evening, the song of some small bird he never identified and never saw clearly went on and on. He used to lie in his bunk when he awoke in the gray light of dawn and listen to the moving strains that wove themselves into all his memories of that time. The singer seemed to represent the voice of the solitary wilderness. He called the unknown vocalist then and always remembered it afterward as "The Loneliness Bird." It was not until years later that he discovered that the disembodied voice that had filled those days belonged to the little white-throated sparrow. How fine it is, in dawns like this, to begin our Trail Wood day with this pure, simple, deeply moving song of The Loneliness Bird.

M A R C H 2 8 . The newspaper I carry back from Kenyon Road today tells of wars in far-off parts of the world, of fiscal troubles in big cities, of the ups and downs of the stock market. But on this day in March, it is none of these news items that takes first place in the interest of our village. Other news—not published with ink on paper but instead written on the sky with sound and motion—is the main topic of conversation wherever neighbors meet. Telephones ring and the intelligence is passed from house to house.

Shortly before nine o'clock this morning, 200 Canada geese, flying north, pass over the village two miles to our south. In one great flock, strung out in a long V with unequal legs, they cleave across the sky. About 300 feet above the ground, their rising and falling wings catching the sun, the far-carrying clamor of their voices a sound of untamed wildness in the air, they drive steadily into the north. Over the white houses, over the sugar maples along the main street, over the high, shining spire of the white church, then out over the open country beyond, above side roads, over Hampton Brook, above our Trail Wood fields and woods, the long wedge of the flying birds pushes on. The story their passing tells is of melting ice on northern lakes, of the sure advance of spring. As one of our neighbors said in the village yesterday, when a smaller flock went through: "Other signs of spring may fail but you can't fool a goose."

After they disappear over the treetop fringe of our North Woods, for how many hours will these migrants forge ahead? For how many miles will they fly before they stop to feed and rest? That depends, to a great extent, on how far the thirty-five-degree isotherm has extended up the map. For it has long been observed that in their spring migration these waterfowl tend to move north in the wake of the advance of this above-freezing temperature. Passing from south to north up the length of Hampton, these homeward-bound birds have presented a kind of visualization of the northward sweep of the invisible isotherm.

They also represent something nearer to the heart. Mile after mile along their route, people lift their eyes to the clamoring birds. And their spirits rise. In them they sense a promise in the sky, they find an assurance that spring—the real spring, the spring of the higher sun, of the first of the wild flowers, of the return of the songbirds—is close at hand, is almost here.

MARCH 29. Here where on so many summer days I have leaned against the rail of this rustic bridge spanning the runoff stream below our pond, Stepping Stone Brook, looking down on activity in the water below, I stand on this March morning. Winter is over. Ice is melted. After the long stillness beneath its solid lid, the water has a crystalline transparency, completely free from sediment. Shoals of black, formed of decaying leaves intermingled with twigs, mark the hollows of last year's sunfish nests. Little puffs and clouds of vivid green algae already ride in the easy flow of the runoff water.

Just as I start on again, I am halted by a movement almost directly below. A small painted turtle, about the size of my palm, comes swimming slowly in and out among the algae clouds. Its body is clean, its colors brilliant. Every part of its shell seems burnished. Even its claws shine. I can rarely recall seeing so beautiful a painted turtle. Not long before it had pushed its way out of its winter hibernaculum. It is the only sign of animal life I see. I am observing its first swimming advance after so many months of torpid inactivity.

While I lean on the rail, the turtle tilts downward toward one of the debris-filled hollows. It pushes its head under the decaying leaves, then claws its way deeper and deeper into the mass of rotting vegetation. I see its hind feet disappear. The sediment it has stirred up drifts away. The painted turtle has returned to a last segment of its winter sleep. By chance I have been standing here on the bridge at the precise time of this short break in its inactivity.

The month of March, so temperamental in its changes from sunshine to storm, has veered to a milder day. By mid-afternoon the mercury reaches fifty degrees Fahrenheit. I see a *Polistes* wasp airborne near the lilac bushes. During these latter days of the month, the hibernating fertilized queens become restless, stirred to activity by any sudden rise in temperature.

A riddle represented by a lonely voice in the sky brings this day to a close. After an evening reading by the fireplace, Nellie and I walk down the lane to the brook and back. There is almost no wind and, except for the murmuring of the stream, the night is very still. We are standing near the break in the wall opening into Firefly Meadow when our ears catch, coming down from the star-filled heavens, a sweet, plaintive, ploverlike call. It is repeated half a dozen times, growing fainter and fading away. What unknown bird of passage are we hearing? What migrant would be in the sky so soon? To us it sounds like the voice of an early black-bellied plover moving north over a starlit land.

Thus we come to the end of a day of three memories—memories of a painted turtle, a wasp beside a lilac bush, an unknown bird winging through the night.

M A R C H   3 0 . Black and white and tawny red swirl and streak above the carriage stone across the lane. Two chipmunks with territories that overlap are locked in a struggle for possession. They leap. They tumble. They roll. They spin in violent whirls of attack and counterattack. Their striped bodies clash and merge and break apart. From one side of the arena to the other the struggle follows its zigzag course. With only the briefest pauses for breath, the little lightweights battle on. They reach the edge of the stone and the violence of their conflict sends them tumbling off together into the weeds below.

A few feet away, under the protection of a highbush cranberry, a cottontail rabbit watches the strife with mildly interested eyes. Rabbits keep out of foreign wars and private quarrels. As the linked bodies of the two fighting chipmunks come crashing down among the dry weed stems, the cottontail makes two leisurely hops and takes up a new position farther from the battle. But the fight is almost over. While the rabbit and I watch as spectators, the chipmunks break apart. One turns tail and streaks away with the chittering victor in close pursuit.

M A R C H   3 1 . March goes out like a lamb—a wet lamb. Heavy rain in the night and mist in the morning and a breeze that veers to the south before noon. The ground is soggy but the air is mild when we set out toward the end of this final day of this transitional month of March.

From the edge of the pond, as the dusk begins to deepen, rises the round, clear, musical call of the tiny frog, the spring peeper, *Hyla crucifer*. First one, then another, then another. To me the spring peepers always sound like creatures of a dawn world, inhabitants of the earth in the first days of creation—so innocent, so round-eyed, so born-yesterday. They give the impression of something tentative, frail, and vulnerable. Theirs seems the voice of innocence. Not born yesterday—born today. Their bright calling—a kind of "Spring! Spring!"—will chant through the nights of the weeks ahead.

It is always a source of wonder that so tiny a creature as this small frog with the markings of a cross on its back, a creature that appears no larger than my thumbnail, the smallest frog we have, can produce so far-carrying a sound. Other peepers join in. At first they all seem to be giving the same, identical call. But as we listen intently we notice little variations. One adds a sort of trill at the end. And while most of the hylas give the single ringing call, the pitch may vary with individuals.

As we turn up Hampton Brook and follow it to the waterfall, we leave the voices of this little band of calling hylas behind. But as we draw near the waterfall, we hear, louder and louder, the great chorus of another pond on a neighbor's land, beyond the Old Colonial Road. The mingled voices of hundreds of peepers rise in a clamor so great we hear it even above the sound of the falling water when we stand within reach of the cataract. And when we come to the edge of the larger pond, the chant becomes nearly deafening. Individually, the calling of the peeper is a frail, sweet, lonely sound. But this vast intermingling of a multitude of little voices rises in swelling waves, a shrill, commingled din that beats against our eardrums.

It follows us far into the darkened woods as we come home. With it this clamor of the mating hylas brings a sudden awareness of all that overpowering rush of fertility, that renewal of life in infinite variety, that is the gift of every spring.

A P R I L 1 .   Today we are off for the North Woods to search for the blue butterfly. Each year as March draws to a close and the earliest days of April arrive, Nellie and I wander along mossy trails, in open glades, down the Old Woods Road, our eyes roving ahead and beside our paths, alert for a glimpse of the small gay insect that for us symbolizes the return of spring. An elfin creature, it flies on diminutive wings of blue tinged with violet. Carl Linnaeus bestowed upon it the scientific name it still bears: *Lycaenopsis argiolus*. Commonly this early butterfly is known as the spring azure.

The dusky-hued mourning cloak butterfly, hibernating as an adult, appears even earlier in the year. We sometimes see it abroad during thaws late in winter. But when we encounter the little azure, the first of our native butterflies to emerge from an overwintering pupa, it seems leading the parade of spring's colorful returning life.

Three times today we are rewarded in our search. Three times we see the butterfly of spring that we are seeking. Once we are not far from Hyla Pond. The second time we are just beyond Witch Hazel Hill. The third we are on the long slope that descends to the ford where Hampton Brook crosses the Old Woods Road. Each time we stand in some sunny glade, watching the small, beautiful creature, immaculate in its newly spread wings, whirling over the path or alighting on the fallen leaves.

All along our way we encounter other humble signs of the season: earthworm castings, tiny dark spiders running over pale mats of last year's grass, the new tunnels of moles. On this day we are in the Time of Little Brooks. Water from the saturated ground is flowing down small temporary streams. They wind among mossy rocks and pause in diminutive pools behind dams of twigs and dead leaves. Some of these streams in miniature have a life of only a few days, others continue for a fortnight or more.

As we are coming home again, we encounter a grouse, a cottontail rabbit, and a woodchuck. They—all three—have outlasted the winter. Spring is here for them, for us. We leave the woods behind and emerge into the open meadow of the Starfield. A quiet elation fills our minds. Spring, spring, spring is here! We have seen the little blue butterflies for another year.

A P R I L 2 .   At the end of my walk today, I lean on the top pole of the meadow barway while the sunset fades. A tree sparrow alights nearby and lifts its voice in its sweet and simple song. Two wild ducks curve over the pasture and over the woods, speeding in black silhouette against the glow of the west. I listen to the faraway barking of a dog. Somewhere off toward Old Cabin Hill a woodpecker on a dead limb sets up a tattoo, hammering with its bill. The shadows lengthen. In its leisurely drift, the calm air of evening brings an April sign of the seasons—the faint, rather pleasant smell of a distant skunk. Then, as the dusk deepens over the lowland meadow to the west, there is wildness in the sky. A woodcock circles up and up, then tumbles down in a falling-leaf descent uttering the twittering song of its courtship flight. Such are the simple country events that bring a quiet pleasure in this hour when our part of the globe is rolling from daylight into dark.

Each of these things occurring around me represents a single frame of an endless moving picture. On an initial encounter, we always have a tendency to imagine that our new surroundings and the events of our new surroundings are permanent, that what we are seeing is established and unchanging. But a truer perspective comes with the passing of time. Nothing—not mountains nor sea nor shore nor rocks—is *exactly* the same on two successive days. The changes may be imperceptible to us but they are there. Lichens are working on granite. Death is replacing life. Leaves are expanding and leaves are wilting. The world is never finished. Everything is going up or going down around us. The things we see most often, the species that are endlessly repeated in individuals, are the successes, the end product of unceasing evolution.

And what has this process we call evolution in store for us? Ideas on that subject have their own evolution. I remember once coming across a book in the New York Public Library entitled *Our Planet, Its Past and Future*. The author, William Denton, in his day was a popular lecturer on geology. His book appeared in 1870. All I remember of it now is its final chapter: "What Will Be the Future of the Globe?" It provided an illuminating glimpse into the outlook of a former day, a time sorely in need of Doubting Thomases, a time of all-too-easy acquiescence with the dictum: "Say it three times and it's true."

Evolution, according to this author, will bring about the disappearance of everything harmful to man. Poisonous plants will disappear and wonderful fruits will take their places. Noxious insects will be evolved out of existence. Evidence: Dragonflies were immense in Carboniferous times and now have become smaller, indicating they are on their way out. Poisonous snakes, tigers, and animals injurious to man will go. Evidence: The Bible says: "There shall be nothing to hurt and nothing to destroy." Land surfaces will increase for man's benefit. Winter will grow less severe and the climate will be modified to the needs of mankind. Volcanoes will die out and earthquake power will be harnessed so quakes will be destructive only in wild areas remote from man. All this will happen, says the author, "when the heirs of the world shall obtain their inheritance. For the world is for man, and whatever wars with his highest well-being must disappear."

"If," he adds significantly, "these various improvements should not take place by the operation of natural forces, they may be effected by the instrumentality of man." There is the rub. For in the years since those words were set down, we have watched "the instrumentality of man" at work and have seen the disastrous consequences encircle the globe. Today we approach the old question "What has evolution in store for us?" with a new complication. Now the imponderable element in our calculations is man himself.

APRIL 3. Fine snow, hardly more substantial than frost, dusts the fields this morning. But almost as soon as the sun rises it disappears. With my field glasses slung around my neck, I head out after breakfast to see what I can see.

The first thing I see is a hunting fox, a red fox mousing among the grass clumps of Firefly Meadow. More than the gray, the red fox is abroad in daylight. For nearly half an hour, partially screened by an old stone wall, I follow its fortunes through my binoculars. It pursues a wandering course down the meadow, beside the pond, up into Woodcock Pasture, across the Starfield, and back to plunge into the deep shade of Veery Lane.

Frequently its color almost matches the tawny hue of clumps of winter-weathered grass around it. Over and over I see it pounce with the agility of a cat, forepaws held together—sometimes apparently with success. Its pointed ears, white in front, coal black behind, are constantly in motion, pricking forward, twitching backward, the black appearing and disappearing. Once in the Starfield it leaps in one direction, then in another in swift succession. Each time it jumps, its bushy tail with its white tip flips upward. For a minute or two at a time, it stands frozen, ears forward, head held high, peering intently down into the grass.

When it comes to the apple tree that leans over the path close to the entrance of Veery Lane, I watch it give a great leap up the slanting trunk trying to snap up a gray squirrel that shoots into the upper branches. The jump is a kind of graceful wingover, a catlike upward bound—contact with the trunk, a quick turn, and the leap down again all in a sudden, graceful flow of movement.

As it wanders on down the path, it starts and stops, turns aside, moves on again, follows a course as erratic as that of a foraging skunk. I see the white tip of its tail wink out of sight as it rounds a

The trail that runs beside Hampton Brook winds among high
clumps of cinnamon ferns as it draws near the waterfall.

(*Above*) False or white hellebore, swiftly growing, swiftly dying, unfolds its leaves in the sunshine of a spring morning.

(*Below*) Out of its winter hibernation, a pickerel frog rests in shallow water as it basks in the warmth of the spring sun.

Bloodroot in a wild garden of spring, above. Below, high water tumbles in a cascade over rocks in Hampton Brook.

A small skipper butterfly clings to a day lily bud. Its ⸍ habit of darting from place to place gives it its name.

Nestlings in a heat wave. Young phoebes, with gaping bills and drooping heads, await return of the parents with food.

A botanical work of art takes form as a bud on a shagbark hickory tree unfolds in the increasing warmth of spring.

Trail Wood, with its house, pond, fields, and woods seen from a
thousand feet in the air (*Aerial photograph by Virginia Welch.*)

turn. The little adventure is over. It leaves me remembering this particular April third as The Day of the Hunting Fox.

APRIL 4. I come to the woods in many moods. The intensity of our awareness differs from hour to hour and day to day. Sometimes our minds are flooded with light; sometimes everything appears in dim twilight illumination. It is as though our moods functioned like the diaphragm of a camera, opening and closing, admitting more light one time, less another. There are days when I follow the trails alert to everything around me. There are others—extremely rare—when I find myself preoccupied, walking almost insensible to my surroundings. Again on other days I am neither alert nor insensible. I am not calloused nor blasé nor dull. Without thinking what I am doing, without conscious effort of observation, I drink in the sights and sounds and smells I find around me. It is so that I walk through the woods today.

Two mourning cloak butterflies whirl in a woodland glade, following one another as though riding an aerial merry-go-round. Redwings call incessantly from the swamp below, two white-breasted nuthatches from the hill above. And from somewhere between the two comes a soft, nodding, soporific sound, the early voice of a mourning dove. And accompanying my every step is the smell of spring, the rich woodland scents of loam and leaf in this time of moisture and growing warmth. Such things I encounter along my way. It seems to me, on this day, I absorb them all, without effort, as a plant absorbs nourishment from the soil.

APRIL 5. My walk this morning is among pussy willows and little fairy pools set amid green mosses where the melting snow has left water collected in our lowland woods. High overhead, somewhere beyond the trees, unseen in the sky, the first killdeer of the year repeats its plaintive, lonely shorebird cry. A new voice for a new season. How pleasant it is to set down all these varied signs of spring!

At the head of its small woodland glen, I stop by Whippoorwill Spring. Its flow of clear, cold water tumbles from an opening in a precipitous bank, dropping into a miniature pool, then flowing away down the valley to join the pond at Whippoorwill Cove. For some time, leaning back against a tree where a plank seat is nailed between two trunks, I watch the twinkle and glint of the sun playing over the fall of the water. Then, after cleaning winter debris from the pool—twigs pruned by storms and the rotting leaves of last year's foliage—I start for home.

Coming out into the fields, I catch sight of two odd acquaintances, incongruous companions from a farmhouse down Kenyon Road. A beagle dog and a large yellow cat are trotting side by side. I often see these two animals sitting close together in the dooryard of the farm. Apparently the cat grew up with the beagle. It knew it from the time it was a small kitten. It seems to fancy itself another dog or its companion another cat. I watch the two— the beagle padding stoically along, the cat, with more dainty steps, keeping pace, now beside it, now just behind.

These two curious wayfarers are more than a quarter of a mile from home. The dog appears a bit embarrassed by the presence of the cat. At times it gives the impression of trying to lose its persistent follower. But the cat is not rebuffed. Whenever the beagle pauses, it rubs itself against it. Now it rubs its back under the beagle's chin. The dog turns its head aside and begins plodding on. It is an elderly dog of placid disposition. In the main, it receives the attentions of the cat much as a growing boy receives the attentions of a small sister tagging along in his footsteps. Across the width of Firefly Meadow and out over the Starfield, I follow the advance of this oddly assorted pair. I see them disappear from sight into the edge of the woods where Fern Brook flows.

APRIL 6.   The live-and-let-live nature of the rabbit. Have you ever considered it? For some time I have been thinking about this aspect of its life as I lean against the old stone wall overlooking the singing field of the woodcock. Something I have seen taking place below me has brought it to mind.

Hopping about in search of lush grass or succulent herbaceous plants, a rabbit exhibits none of the weasel's bloodlust, none of the fox's predatory intensity. Its movements are relaxed, its attitude inoffensive. It poses virtually no threat to its fellow creatures.

When I stood in almost this same spot looking down in the gathering dusk on this same Woodcock Pasture a few days ago, I saw a large black and white cat that had come down through the woods from a house on the ridge to the west. It was flattened in the grass, intent on the movements of the woodcock parading back and forth on the ground before taking wing. The bird was wary. It recognized a predator. It was on guard. When it alighted after its next aerial song, it came to ground in another part of the meadow.

But tonight, as the robin ends its sunset song and the twilight deepens, I have seen a different creature in the grass, this time even closer to the parading woodcock. It is a feeding cottontail. It hops about, sits up from time to time, nibbles here, nibbles there. It ignores the bird and the bird ignores it. Each time the woodcock ends its plunging descent from the sky, it alights at the spot where it ascended, hardly more than a dozen feet from the cottontail. Instinctively it recognizes the inoffensiveness of the rabbit.

So this sixth day of April comes to its close. First the sunset voice of the robin. Then the peaceful feeding of the cottontail. Then the skysong of the woodcock. Then night.

A P R I L 7 .   Two hawks, redtails, wheel in a climbing spiral over the western meadow. With only an occasional wing beat, they soar up and up, riding on an invisible elevator of ascending air. Watching their slow spin against the burnished blue of the sky, we see their revolving forms grow steadily smaller, diminish to crow, then to robin size.

Gradually, one above the other, turning continually, the two birds move away toward the south. It is the lower of these two buteos that holds our special attention. Something dark swings below one of its legs. Nellie and I examine it through our field glasses. The trailing object is a leather strap or thong. Apparently, while this hawk was being trained by some amateur falconer it escaped and returned to the wild with its jess, or leather holding strap, still dangling from its leg.

The predators sink to fading specks in the sky. We lower our field glasses. Losing their magnifying effect, our eyes return us once more to the familiar proportions of the world around us.

This abrupt transition revives a memory of the youthful experience of a friend of mine, John K. Terres, who in later years has added valuable books to the literature of ornithology. When he was in the first flush of his enthusiasm for birds, he once told me, he spent a summer working with a surveying crew. During every lull in this employment, he monopolized the transit. Looking through it as through a telescope, he studied the enlarged images of the thrushes and towhees along the fringe of a nearby woods. Another friend, when he was a small boy and unable to afford field glasses, used to climb into treetops each spring and fall. Lying motionless among the upper branches, he familiarized himself with the field marks of the migrating warblers, observing them close at hand.

At the opposite end of the scale from these two memories is a third. It concerns a wealthy physician whose hobby was photographing wild birds. He assembled elaborate equipment and took innumerable pictures. But always they were of perching birds. His interest lay not in the birds themselves—he cared nothing about their songs or habits or life histories—he was interested only in getting as many pictures of perching birds as possible in order to aid a taxidermist friend mount his stuffed specimens in more lifelike poses.

From how many widely varying points of view, on how many planes of approach, does the reaction of human beings to wild birds manifest itself!

APRIL 8. A landmark in our South Woods, when we came to Trail Wood, rose beside a red maple tree—an immense and ancient wild grapevine. Its main stem, a few feet above the ground, was larger than I could reach around with my two hands. I measured it with a steel tape. It showed its circumference was fifteen inches. Ascending the maple, this vine spread in an interweaving network of smaller vines throughout the tops of the surrounding trees. The total length of all the sinuous coils in this canopy must have been at least 1000 feet. This patriarch vine now is gone. But its memory is retained in the Big Grapevine Trail, the name we gave to the path that threads among birches and maples above the slope where seven springs send seven rivulets trickling down the incline to unite in the narrow lowland of Seven Springs Swamp.

Each time I come this way, I notice on a dozen or more of the birch trees the flat-bottomed, hoof-shaped shelves of the fungus *Polyporus betulinus*. They grow on both dead and living trees. Deer eat them when they are young. In early times pioneers gathered them for fuel. When dry, they burn with a white flame and hold fire like punk. The upper surface of each is smooth, tough, corky. The lower surface is riddled with a maze of small holes. Each is the opening of a tube in which spores mature. Once when I broke free one of the shelves and brought it home and placed it on a dark sheet of paper overnight, it imprinted its form in white dust, the dust of the spores shed from the innumerable tubes.

On this day I come to the South Woods carrying a red carpenter's level. Whenever I have seen one of these projecting fungal shelves, its flat bottom has appeared parallel with the ground, its spore-shedding tubes pointing directly downward. When a birch is tilted, the fungus grows at an angle to the trunk, still lined up with the ground as though adjusted with a plumb line. When *betulinus* appears on the prostrate trunks of fallen trees, its shelf projects out parallel to the trunk, its bottom still lined up with the ground below. A botanist once tried an experiment on a fallen birch. He slowly rotated the trunk over a period of time. The result was a spiral of the fungus plates extending up the length of the prostrate tree.

To my eyes, when I walk along the Big Grapevine Trail, the shelves always appear to extend from the trunks of the trees with their lower surfaces parallel to the ground. But are they really parallel? To find the answer to that recurring question, I have brought along the carpenter's level. Among the vertical trees, I begin placing it against the lower side of all the fungi I can reach. I watch the bubble come to rest. Time and again it halts its swing within the centered lines. Almost always the flat spore-shedding bottom is exactly, or almost exactly, level. But, as in most things in nature, there are exceptions. *Polyporus betulinus* is not infallible. Here and there I come upon a shelf tilted up or to one side ever so slightly. But the general rule holds true. Wherever the birches grow throughout the woods, the shelves of most of the fungi are pointing their spore-openings directly toward the ground.

APRIL 9.   Thinking of the birches and their fungi that I examined yesterday along the Big Grapevine Trail of the South Woods, I reflect as I walk along on the difference between the viewpoints of the forester and the naturalist. To the forester, the tree is all-important. To the naturalist, the enemies of the tree— the fungi, the boring beetles, the carpenter ants, the excavating woodpeckers, the *Tremex* larvae—all the living host that surrounds and preys upon the tree—the life of each, all its curious habits and its intuitive wisdom, the amazing delicacy of its senses—these, too, are of equal interest to the naturalist. His great concern is the rounded whole, the life of the tree and the life that surrounds the tree, the place of each in the intermeshing, overlapping, conflicting roles that combine into the vast interplay of aims and efforts that make up what is commonly called the balance of nature.

APRIL 10.   Yesterday the thermometer was up; today it is down. Yesterday the sky was far away, shining, burnished blue; today it sags low, dragging sluggish clouds, heavy and opaque, just above the treetops. Before dawn the April rain begins. Hour after hour it has been soaking the meadows, dimpling the pond, drenching the woods, lifting the level of the brook, widening the waterfall. Now it streams down my raincoat as my rubber-booted feet slosh along hillside paths that have become running brooklets in the downpour. Sunshine yesterday, rain today—so the meadows gain their green. I find myself repeating the words: "Rain, rain, April rain! Rain and green grow the grasses-o!"

Circling the pond on the muddy path at its edge, I see tiny "tackheads"—minnows of the golden shiners—swimming in schools beside Turtle Rock, where turtles sun themselves on August days. Close by I notice submerged curtains of early-forming algae, diaphanous, slowly waving in the water in pulsations, shimmering in a green aurora. Calling back and forth while keeping apart, two white-breasted nuthatches, in a kind of solitary sociability, are "yanking" among the wet trees of Azalea Shore. A little later, as I top the rise and reach the open fields, I see a red-shouldered hawk sail the length of the Starfield and curve away down the edge of the woods. Even on this somber afternoon, even in this chill rain, my brief encounter with nature has lifted my spirits. It seems to me that I need contact with the earth as much as any herb or tree.

As I enter the North Woods, great drops that have collected on the ends of twigs splash down on my dripping hat. Breathing in the moist air, with the murmur and spatter of the rain in the woods all around me, I cross Fern Brook.

I have taken only a few steps beyond when two sounds reach my ears. The first is the soft warbling of bluebirds flying overhead—bluebirds in the rain. The second is a great hullabaloo that breaks out in the meadow I have just quitted. Our neighbor's beagle—this time with a smaller dog as its companion—has caught the scent of a cottontail. I turn back to the pasture edge to watch the chase.

Across the field I glimpse the two dogs rushing through the wet grass, dashing this way and that in a frenzy of excitement. Far ahead of them, lifting in great bounds that give it a clear view of its pursuers, the cottontail is adding distance with every leap. The chase, accompanied by much deep baying and high-pitched yipping, traces a great circle over the meadow. It ends at a large pile of wood heaped near the sheds. Here the cottontail finds sanctuary. Here the trail disappears. But here the clamor continues. I turn back on the woodland path. The sound of the frustrated dogs carries to my ears for a long time as I range on through the soaked and dripping woods.

APRIL 11. Water that forty-eight hours ago drifted as vapor in the sky, that twenty-four hours ago fell on the land as rain, now rushes down the rocky bed of swollen Hampton Brook. Close to where I stand it cascades over rocks that on a summer day last year I piled in a low barrier across the stream. For some time I remain motionless, looking down, almost hypnotized by the slide and tilt of the torn water, the swirl and shift of the glistening lines of foam.

Then I concentrate on one smooth rock. Across its flat surface the tumble of the water for a moment levels out. It ceases its churning, streams in a smooth transparent flow to be shattered again by a boulder just below. This one spot is like a clear window opening into the heart of the cataract. And peering through it, I glimpse a form of animal life triumphant, myriad creatures that

find in the midst of this crash and foaming flow their most congenial home.

All across its surface, the rock appears clothed in patches of
shaggy brownish moss. Several other rocks, vaguely seen in the
smother, support similar close-packed masses. Each is formed of
thousands of minute, waving, cylindrical forms of animal life. All
are anchored to the rock at their lower ends by round sucking
disks fringed with tiny hooks. Behind the head of each is a second
similar disk. In changing its position, the aquatic larva uses these
two disks alternately, looping along underwater like a measuring
worm on a twig. Should it lose its hold and be carried downstream, it still clings to its anchorage by means of a thin silken
thread spun from its salivary glands. In this way it may advance
safely from rock to rock.

Again and again I return to watch these multitudinous larvae
of the *Simuliidae*, the blackflies, through the window of the
clearer water. They are found only in the swifter current, where
the white water of cascades provide them with the extra oxygen
they need. Hour after hour they wave back and forth, sweeping
two fanlike organs through the water, straining out the infinitesimal bits of food, mainly algae and diatoms, on which they feed.

Peering down, I observe them in the time of their food-
getting. What happens after they have come to the end of their
larval life and spin their boot-shaped underwater cocoons, open at
the downstream end, I have never been fortunate enough to observe. Like the larva, the pupa breathes through tracheal gills. At
this stage in the life of the blackfly, the rocks of the rapids are carpeted with hundreds of cocoons. Within each, as the transformation into the adult nears completion, a bubble of air forms. Then
the cocoon splits and the blackfly, riding within the bubble,
shoots upward through the tumbling water, pops to the surface
amid flying spray, and instantaneously lifts into the air on its tiny
wings. Soon afterward, the mated females begin laying their eggs
in vast numbers on rocks and objects washed by the spray of
rushing water. At other times they place them on wet grass or
leaves dipping into the stream.

If this were the end of the story of the activity of the blackfly,
Trail Wood life in late spring and early summer would be hap-
pier. Like the females of the mosquitoes, the females of the black-

flies are drinkers of blood. They swarm around us in the bright sunshine, kept only partially at bay by the insect repellent we rub on our hands and faces. Whenever they breach these defenses we feel nothing until the meal is done. Then follows swelling, a burning, a stinging, an itching that may last for days. Our species is the "Adirondacks blackfly." The discomfort it inflicts on us is also shared by both domestic and wild animals.

APRIL 12. Within sight of the house, a log-cock, a stump-breaker, a cock-of-the-woods—a pileated woodpecker—has been at work. Unknown to us, unseen by us, unheard by us, this largest woodpecker of the north has littered the ground with chips. Eight feet or so above them, at the edge of the North Woods beyond the Starfield, a long gash shines out in the side of a decaying tree. Half a foot deep and three feet long, it reveals where the black and white and red bird has chiseled into the wood in search of hibernating masses of carpenter ants. Most often in winter and early spring such holes are hewn into the trunks of both dead and living trees. I examine the cavity carefully. I run my hand along the ragged sides. Nowhere can I discover any evidence of carpenter ants or their tunnels. With all the ax stroke of its descending bill, with all the chips sent flying, the pileated woodpecker has labored in vain.

Farther in the woods, as I am following the windings of Hampton Brook, a scrabbling sound above my head stops me. I look up and glimpse the whirling forms of two gray squirrels in a wild twisting, leaping chase through the treetops. I follow them with my eyes until the vanquished disappears and the victor returns. "He who fights and runs away," says the old jingle, "lives to fight another day." But I suspect this is the tale told by those who have run away. It is a story the timid never tire of telling. The truth more likely is that he who fights and runs away lives to run away another day.

Now, with my walk through the woods long ended and with darkness settled down, I sit by the fireplace at the end of this April day, rubbing dressing into the leather of my tramping boots. It is a scene from the past, this simple occupation on this night of advancing spring. It occurs to me that it might be taking place on any of the 170 years of nights that have slipped by since this house was built five years after the eighteenth century ended.

APRIL 13. In the night the wind blew strongly from the south. This morning most of the tree sparrows—those companions of ours in the time of the winter drifts—have begun their long flight to the north. Only two or three laggards remain behind for a few days more. When they are gone, the rest of spring and all of summer and part of fall will pass before we hear again the sweet tinkling chorus of their sunset song.

At the top of Firefly Meadow, Nellie and I pause on our way to the pond. We see crows going over the wet fields. We see mallards trailing their V's across the still water. We see the liquid mirror of the pond reflecting the clouds of red that have enveloped the swamp maples with the opening of their flower buds.

As we descend the slope, we hear the wild laughter of a homecoming flicker. We hear the bright little call of a goldfinch overhead. The males of these finches are already in full breeding plumage although a quarter of a year will go by before their mating and nesting time arrives. But the loudest sound we hear comes from one of our smallest birds. It fills all the hours of these days of lengthened light, the ringing "hear-hear-hear" of the tufted titmice.

Everywhere we go on this morning we are surrounded by tiny dark satellites whirling about our heads. The earliest blackflies are on the wing. And where the woods approaches the pond we notice the unfolding of the buds of the spicebush, presaging the filmy mist of bright yellow-green that soon will envelop the understory among the trees as the minute flowers burst open. The coming of the blackflies and the blooming of the spicebush arrive each year together.

APRIL 14. Under my eyes, as I stand here in the woods beside this placid pool, close to the spot where I saw the gray squirrels race through the treetops, I glimpse something new, something rarely seen, something I cannot recall ever having observed before.

On the surface film of the pool eight long-legged water

striders skate about or rest quiescent for minutes at a time. A moment ago there were nine. Just as my eyes focused on this ninth insect, I saw it unfurl little wings and become airborne. No longer did it ride on the invisible surface film. Instead it rode on something far different but just as indiscernible, the invisible air. Where it went I cannot say. In an almost instantaneous action it took wing and darted away. Only by chance had I happened to be looking at it at this precise moment when it abruptly transformed itself into a creature of the air.

Of all the people who watch water striders floating or darting about in sudden rushes as they ride the surface film of ponds and streams each year, I suspect that few are aware that these insects possess wings and are able to fly. It is by air that they travel to new water when their aquatic hunting grounds dry up or become overpopulated. Once I found a water strider inside our house at Trail Wood. It is still a riddle how it got there but obviously it had traveled on its wings.

In this small and momentary adventure beside this quiet woodland stream, in seeing this sudden change from a creature of the water film to a creature of the air, I am at long last experiencing the same luck that Nellie had at the time when our pond was being dug and held only spreading spring-water puddles on the bottom of its bowllike bed. There she too, by chance, had seen a water strider abruptly spread its tiny wings, abandon the surface film, and launch itself into the air.

APRIL 15. Beside stone walls, in weed tangles, along Hampton Brook, in the woods, in wide meadows under the open sky—everywhere Nellie and I turn on this sun-filled mid-April day, we walk amid signs of onrushing spring.

Song sparrows sing from bush tops. Redwings flash the scarlet of their epaulets when they land. Earthworm castings dot the bare open spaces. Mole tunnels meander under the meadow grass. Pussy willows, no longer silver, shine with the gold of ripened pollen. Cowbirds spread their wings and posture in mating performances. Quail call from old stone walls. Fish break the sheen of the pond. Near the waterfall, the early-blooming chickweed unfolds the minute petals of its diminutive white flowers.

And along the warmer southern side of the lowland walls I find the false hellebore spreading wide the ribbed apple-green of its leaves.

On such a day as this you can feel spring, you can see spring, you can smell spring, hear spring, taste spring. You respond with all your senses to the multiform manifestations of change into which the awakening season plunges you.

APRIL 16. In an arrow-straight course, cawing as it flies, a crow passes over me. Other crows are laboring over the pond, over the lane, over the Starfield—all heading in the same direction. Their destination is marked by a raucous uproar, the harsh commingled cawing of many birds. I glimpse a dozen or more black shapes swirling in flight above the treetops to the west. Somewhere near Old Cabin Hill, the crows are mobbing an owl.

Ten miles from here, one recent year, two young men were fishing in the Natchaug River when they heard an even greater tumult in the neighboring woods. Their approach dispersed the clamorous throng of excited crows. The birds had discovered a great horned owl on its nest. This large bird of prey, a creature as fierce as a wildcat, had been pecked to death by the crows. It might have escaped by flight but, rather than desert its two nearly fledged young, it remained, bearing the full assault of the attacking storm of birds. The two young owls, perching on a limb near the nest, were rescued and turned over to a conservation officer to be cared for until they were ready for release. They, too, would have been killed in a few moments except for the timely arrival of the fishermen.

I hurry through the woods toward Old Cabin Hill and the center of the clamor. As I draw nearer I slow to a silent, stealthy approach, hoping to observe the activity of the attackers or perhaps even catch a glimpse of the harassed owl. But other eyesight is keener than mine. Suddenly a change occurs in the cawing. A new note enters in. I have been sighted among the trees—a man, a greater menace than an owl. The din of the birds moves away. It dissipates as it goes, subsiding into a ragged cawing of single crows.

Through a long succession of centuries, this clamor of the crows has been familiar to the eastern forests of the New World.

Heard at daybreak, the harsh, excited cawing of the black birds when they have discovered an owl is a sound as wild as the howling of wolves, a sound as ancient as the days when only the Red Men inhabited these New England hills.

APRIL 17.   Pausing in the sunset halfway across the pasture on my way home from the woods, I listen to one of the first robin songs of the season. It descends from the tipmost branch of the highest hickory beyond the house, from the identical limb, then bare, now stippled with the green of unfolding leaves, where a goshawk from the north once rode in the zero winds of January.

Characteristically, the voices of most thrushes flow and sweep like the music of violins or woodwind instruments. But not that of the robin. Instead it suggests to me the music of a piano. Loud and clear, the notes are pounded out. They come as separate sounds. Nor is there, in the voice of the robin, any of the musing, reflective quality of its relative, the hermit thrush of the dark north woods. Nor is there any of that soaring spiritual intensity inherent in the organ tones of the evening wood thrush.

But how filled with good cheer and well-being is the voice that now carries across these sunset fields. It is a sound of health and energy, of courage and confidence. It is a voice particularly in accord with this season, this time of optimism, of returning life, when all seems well. It is the cheerful song of the spring.

APRIL 18.   Just upstream from the bridge where a single immense slab of rock has supported the traffic of our lane for more than a century and a half, I spend the morning adding stone after stone to the rough dam over which water cascades below a growing pool.

Half a dozen times when I lift some flat stone from the wet ground beside the brook, light flickers in little glints along the moist, gleaming body of a salamander. As each wriggles away, the protective coating of mucous secretion that covers it reflects what it has rarely if ever reflected before—the full glare of the sunshine. Each time I replace the rock carefully, remembering the admonition to amateur naturalists once printed in a conservation leaflet: "If you turn up any stones in your quest, put them back as you found them. To some creatures this is home."

Seemingly out of a far older world, the soft-bodied salamander often is surprisingly long-lived. In captivity it has been known to survive for more than twenty years. In the wild, safe beneath its sky of rock, this primitive creature awaits the coming of darkness to begin its wandering in search of the slugs, snails, worms, and small insects that make up its food.

I continue my leisurely work in the warm April sunshine, my mind busy for a time thinking of all those other forms of life that also find safety under skies of rock—the black field crickets, the teeming colonies of small ants, the varied spiders of the out-of-doors. For many a diminutive and humble creature in pastures, in woods, beside streams, a home protected by a rock means safety. To be a rock lifter in our wanderings is to glimpse hidden life, other little worlds, tiny realms removed from the life we know.

Toward noon, I wander upstream to the waterfall. Unwelcome companions advance with me. More blackflies have emerged from the white-water portions of the rocky brook. Dark and minute, they pass and repass before my eyes. Remembering how Tristram Shandy's father argued that the Almighty had shaped the human nose for the wearing of spectacles, it occurs to me that if blackflies were humans they would believe that I was up on earth just to supply blood for them.

APRIL 19. Once more I am back beside the brook, adding rocks to the dam. I have been working for some time when I notice a gray squirrel. I see it stop its search for food among the fallen leaves. In sheer exuberance and well-being, it leaps, whirls, races—apparently just for fun, just for the joy of activity. It is bright and sleek. It is in the springtime of its life and the springtime of the year. Oh, to feel like a young squirrel on an April day! I stand envying the little animal, envying all the manifold forms of life that with the pinch of winter behind them on this day face an easier time of year with energy superabundant.

I encounter this in another form some time later, when I stand again beside the waterfall, my ears filled with the ever-varying song of the plunging water, my eyes snared in the maze of glittering lights playing over the foaming cataract. I become aware of dark, slender shadows rising and descending in the midst of the water. Moving to the edge of the falls I look directly down. The shadows are small dusky fish, perhaps four inches long.

Again and again, like miniature salmon leaping the falls of the Columbia River on their return from the sea, these minnow-sized fingerlings hurl themselves upward against the plunge of the falls. Each time, after rising a foot or so, they are dashed back amid the shining bubbles that whirl away in the seethe below the waterfall. Once a larger fish, five and a half or six inches long, shoots upward, cutting its way through the descending water, the trajectory of its leap carrying it more than halfway to the top. Then it arcs downward again. It is, I think, a young brook trout that has worked its way up this tributary stream from Little River. The smaller fish appear to be dace but I cannot be sure. Each, in the oxygen-surcharged water below the falls, is finding, just as the salmon find below the greater cataracts, a source of sudden reserves of energy.

On this day, as when observing the flight of the water strider, I have had the good fortune to be at a meeting point in time and space that has enabled me to be the spectator at this dramatic expression of the exuberance of spring—that sense of well-being that has come to the fingerlings of the brook, the squirrels in the woods, the salamanders beneath their rocks, and to the man who walks beside this New England stream feeling a part of all that occurs around him.

A P R I L 2 0 .   In the light drawn across the landscape, the rays of the rising sun grow stronger. They cross the valley of Little River, probe among the bare treetops above Hampton Brook, run along the ridgepole of this white cottage under the sheltering hickory trees and speed on to highlight the flank of the steep ridge to the west. This is, I suppose, like other dawns coming to other clear April skies. But to me, it has a special quality of its own, a special beginning-of-the-world freshness and beauty.

I remember a sentence from one of the books of Leo Tolstoy that I read long ago: "All that the human heart contains of evil should disappear at the contact of nature, that most immediate expression of the beautiful and the good." Nature is, demonstrably, the great expression of the beautiful. And what of the human heart of which Tolstoy speaks? Evil it may contain. But it also contains something else, something not found in nature—compassion for the individual, for the unfortunate, for the one with his back to the wall. The element of compassion seems almost an invention of man. Where is it found in nature? The weak, the sick, the injured are natural prey. Except for parental care for the young, except for help afforded to mates or to family or to close-knit groups, creatures of the wild are on their own.

The warm sun shines down. The reviving rain falls. The nourishing soil feeds the growing plants. A world of beauty—of rainbows and sunsets—spreads around us. Everywhere we look, we see the brightness, the wonder, the interest of nature. But we look in vain for a core of kindness, for evidence of goodwill toward the individual.

Man, for all the evil he has contributed to the world, has also contributed a larger conception of fellow feeling, a dimension of goodwill previously unknown in nature. A world with augmented compassion in it would be a world more nearly in accord with our ideas.

Yet, as I walk on, I have second thoughts. I remember the subzero dawns of January. I remember the small birds, balls of fluffed feathers, sitting on their feet, alive and alert, surviving the bitter wind. Nothing tempers the storm to the weak creature. Not by tenderness is the strength of the world produced. If we could shatter all creation and rebuild it imbued with increased compassion, we might have—from our viewpoint—a world improved. But nature's world has been functioning since timeless time. It has endured. On this earth, nature represents the greatest example of success on a gigantic scale. It works. And who can say with certainty, who can be sure, that this different world, a world of greater kindness and compassion envisioned by man, would work as well?

APRIL 21. One-third of spring now—so soon—is gone. In mild and sunny weather this morning I spend another hour adding more rocks to the low dam where I watched the playful squirrel and peered at the blackfly larvae and uncovered the salamanders. A kingfisher clatters up Hampton Brook. It spies me and veers in a wide detour. And all the time I work and all the time I stop working to watch some small activity around me or to investigate something new in the shallows along the brooksides, the clear voice of a traveler come home to Trail Wood, a towhee by the stone wall bordering the lane, fills the minutes with its call of bright good cheer.

Another returning bird, a robin, has sung for several days, clinging to one of the tipmost branches of the apple tree below the terrace. We recognize it, sight unseen, as soon as the first "tea-leaf" of its song reaches our ears. It stands out among the mounting chorus of the birds because its tones are almost devoid of melody. Its song is thin and tinny. It lacks the robin's usual lusty, rounded notes. Yet, undeterred, the male sings on and on. It exhibits no sign that it realizes its performance is inferior in any way. It is responding to a primal urge of its kind. Spring has arrived. The time of the singing of birds has come. The satisfaction received from such a performance appears to have little or no relation to the results in terms of avian melody. The simple fulfillment of the urge is its own reward.

APRIL 22. In frosty-gray patches, reindeer lichen spreads its dry, crisp mats over the sparse soil of the ridgetop that rises beyond the southern border of the pond. They provide the area with the name we have given it: Lichen Ridge.

Crunching across this brittle footing, Nellie and I examine the more open ground between. We are searching for the burrows of the wild *Andrenid* bees. We encountered them here in April last year, and the year before. And we find them here again today.

In almost the same areas where we saw them before, we walk among tiny circlets of excavated soil ringing the round-eyed openings of the insect tunnels. Looking about us in one place, we count more than fifty burrows scattered between two rafts of lichen.

A year ago, as we wandered among these April bees, Nellie scooped up one of the dark little burrowers in a glass vial. Expanding under our peering eyes as we examined it under a fourteen-power pocket glass, we saw the thin yellow stripes running along the brownish-black abdomen and the dense pile covering the thorax. Before we released it, we also noted that characteristic that distinguishes the *Andrenidae*, the shallow triangle formed by the three ocelli on the head. To know the bee belonged to this group was enough for us. Remembering Frank E. Lutz's observation about this "large and confusing genus," we gave up trying to carry our identification further.

At several of the burrows we glimpse the flat heads of the insects blocking the openings. They remain unmoving. I tap my foot. All the faces disappear. Even slight vibrations in the soil supply an early warning system for the mining bees.

If our gaze could plunge beneath the surface of the dry soil, we would find, near the bottom of each tunnel, a series of branching galleries. The walls of these chambers are carefully smoothed and in some instances are lined with a protective layer of mucus. As soon as they are completed the bees begin stocking them with pills of pollen. For, like honeybees and bumblebees, these smaller insects have pollen baskets on their hind legs. In the midst of the pellets of food the female deposits her eggs. By midsummer the larvae have hatched, completed their growth, and transformed into adults. Some of the fuzzy-bodied bee flies, the *Bombyliidae* that hover, dart about, then hover again, spend their early life as parasites feeding on the pollen stored up in the tunnels of the *Andrenid* bees.

Although usually these insect miners nest close together, digging their tunnels in clay banks or dry soil, each has its own individual burrow. The mated females overwinter, in some instances in the same burrow in which life began for them. In many regions, these wild bees are important pollenizers of cultivated plants.

The small holes of the *Andrenidae* we see perforating the ground of the ridgetop on this April day are familiar over most of the eastern United States and as far west as the Rocky Mountains. Here at Trail Wood the tunneling bees are close to the eastern limit of their North American range.

APRIL 23. I wonder. Has an owl, slipping through the darkened woods, ever dropped in a silent plunge and picked from the air a flying squirrel tobogganing from tree to tree? It is entirely possible. But so remote are the chances of anyone being at the right place at the exact moment on a night of sufficient moonlight or starlight to see what is happening that probably in all history no man has ever seen such an event take place. However, if it is possible, in all likelihood it has occurred—even if unseen by human eyes.

This train of thought runs through my mind as Nellie and I follow, in the dark of the moon, the path that encircles the pond edge. Our two flashlights pick out fragments of our surroundings as we advance—last year's sodden leaves in a patch of mud along the path, a clump of tousled grass, a small spider web among the cattails, its silken mesh dense with imprisoned gnats, tiny wraiths so pale they appear pure white in the rays we turn in their direction.

Along Azalea Shore, down Stepping Stone Brook, at the edge of Whippoorwill Cove, spring peepers—the "peep frogs" of country folk—lift their clear little voices in a chiming chant. Endlessly, tirelessly, on and on into the night, their chorus repeats the same refrain: "Spring! Spring!" At intervals, lifting above this batrachian clamor, comes another sound of the season. It is the musical, fluttering call of the peeper's larger relative, the gray tree frog, *Hyla versicolor.* The calling grows louder as we near the leaning wild apple tree that is rooted close to the water's edge on the northern side of the pond.

I run the beam of my light up the mottled trunk. It reveals —like two thicker fragments of bark a foot apart—the almost perfectly camouflaged bodies of two singers clinging with padded feet to the tree. A plane drones overhead in the night. All the spring peepers and all the gray tree frogs redouble their efforts. Even in midday, we notice how the sound of a plane moving across the sky above them sets these frogs to calling.

Watching in the light of our double beams, we are fascinated, as always, by the throat sac of the gray tree frog. It—as does the throat sac of the peeper—expands like a shining balloon. Then when the end of the call is reached, it collapses and disappears suddenly as though the tree frog has swallowed it. "Bubble-gum frogs" is the name a friend of ours once applied to these batrachians. Again and again we observe how the creature's sides draw in as the sac swells, distending farther and farther, vibrating with the intensity of the fluttering call, then disappearing in its sudden contraction.

During a momentary lull in the serenade around us, we catch

the forlorn little voice of a lone peeper far above us, somewhere among the upper branches of the apple tree. Rarely have I heard a *Hyla crucifer* call from so elevated a perch. Most often we find them in the grass at the water's edge.

Across the pond, when we switch off beams, we see starlight reflected in the black mirror of the water. A fish leaps, sending a circle of ripple rings hurrying over the surface, setting the images of stars and constellations tumbling and gyrating where they pass. Once we hear a low splash and then make out a long V expanding in the wake of a swimming muskrat.

We come slowly up the darkened slope to the wall beside the lane. As we stand looking back we become aware of a faint sound around us, a small rustling and crackling among the fallen leaves. We switch on our beams. Earthworms, stretching out from the mouths of their burrows, are retrieving and drawing underground fragments of leaves, foraging for food under cover of the night.

APRIL 24.   As though sitting in the balcony of a theater, we look down the slope of Firefly Meadow this morning, watching a small drama being enacted at its foot. It is a mystery play and we are mystified by what we see.

Spectators unobserved, we remain motionless, largely hidden, leaning on the stone wall under the hickory trees. Below where the slope of the meadow merges with the lowland woods, a male grouse has emerged from among the trees. With its ruff raised, its body lifted to its full height, it stands facing an alighted crow. The crow changes position. It walks about. It turns away. It swings back toward the grouse again. That woodland bird remains rigid, tense, vigilant.

This continues minute after minute. At last we see the grouse gradually begin drifting back out of the field toward the woods. Its ruff lowers and it slips quietly away among the trees. The crow waits a few moments longer, then flaps heavily into the air. We change our position and discuss what we have seen. How had the crow and the grouse come face to face? What had produced their unusual confrontation?

In recent days we have caught a low-pitched, reverberating sound, accelerating in tempo, rising to a muffled drumfire at the end—the sound of a cock grouse on some log in the woods drumming with its wings to attract a mate. It is a sound that can be heard half a mile away. For the ruffed grouse, this is the equivalent of the flight song of the woodcock or the woodpecker's rapid-fire tattoo xylophoned on some dead and hollow limb. It sends forth an invitation, one of the many being broadcast in various ways by wild creatures in this mating season of spring.

At this time of year the male grouse is especially belligerent. Twice in the same place a friend of mine driving slowly down a rutted woodland road had his light truck attacked by a grouse. Each time it raked along the hood and top of the machine in repeated onslaughts. The Staten Island naturalist, Howard H. Cleaves, once recalled another instance of the kind in which a cock grouse, apparently aroused by the noise of the motor, faced a farm tractor with lowered head and ruff extended. The pugnacious bird even flew up and fearlessly rode along on the back of its mechanical adversary.

It is this combativeness of the male grouse in this season of the year that appears to provide the most probable explanation for the puzzling confrontation we have observed.

A P R I L   2 5 .   A faint sweet perfume drifts across the fields this morning. For days now I have detected the same delicate fragrance riding on the air whenever I have stepped out into the late-April dawn. It arrives from the lowland woods to the north and I suspect it comes from the blooming of the swamp maples. Looking across the fields, I see the treetops billowing up in filmy clouds, flushed with a rich tinging of red. Clustered on hundreds of thousands of twigs, small masses of minute flowers clothe the maples. In a close-up view, crowded together, they suggest tiny explosions of brilliant fireworks stilled in the midst of a display.

More than once Nellie and I have observed male purple finches feeding among these masses. The two seem to merge together. The color of the flower clusters and the little songbirds match almost exactly. Both the petals of one and the plumage of the other exhibit the same tinting of raspberry-tinged red. We always notice how the purple finches land in a tree with a flourish, a sudden sweep or curve as though they have made up their minds at the last instant.

When first we saw these birds poking among the maple

flowers, we assumed they were hunting for insects or getting nectar. However, as we studied them through our field glasses, we
noted that they were tearing the blooms apart. But we never saw
a petal fall to the ground. The finches apparently were feeding on
the flowers themselves. At various other times at Trail Wood we
have encountered birds that eat flowers. One May day, a cedar
waxwing alighted among the apple blossoms of the tree below
the terrace. We saw it hopping about, rapidly tearing off and consuming the pink and creamy-white petals. House sparrows are
notorious for the destruction they cause among the blooms of
various fruit trees. Later in the season, orioles sometimes rip
apart Nellie's day lilies.

Of all the species we have observed feeding among flowers,
the one we probably will remember longest appeared in a setting
of special beauty above us one spring day as we followed the narrow trail among ferns that runs along the base of Old Cabin Hill.
A wild cherry tree, its twigs laden with panicled flowers like clusters of white foam, arched out over the path. And all around us
the ground was littered with fallen blossoms. In the tree, its black
and white and rose-red plumage moving among the flower
masses, a bird as large as a small robin hopped from limb to limb.
For a long time we watched this male rose-breasted grosbeak in its
feeding. Unlike the finches, it was not consuming the petals
themselves. Instead, we saw it pick off the individual flowers with
its heavy white bill, nibble at them, then drop them. From each
it appeared to be obtaining a minute quantity of the concentrated
nourishment of nectar.

APRIL 26. With a small book in my pocket, I turn down
the winding trail that leads to Whippoorwill Spring. It is about
ten thirty in the morning and the day grows warmer. A little
azure-blue butterfly—that delight of the early spring—dances
along the path before me. Beside an oak tree on the hillside I lean
down to enjoy the deep red of a trillium in bloom. A towhee,
scratching among last year's leaves, calls from the ridge above. So
spring sweeps on. Now it is nearing May.

My path leads me along the steep northern side of a small
woodland glen cushioned with the green of mosses, crowded with
the lush growth of skunk cabbage, and, at this season of the year,

clouded with the pale yellow blooms of the spicebush—the spice-bush that, with its flowers after winter and its golden autumn foliage before winter, adds to our woodland beauty in both spring and fall.

The secluded glen below me is shaped like a slender arrow-head pointing west. At the point of its V, the sparkling water of the spring gushes forth from the almost vertical face of the valley's western end. It falls a foot or two into a rounded, mossy, minia-ture pool, shining in the sun as it descends. Then the water drains away, trickling in a shallow flow among the skunk cabbage and the spicebush to end its leisurely descent in the pond below.

Close together on the hillside overlooking the spring grow two oak trees. Between their twin supports I have nailed a plank to form a bench. After catching and drinking a handful of cold, re-freshing water from the spring, I sit on this woodland seat looking about me for a time, listening to the calling of the towhee, the soft crooning of the wind among the small, newly unfolding leaves, breathing in the moist, scented air of the glen.

Then I extract the book I have brought along. It is a compact edition of the old classic *The Canterbury Tales* by Chaucer. For the better part of an hour, I dip into the pages, reading here, then there, finally closing the book. The pull of the present, the beauty of this small, secluded spot, the events of the living spring draw back my attention. Now, this very minute, all around me, "that Aprille with his shoures sote" has had its way. Now, this minute, as in Chaucer's ancient time, "smale fowles maken melo-dye."

The melody will increase day by day. How hard it is not to speak at this time of year of the happy birds of spring! For their singing reflects our mood in these days of returning life and light. Our impression is that they *are* happy. Scientists—rightly so—scoff at bald statements that refer didactically to birds singing happily on the boughs. To say this is, admittedly, to say more than we know. How can we *prove* that the singing bird is happy? But, for that matter, how can we *prove* that the singer is *not* happy? So far as I am concerned, it seems to me that the weight of evidence is on the side of happiness. If it cannot be proved one way or the other—without getting inside the bird itself—the singer's elation and gladness in the days of its song can be *as-*

sumed with much justice by observation, which, in fact, is how we come to the conclusion that people we see are sad or happy.

APRIL 27. The breeze wanders over the meadow and drifts along the edge of the woods and probes among the trees. And everywhere across the winter-trampled woodland floor, among the twisted roots of the red maples, hundreds of white or pink-tinted wild flowers bob and dance on slender stems. Toward the end of every April we find them here—sprung from their horizontal, perennial roots—the windflowers, the wood anemones. In Greek mythology, Venus created them from drops of blood of the slain Adonis. A thousand years ago, the Saxons named these early blooms of spring "flaw-flowers"—flaw meaning gust—because they wave in every slightest gust of air. More modern are such common designations as wood flowers and nimble weed.

Nellie and I sit on a log among these dancing flowers that Carl Linnaeus, nearly two centuries ago, listed as *Anemone quinquefolia*. The second part of this scientific name refers to the deep-green, five-parted leaves. Above these leaves a single smooth stem bears aloft a single flower. Here, around us, the stems are four or five inches high; in more favorable locations they may lengthen to as much as ten inches. The frail-appearing blooms, nearly an inch across, consist of from four to nine petallike sepals surrounding a cluster of yellow stamens. We watch flies and early bees carrying the pollen from bloom to bloom.

In the sequence of their appearance, this wild flower is preceded by our other anemone, the rue anemone that clusters together in more open, sunlit areas. We encounter it first each spring along our lane and at the borders of our pastures. This anemone is one of the longest blooming of all our early wild flowers. Both anemones, wood and rue, are close relatives of the pasqueflower. All are members of that huge botanical assemblage, the buttercup family.

Along paths that lead deeper among the trees, the wood anemones grow fewer. Then in open glades we see them multiply again, especially where the soil is damp. But it is the same each year: always along the southward-facing edges of our woods we enter among the largest concentrations of these wild flowers.

Even if no other means of distinguishing between our two

anemones were available, the end of such April days as these would bring a sure method of recognizing which is the wood and which the rue. When the light commences to fade, the flowers of the wood anemone begin to close while those of the rue anemone remain open. It is sunset when we come home from our woodland walk. The windflowers, where we sat a couple of hours ago, still wave in the lessening breeze. But the blooms that nod at the top of their slender stems are smaller. Already the wood anemones are closing their flowers for the night.

A P R I L 2 8 .   In watch spring coils, the gray-green and scaly fiddleheads of the Christmas ferns rise above the prostrate masses of the last year's fronds. They dot the steep hillside above me. They cluster beside this narrow trail that leads north past Hyla Pond.

Spring sunshine floods among all the earliest green of the expanding leaves. It highlights far below me, as I round Witch Hazel Hill, diminutive Hyla Rill, weaving amid the mosses of the valley in its short flow from Hyla Pond to the greater pond outspread behind the beaver dam. I stand leaning against the smooth blue-gray bark of a hornbeam tree, looking down. Little flashes of brilliant red explode among the lower trees. Each represents a red-winged blackbird flaring open its scarlet epaulets as it alights.

No longer do the male redwings clothe the treetops with their descending flocks. No more does their combined excited calling swell into a surging torrent of sound. Now they are spread out across the lowlands. The social birds of a few weeks ago are fiercely individual, fiercely competitive now. They have staked claims and are defending nesting territories. A kind of centrifugal force has set in, pulling them apart, where a kind of centripetal force held them together before. In a sudden reversal, the redwings have become combative and possessive. And in this defense of their chosen nesting sites, the color I glimpse in tiny flashes among the trees below me plays an important role.

A few years ago, an American ornithologist conducted experiments to determine the practical value of the scarlet wing patches flaunted by the landing males. On a number of redwings he dyed the feathers of the epaulets and their yellow margins the same black as the rest of their plumage. When he released these birds, he observed they could defend their territories no more than half the time—territories they had successfully maintained before. In contrast, other redwings still able to exhibit the brilliant hue of their wing patches held their ground at least ninety percent of the time. The conclusion of these tests is that the use of the eye-catching red of the epaulets is of fundamental importance to these birds in defending their territories against other males.

APRIL 29. Hellebore Crossover is the name we have given to this short connecting path along the edge of the beaver pond in our far North Woods. Running close beside a wall below the long drop from Witch Hazel Hill and the hornbeam where I stood watching the redwings yesterday, it is a natural trail, a wild trail, among the oaks and red maples. Deer used it before us, and the wall that parallels it forms an elevated pathway, paved with flat stones, along which the red squirrels scamper.

Now, as April draws to a close, the lush, dense stand of false hellebore, the white American hellebore, *Veratrum viride*—that plant of numerous names: poke root, earth gall, Indian poke, bear corn, Devil's bite, tickle weed, itch weed, and poor Annie—is at its height. Swift to grow and swift to fade, it is a denizen of swamps and wet woods from New Brunswick south to Georgia and west to Minnesota. In the Adirondacks, it ascends the mountains to an elevation of 4,000 feet. The thick, poisonous rootstocks are perennial.

On this morning we see the hellebore, the wall, the trail through a veil of attendant motes—blackflies that we repel by liberal applications of insect repellant. The leaves of the light-green plants rising from the moist soil are wide and large, clasping the stem, roughly oval and heavily ribbed. They become narrower as they approach the top of the stem. And surmounting all, at blooming time, are the many-flowered panicles with the small yellowish-green stars of the blooms densely packed together. From them come the "bear corn"—the seeds very flat and broadly winged.

Hellebore and skunk cabbage are neighbors. But the skunk cabbage requires more water, often grows in water, while the hellebore rises from damp soil rather than from permanently flooded areas. In the morning sunshine on this spring day, the almost translucent leaves of both plants glow, luminous in the back-lighting. We loiter for a time on this path, advancing slowly, en-joying the simple beauty of these precocious growths of spring.

There is much to be said for such simple enjoyment of simple things. I remember all the natural history organizations to which I have belonged and all the organizations to which I still belong. I recognize their contributions in distributing knowledge and carry-ing on beneficial programs. But I remember how each probably began when two or three or more people got together with the idea of furthering some branch of observation. Officers and by-laws and programs of action sprang up in the wake of their simple enjoyment. Hardly do two people become interested in hellebore or skunk cabbage, it seems to me, than they set about forming a National Hellebore Association or a Skunk Cabbage So-ciety of America—electing one of them president and the other secretary-treasurer. Soon there are dues and regular meetings and annual conventions. And all the while what each really de-sired was the enjoyment of the thing itself.

A P R I L   30. The last of the April showers ends soon after daybreak. The robin sings on its high limb in the apple tree sur-rounded by dripping leaves. A towhee by the wall scratches soundlessly among the soft wet litter. In small gleaming drops, collected water shines out across the dark green umbrellas of the mandrakes and glitters on light blue petals where dog violets are in bloom. Wherever we go across the meadow we leave a darker trail in the wet grass.

When special friends come to see our paths at Trail Wood, we always hope for sunny weather. But when we are alone, I find more and more as I grow older that I fret less about the frequent changes in our New England climate. I try to accommodate my-self to it. I strive to be more philosophical about the weather. If I could do this entirely, I would be as philosophical as a woodchuck or a frog.

As the hours pass, the day warms up, the sun shines, the grass dries. Once, when we are among the fields in the late afternoon, I catch a sound that shifts the scene to a hot summer day on a hill-side in the dry far North Woods. It is a drawling "pee-a-wee." Can it be the voice of a wood pewee, returning weeks ahead of its normal time? Nellie says no. She has spied the source of the sound. It is the same bird she heard giving the same call after the pewees had left for the south in autumn—a starling imitating a pewee.

Sunset comes with those sweet, unpretentious strains we never tire of hearing—the song of a little field sparrow riding the crest of a low bush in Firefly Meadow. It goes on and on. So this final day of April comes to its end. Another dawn, another month. Tomorrow it is May.

MAY 1. April is promise; May is fulfillment. May is the time when everything seems happening, when life rises to a peak. May is the birdsong month. May is the time when we rebel most of all against routine, when we want the largest margins to our lives, when we desire to wander through the woods and over the fields completely free—as John Muir and Henry Thoreau were free—of all entangling engagements. Perhaps the walks of this month should be headed: Sauntering Through May. May in spring and October in fall—these, for the average person, are the two best loved among all the twelve chapters that make up The Book of the Year.

Before sunrise today the season slides into bounteous May. And what do I see—morning, noon, and night—walking through this initial day of this favorite month of spring?

Morning. A red-shouldered hawk drops down from a tree to a rock at the edge of Azalea Shore on the southern side of the pond. It wades out into the shallows and for several minutes I see it through my glasses splashing and shaking itself as it takes a bath.

Noon. The oriole is back in the hickory tree. Among the upper branches, still unclothed in leaves, I catch the male's black and white and orange plumage flashing in starts and stops, the contrasts brilliant in the midday sunshine. I hear the clear, flutelike whistle, the same rich tones that have been part of each year since we came to Trail Wood. Two or three days before the usual time of its arrival, the bird has returned from its long flight to the south, and then, in season, to the north again. It has come home to the identical nesting tree where, a year ago, its mate wove a deep pendent pocket nest anchored near the tip of one of the highest limbs.

Night. Close to the edge of the pond, when dusk is merging into dark, I come upon a lone spring peeper. It is the sole occupant of a small and shallow puddle. There it is master of all it surveys. I hold it in my flashlight beam. I watch its shining throat sac expand into a round balloon, collapse, then expand again as it reiterates its carrying, piping little call, inviting any female within reach of its voice to join it in this, its private lake in miniature.

M A Y 2 .   Ninety-nine years ago, on a September day in England, Richard Jefferies entered in his nature diary a single sentence. It consisted of only three words: "Fine beautiful wasps." A quarter of a century before, Henry Thoreau, returning from a walk on a spring day in Concord, recorded his observation of greatest interest in an even shorter entry, a single hyphenated word: "Mouse-ear." Standing here among the clumps of juniper and looking across the pond toward the distant house, I jot down the most important event of this day in a sentence almost as succinct: "Swifts return to the chimney."

M A Y 3 .   Sitting on this old log beside the twisting course of Hampton Brook, I have been watching the treetops, observing a small band of warblers darting about, endlessly in motion as they feed among the filmy clouds of newly expanding leaves. Spring sunshine illuminates the woods. The air is warm, filled with the primeval smells of the earth in May.

And all the while overhead, among the upper twigs and branches, I glimpse the bright little bodies of the warblers, the blue-winged, the chestnut-sided, the black and white, the yellowthroat, and the prairie. They start and stop, dart and flutter. Their colors catch the eye. Their clear, carrying, emphatic little voices fill the woods. They are active life in its most visible form on this sunny morning in early May.

For them the uncertainties of migration are almost over. For them the climax of the year, life at its best, the mating and nesting season, is close at hand. As I follow them with my eyes among the stippled green of the treetop foliage, I am remembering other warblers and other springs, faraway springs, springs in the valleys of the Great Smoky Mountains; among live oak trees—bent and sculptured by the sea winds—along the lower coast of Texas; in sheltered woods behind the rampart of the lakeshore dunes in northern Indiana.

Mine, I realize, is an emotional as well as a visual memory. It is often concerned with moods and atmospheres, scenes and appearances, shifting tints and fragrances, lights and shadings. Even when I was very small I remember standing entranced as a sunset altered minute by minute. At the time I felt: I want to remember this *always*. I want to be able to recall this *very* sunset just as it is now whenever I desire to see it again. Even in those early years, I was acutely aware of how time is ceaselessly sliding away, streaming to the rear, carrying with it moments that have affected me deeply. Always I have wanted to salvage such moments from the rush of time, to preserve them permanently in memory, to enjoy them again and again in retrospect. I have been a lifelong hoarder of memories.

Yet sitting here, warmed by the sun, watching amid the early green of the treetops these vivid sparks of life, these little birds with their intense activity, their prodigal expenditure of energy, seeing them in this sun-filled, food-filled woods of spring, I know that the dwarf, Gimli, in J.R.R. Tolkien's *The Fellowship of the Ring*, was speaking truly when he said: "Memory is not what the heart desires." It is not the other springs but this living spring that the heart desires. It is not remembrances of the past but reality for today and tomorrow and tomorrow, life and more life, the living birds, the living trees, the sunshine of this very moment. Only those deep in pain or burdened with unbearable sorrow can fail to agree with W. H. Hudson's final words in *Far Away and Long Ago*: ". . . I could always feel that it was infinitely better to be than not to be."

In a little while most of this band of northward-moving warblers will be gone from these woods. In a little while longer they will be gone from the earth. Their species will continue on but they as individuals are—as are all the forms of life the world contains—transients. This the rational mind sees, realizes, accepts. The brain makes peace with the facts. But the heart still longs for the impossible, for more and more life, for life unending. It wants the warblers to stay with us, to fill the woods with their bright colors and bright songs forever. Yet even as our hearts long we are aware that only in the unreality of some fairy tale is such longing ever realized. And so we come back to memory, memory that brings alive the life of other years, that expands the boundaries of our enjoyment. The life we have loved lives on in memory. It is through memory that the mind goes part way in satisfying this longing of our hearts.

MAY 4. Up until yesterday we have had one resident water snake in our pond, a rather pretty banded reptile, easily alarmed and always disappearing swiftly at my approach. In the morning yesterday, I saw it had been joined by a stranger. By noon a second newcomer had made its appearance. By evening a third had arrived.

Now, as Nellie and I examine the pond edges in the morning light, we discover the number has risen to five. While we watch, we catch sight of one water snake after another swimming steadily toward Summerhouse Rock. There they haul themselves out and lie on its flat surface. Soon we are looking through our field glasses at a mass of wriggling, intertwining serpents. This May our pond is a rendezvous, a mating place, that has drawn water snakes from the surrounding area. One, I believe the female, is thick bodied, older, and almost black in color. Most of the others, probably the males, are younger, brighter, more reddish in hue. Numerous males are trying to mate with one female. But they do it without any evidence of fighting among themselves.

Why have they selected our Trail Wood pond for their rendezvous? How have they all arrived at this one rock? My guess is that they have followed scent trails left behind by the female. For a good part of the day, this mating tangle of sinuous bodies remains on the rock. During long periods all the water snakes are still. Then a general wriggling and changing of position sets in. By evening the May encounter is at an end. The aquatic snakes begin to scatter. In the next dawn we will see only one, the resident that makes its home along Azalea Shore. I wonder where the others will go, how far they came for this tangled assemblage beside our pond, where the female will give birth to the living young that will number anywhere between sixteen and forty-four. Life for these creatures is precarious and usually short. Both when young and as adults they fall victim to numerous predators.

At Trail Wood such northern banded water snakes are relatively rare. I recall seeing only one this spring away from our pond. It lay sunning itself on the grass beside Hampton Brook. As I looked over the low wall of the bridge, it saw me. It slid into the water, sank to the bottom, slipped under a flat rock, then lifted the forepart of its body vertically a foot or more, bringing its head just above the surface of the stream. Here, while minutes passed, it remained anchored. It looked at me; I looked at it. It tired first. Lowering its head again, it wriggled from beneath the rock and, remaining submerged close to the bottom of the brook, undulated away downstream, weaving among the stones, hurried along by the current.

MAY 5. On this May morning, I have been following the ups and downs of a tree-climbing woodchuck. I first see it foraging amid the long grass of a small open space at the foot of an apple tree now in the last years of its long life. Its trunk, rising like a long hollow shell for the first six or eight feet, tilts sharply toward the south.

For a time the animal feeds hurriedly, stretching out its neck, swinging its head from side to side, nipping off and chewing rapidly mouthfuls of the blades of the lush spring grass. Hardly a minute goes by without it lifting itself to its full height to make a swift survey of its surroundings. Then it drops back and resumes its headlong feeding. In time, its hunger growing satisfied, its hasty grazing slows down, grows languid. I see it several times look up along the leaning trunk. Adventure seems on its mind. Then it comes to a decision. It scrambles awkwardly onto one edge of the troughlike trunk and commences hitching itself up the incline.

Every few feet it pauses and sweeps the scene around for danger in a reconnaisance sometimes rapid, sometimes prolonged. Even at Trail Wood, where the woodchuck is not persecuted, it is one of our wariest animals. As soon as it emerges from its burrow it is eternally on guard.

Late in summer I have seen woodchucks struggling vainly to climb bushes to reach berries. A friend of mine once saw one of these heavybodied animals among the topmost branches of a high tree feeding on bunches of wild grapes. But, on this spring day, the climber in the apple tree gives the impression of seeking a break in routine, a little adventure above the ground, the enjoyment of a wider view than its earthbound existence customarily affords.

I watch it reach the lower branches and hoist itself laboriously higher into the tree. Sometimes it slips, flounders, drops back a perch or two. Then it begins ascending again. Halfway to the top, at its highest point, it stops and looks around for a long time. Then, headfirst, it commences descending again. It reaches the lowest branches, leaves them behind, and disappears. During the rest of its descent it scrabbles downward out of sight, hidden within the hollow trough of the trunk. At the bottom it pops up, gives a little jump, and is back on the ground. Its high adventure is over. Once more I see it fall to snatching mouthfuls of grass, bolting down food, resuming the hurried feeding that is the main work of the woodchuck's day.

MAY 6. The alarm clock of spring has sounded. Everything is awakening with a rush. The shadbush, even before its leaves are fully unfolded, fountains up in the white cloud of its loosely clustered flowers. High-bush blueberries are spangled along their twisted branches with the minute dangling bells of their blooms. Cardinals whistle; flickers call; bluebirds sing. Back and forth low among the meadow tussocks, overwintering queen bumblebees drone, drifting slowly in their search for underground nesting sites. A newly returned brown thrasher picks rapidly among the blades of a grass clump. When it flies away, I investigate. The clump is alive with small red ants that have poured from underground into the spring sunshine.

Those low-growing relatives of the bedstraws, the simple bluets with their four-petaled flowers of pale blue or white, each with a yellow eye at the center, spread in patches and drifts along the meadow path. From a distance their dense stands resemble dewy cobwebs spread across the shorter grass. Once in the midst of such a stand I come upon a glimpse of beauty in miniature, a fairylike scene where a tiny spring azure butterfly has alighted for a moment on one of the little flowers of the bluets.

And everywhere—beside the brooks, along the edges of the woods, across the open fields—on this day in spring, I hear the singing of the birds. A robin in a treetop, a field sparrow clinging to a dry mullein stalk, a song sparrow riding the tip of a small cedar tree—I hear them all. Again and again, from along the course of Fern Brook, rings the loud, emphatic song of the Louisiana water thrush. I pause in the pasture listening. I hear a phoebe and then a towhee. Somewhere in the distance a bobwhite reiterates its name. I catch the rippling warble of purple finches, the clear whistle of a tufted titmouse, and now, as I stand still listening, the clatter of a kingfisher over the pond.

I return home at the close of this day of onrushing spring wishing I could experience it all over, know it again on a thousand days, see the same scene unchanged, hear the same birds sing, stand among the same wildflowers just as I stood today. But I know—as I knew in watching the warblers a few days ago beside the bend in Hampton Brook—that this can never be. But it is something to have the memory set down in a journal, to have it preserved on a printed page. In nature, everything flows. All is change. In truth, we never cross the same river twice. But the printed page does not change. It is the river we can cross again.

MAY 7. In swift, tight circles, the pair of broad-winged hawks mounts into the sky over Monument Pasture. We watch the soaring birds grow smaller as they climb. We hear the shrill, receding, whistled piping of their calls. We see them pass and repass. We see them make dives and swerves that are parried by instantaneous tilts and veers. Unaware of passing time, Nellie and I stand gazing up at this skyborne exhibition of the wild ecstasy of the nuptial flight.

With scarcely a wing beat, the two hawks spiral upward within their column of ascending air. Each minute they leave farther behind the heap of flat rocks that in long-ago spring and summer days in this pasture the hired man named Hugh shaped into the form of a monument to himself.

Some days ago, our broadwings came home, home to their old nesting area in a secluded part of the Seven Springs Slope in the South Woods, home from South America after a round trip of thousands of miles over land and water. Since then the woods have echoed with the "whee-ooou" of their mating call, repeated in the air and from perch to perch. Before long we will see, in the high fork of some large oak or maple, a stick nest growing in size.

To it the birds will add, at intervals, twigs and small branches with the green foliage still attached. In the weeks that follow, we will encounter the mated birds frogging along the brooks, hunting chipmunks in the woods. We will see them searching for prey on the wing or perched motionless on the lower limb of a tree, peering downward for some slight sign of movement below.

Called the gentlest and tamest of the hawks, these woodland broadwings are, in the main, peaceably inclined. They rarely attack anything except the prey on which they live. Now, looking up, we see in the swift spiraling of the birds high above us another sign of how far the season has advanced.

M A Y 8 . Have you ever watched a pill bug change its skin? Until this mid-afternoon in May, if anyone had asked me that question I would have had to answer no. But in the past five minutes my reply has changed to yes.

Amid a moldering pile of logs, I have surprised one of these small and primitive creatures. A terrestrial isopod crustacean less than half an inch long, it suggests a prehistoric trilobite in miniature. Because of its ability to roll its brownish-gray body into a round pill-like ball, in the manner of an armadillo, the scientific name of its family is *Armadillidae*. It is encountered most often under bark and stones and amid moist and rotting wood. It is a scavenger in its feeding, consuming organic matter, both plant and animal.

The thing that first catches my eye as I am passing by is a small spot of white on the dark surface of a decaying log. It is the almost translucent shell of chitin out of which the pill bug is gradually creeping. This exoskeleton—the hardened outer coating or skin that takes the place of an internal framework of bones in the body of these creatures, as among all the hosts of the insects—has split at the front. Gradually the forepart of the pill bug's body, clad in the new, still soft integument, emerges from its old external skeleton. The creature moves a little, then rests for a time. Then it hitches ahead a little farther. During the rests the shell of new chitin rapidly hardens.

This sequence I am watching on this May day amid these rotting logs has recurred during unnumbered millions of years. Before insects appeared on earth, these more primitive forms of life were shedding their exoskeletons just as I see today. When the transformation is complete and the pill bug moves away on its seven pairs of walking legs, it leaves behind it a little ghost pill bug clinging to the log—a white discarded mold of its small body, perfect in all its parts.

MAY 9. My walk on this May morning is The Walk of the Seven Stops.

Stop 1. I stoop to tie my shoelace on a woodland trail and in consequence see something I otherwise might have missed. It is a small cushion of shaggy moss growing on decaying wood. The leaves, all turned to one side, suggest scythes or brushes. Children have fancied they resemble duck heads and soldiers with lances marching to war. The common name for this primitive plant is broom moss. Its scientific name is *Dicranum scoparium*.

Stop 2. The beech leaves are off their twigs at last. All winter on this sapling tree beside the trail, the pale tan flags of the slender, last year's foliage have fluttered in the wind. Now the swelling of the buds has loosened their grip upon the twigs.

Stop 3. At the edge of the woods I pause to listen to the clear call of a little tufted titmouse. It is repeated endlessly. The sound comes to my ears with small variations. Sometimes it is "Cheer! Cheer! Cheer!" sometimes "Chew! Chew! Chew!" sometimes "Hear! Hear! Hear!" At other times it is more nearly like "Year! Year! Year!" In whatever form it arrives, it rings out among all the bird songs around me. I remember an experienced field ornithologist who once told me he had found that he saw forty-five percent of all the species he would encounter on a given day in spring during his first hour in the field.

Stop 4. Out in the meadow I look up. High above me two red-shouldered hawks spin in an updraft. Just as I get my glasses focused on one of the soaring birds, it sweeps back its wings, tilts steeply downward, and, like an arrowhead, streaks in a long plunge toward the earth. I follow it down and down. I see it near the ground, open its wings, check its descent, and begin climbing upward again. A hawk sporting in the air of spring.

Stop 5. Another hawk, one of the broadwings, goes beating across the field low above me. Looking up, again through the round, magnifying windows of my binoculars, I see it, as it passes by, give a little flutter to its tail as though it has been bitten by a parasite.

Stop 6. In the yard below the terrace, one of our cottontails nibbles tender new grass, washed by rain in the night. I watch it nip off a blade and chew it rapidly beginning at the lower end, the blade growing shorter and shorter until the tip disappears. A bluejay flies into the apple tree with a raucous note of alarm. I notice how the rabbit instantly sits up, its head held high, looking around, tense and ready to leap. It is an animal tuned in on all the warning sounds around it.

Stop 7. One last pause before I come indoors, a pause to watch a white-breasted nuthatch at a feeder still stocked with sunflower seeds. I see it pick up a seed, discard it, pick up another seed and discard it. I begin counting. It discards twenty-eight seeds before it chooses one to its liking and flies away.

The Walk of the Seven Stops. On every trip afield, it is the halts, the pauses, the moments when activity ceases that mark encounters of special interest.

MAY 10. Pure, sustained, musical—the most beautiful of all our batrachian sounds—the song of the American toad fills this twilight. Rising and intermingling, the clear, trilling, overlapping chorus is the dominant music of the night. It ascends from all the shallows of the pond where, from our woods and fields, the toads have congregated for the mating of spring.

As long as I can remember I have heard this song unchanged. It rose in the dusk of earlier days as moving as it reaches my ears tonight. It is the same today as it was in the year 1900 when, beside the Hudson River, John Burroughs recorded in his journal this superb description of the sounds I hear: "In the twilight now the long-drawn trill of the toad may be heard—tr-r-r-r-r-r-r-r-r-r —a long row of vocal dots on the dusky page of the twilight. It is one of the soothing, quieting sounds—a chain of bubbles, like its chain of eggs—a bell reduced to an even quieting monotone."

When we descend with flashlights to the edge of the pond, we find the mating toads and the long convoluted strings of the extruded eggs all along the shore of Driftwood Cove. Only a few days will pass before we will look down into the shallow water here and see the bottom strewn with tiny tadpoles.

It was in this identical spot where we are standing that in the May of another year I watched something terrible and wonderful taking place amid the sunken driftwood of the cove. Here the toads were mating, the smaller male grasping the female firmly with its forelegs. Unmated males trilled at the pond edge. As I watched on that day, in sunshine after rain, I saw five pairs mated in the space of a square yard. Several were partially hidden under floating refuse left by a muskrat's feeding.

Into the midst of these mating toads slid a water snake about two and a half feet long. It eased its way close and opened wide its mouth. Almost gently it slipped it over the head of one of the females. The male clung unmoving in place. Then the snake slowly, steadily began the process of swallowing its prey, in this case two victims at once. At last only the feet of the doomed toads extended outside. Then they disappeared.

But all the time, even as she was being devoured, the female continued to extrude the long, swirling, black-dotted, gelatinous strings of her eggs. The fertility of her kind was triumphing even in death. As long as the last moment of her life remained, she continued the immemorial work of reproduction. Two toads were disappearing while hundreds, in egg form, were being released into the water.

MAY 11. The song of a Louisiana water thrush beside Hampton Brook, of a scarlet tanager in the higher woods. Bluets in the meadows, bluets along the mossy edges of the Old Colonial Road. A flicker hammering on a stub, excavating a nesting hole, near Stepping Stone Brook. Tadpoles, with lashing tails, plunging into the sanctuary of drowned grasses at the edge of the shallows. A spotted sandpiper teetering on Summerhouse Rock. Black and gold, a huge bumblebee working among the tiny bell flowers of a barberry bush, painstakingly extracting infinitesimal droplets of nectar. And everywhere around the house the perfumed air, the rich fragrance of lilacs in bloom.

A gentle mizzling rain begins drifting down—a pasture rain, making the meadows lush. I come home, cross the yard to the kitchen door, and, looking down notice the toes of my walking shoes are white, as though covered with snow. The wet petals of fallen apple blossoms have plastered themselves against the leather.

MAY 12. Snow fleas in May! This is something I never expected to see. Yet at the base of a large ash tree, where at its eastern edge Firefly Meadow drops down to Hampton Brook, I come upon a host of the infinitesimal springtails concentrated in one small area. They pepper the surface of a square foot of moss and leaf litter. I sit down beside them, lean on an elbow, peer closely. A constant drift of movement sweeps across the mass. In waves of seemingly aimless motion, the wingless jumping insects stream this way and that. They suggest grains of dust stirred by a breeze. Always before when my eyes have encountered these almost microscopic insects of the order *Collembola* I have been surrounded by ice and snow. Always before I stood in the midst of a winter thaw. Now both they and I have attained a more enduring thaw, the great thaw of the springtime.

MAY 13. There is a resting place in the woods at the top of the slope to the west of the pond, one of several short planks I have nailed between adjoining trees to form trailside benches for two. Here Nellie and I have been sitting for some time, each leaning against a different treetrunk. In an unseasonable ascent, this afternoon the mercury has shot up to eighty-seven degrees. In this sudden heat, while our bodies remain inert, our eyes rove about exploring the woods around us.

Not far away the mound of a wood ant colony has formed a landmark ever since we first came to Trail Wood. On this day we see its summit has been torn apart. Yellow-shafted flickers and ruffed grouse have been at work—the flickers probing with their bills for ants, the grouse dusting in the soft material the insects have mounded up and that the woodpeckers have disturbed. Bowllike craters record where the larger birds have twisted and scratched and fluffed their plumage and thrown the fine material about in their assault on body lice.

Nearer at hand, only a few feet away, a fluttering motion arrests the wandering of our eyes. On the ridged bark of an oak tree, a pale tan forest moth with a wingspan of only about three-quarters of an inch is struggling to become airborne. It appears trapped, in some way anchored to the tree. The tip of one of the insect's wings is bent as though caught in a crevice of the bark. Then we notice what holds it fast. A black carpenter ant, with its six legs braced in the depression, has its jaws clamped on the wingtip. The desperate struggling of the little moth is insufficient to pull it free. With a forefinger I push it gently to one side to obtain a better view of the ant. The ant lets go and the freed moth, one of the innumerable woodland *noctuidae,* springs into the air and, apparently undamaged, flutters away among the underbrush.

Later on, in slow motion in the heat at the end of the day, we follow the curve of the pond on our way home. Thunder, still far away, rolls and mutters along the western horizon. Halfway down Azalea Shore, where seepage from a small spring has collected into a shallow pool three or four feet across, we bend down to examine a whitish mass of salamander eggs, probably those of the common spotted salamander. The cluster suggests a smooth, rounded piece of ice spongy with air.

Our last stop before climbing the slope to the house is beside Driftwood Cove. The toads are still trilling but now singly, scattered soloists instead of in the great commingled chorus of so short a time ago. Already some of the earliest deposited of the convoluted strings of eggs are hatching, dropping to the bottom a new generation of toads that now lie scattered—tiny black commas printed on the bottom of the shallows.

MAY 14. Here, downwind from an old apple tree, I check my walk in the midst of a shower of blossoms. Petals stream toward me, swirl around me, scud past me. They unroll in a thin carpet of white over the green of the grass.

For a week now, fruit-bearing trees along the walls and near the brook and scattered through the woods—chokecherries and wild plums and seedling apple trees—have reached the height of their blooming. Our lone pear tree, living out its life rooted close beside the north wall, has lifted in a towering, foaming fountain of white. Rounded apple trees, remnants of an earlier orchard, transforming from glossy green to shining pink-tinged white, rise in billowing clouds of blooms. At the foot of my Insect Garden, down the slope toward Hampton Brook, one of the oldest of these apple trees, perhaps a centenarian—the tree the woodchuck climbed—is clad in blossoms of half a dozen varying kinds. I see one branch with noticeably smaller blooms, another with blossoms more deeply flushed with red. Each represents some long-ago graft that culminated in this single tree whose branches yield the colors and flavors of diversified fruit.

Surrounding all these blooming fruit trees, during all the hours of sunshine, the air hums, swept by the blurring wings of honeybees weaving in and out, gathering nectar, collecting pollen, fertilizing the blooms. But now, near the close of blooming time, in the gusts of this morning, the petals are stripping away. They are strewn downwind in languid showers; they are flung in sudden, short-lived blizzards. Like snow in spring, they whiten the grass.

Swiftly the blossoms of springtime spread their petals, swiftly they fall. How soon their beauty goes. A week is a long time in the life of some spring flowers. For the fruit trees, much depends on what happens during the space of comparatively few days at blooming time. A sustained rainy spell, a time of violent winds, hampering the work of the honeybees, reduces pollination. The fate of the year's crop is changed by the time of the winds, the time of the frosts, the time of the rains in May.

MAY 15.   It is two o'clock on this May afternoon. As I start for the Brook Crossing in the Far North Woods, I pause to watch a chickadee. In its quick, darting flight it lands on a stone wall where Nellie has scattered a handful of sunflower seeds. Instead of snatching up a seed and dashing away, as is the custom of the chickadees, this bird spins in a complete circle. In some accident of flight, after surviving cold and gales and freezing rains throughout the winter, it has lost the sight of one eye. It can see only to the left.

The broken wing, the sightless eye—these are two of the greatest catastrophes that can befall a bird. Yet this particular chickadee has been appearing day after day for a week or more. I notice that it is continually on guard, alert for danger. By watchfulness, making the most of its one good eye, it compensates as best it can for its immense handicap. By whirling in a complete circle, making a swift survey of its surroundings, it provides the only possible substitute for full vision.

But this afternoon, as I stand watching it, I notice it is particularly on edge. It whirls once, twice, three times, four times before it is satisfied. At last it snatches up a seed and streaks for the protection of the branches of a nearby tree. As it flies, it holds its head far to the right, insuring thus a greater visual sweep for its one undamaged eye. Perhaps the little bird has had some recent narrow escape. Its nervousness seems to grow. It increases the number of revolutions on the wall, whirling at intervals of only a few seconds, before it feels sufficiently safe for the momentary distraction of picking up a seed. On its last trip before I turn away, it alights and rotates, carrying its good eye through a full 360 degrees, one time after another in rapid sequence until it has turned thirteen times. Only then does it feel transiently secure. By such extreme vigilance has the handicapped bird survived in a hostile world.

Watching it, I remember a far different creature, a bullfrog, we used to encounter one summer beside the pond. In some accident it, too, had lost the sight of one of its eyes. Each time we came along the path and it sensed our approach it, like the chickadee but with movements more cumbersome, turned in a complete circle, giving it a panoramic view of its surroundings and whatever dangers they contained.

MAY 16.   I turn aside to visit an ancient apple tree. And life circles back to an earlier time.

Perhaps four feet above the ground, a large knothole stares out round-eyed. It opens into a pocket, a steeply descending cavern eroded into the heartwood. At its bottom a pool of collected rainwater glints in the dim illumination. Years ago, just such a teacup-sized lakelet in the heart of an apple tree on a hillside above a cattail marsh on Long Island introduced me to a strange, pallid creature fit for Wonderland. It breathes through its tail while it walks underwater. This tail—a hollow tube, extensible like a periscope—enables it to move about in the manner of a deep-sea diver while it scavenges for food among the decaying material at the bottom of the pool. This diver with its breathing tail is the rattailed maggot, larva of the Syrphid fly, *Eristalis tenax*.

I dip the beam of a small flashlight into the cavern, and I see, just as I had on the Long Island hillside of my first Insect Garden—the garden I wrote of in *Near Horizons*—the surface of the water is dimpled. Each dimple is formed by spreading hairs at the tip of the breathing tube where they rest on the surface film.

Over the rough texture of the decaying wood rising beside the water, small dark spiders run and halt and run again. Pale trilobite-shaped sow bugs wander in slow motion. And on the surface of the water I note a mosquito with striped legs floating close beside a raft of eggs, shining white. On this day it is neither the rattailed maggots nor the dark spiders nor the sow bugs nor the mosquito that holds my interest. It is another inhabitant of this microworld within a knothole.

With its forefeet anchored to the wood at the edge of the pool, with only its eyes and nose projecting above the surface, the gray tree frog that I have come to visit remains undisturbed. The only sign of life I see is the palpatation of its throat, sending little ripples running away over the surface of the knothole lake.

Here, as nearly as it is possible, I have made friends with a tree frog. It lets me come close. When it is resting at the mouth of the cavern, it lets me peer around it into the decaying interior of the tree. It even endures stoically the indignity of having me tickle it under the chin with a forefinger as it clings with its great adhesive pads to the wood, half asleep in the heat of the day. At such times it looks like a little gray owl sitting in the entrance of its nesting cavity. In how many other places among the decaying orchards of New England, I wonder, is there such an old apple tree, such a knothole cavern in its trunk, such a miniature pool of stagnant water cupped within—and a gray tree frog that has made it its home?

MAY 17. This is written in the quiet of the late afternoon woods a quarter of a mile from the knothole cavern home of the gray tree frog. For some time Nellie and I have been wandering among the glacial rocks of the wooded Boulderfield beyond Old Cabin Hill hunting for the site of our only fragile fern. We find the spot again, close to the last remnants of the fallen trunk of a chestnut tree. Now, resting on one of the moss-cushioned rocks, my elbows on my knees, I find my mind busy with memories of three small events that stand out from the hours of this warm May day.

The first concerns a bird—a rose-breasted grosbeak. We were standing this morning in a small opening at the top of the Seven Springs Slope south of the Big Grapevine Trail. Here rises the largest witch hazel at Trail Wood. Its straggling growth ascends to a height of twenty feet. The biggest of its spreading trunks has a circumference of more than a foot. While we were examining it, the grosbeak drew near. It fluttered from perch to perch, at each pause filling the woods with its rich, slurring song. Progressing thus, it made a complete circuit of the edge of the clearing. So our day began with a grosbeak singing a ring around us.

The second memory concerns an insect—a butterfly. We were coming home toward noon when Nellie noticed a tarnished orange and black monarch butterfly sailing in leisurely, drifting flight low over the weed tops. It is the first monarch we have seen this year. Where had it emerged from its chrysalis? How far had it flown? What had been its history? We watched it ascend Firefly Meadow and meander away over the lane, across the yard, above the Starfield. The last we saw of it, it was still working unhurriedly into the north, winging its way in the vanguard of the reverse migration of spring.

The third memory relates to two small birds, a stone wall, and an apple tree. Early in the afternoon, not far from where the monarch passed above the wall that forms the northern boundary of the yard, color and movement arrested our attention. Clad in all the brilliance of its breeding plumage, a male purple finch was performing, not for our eyes but for the eyes of a female perched in the top of a nearby apple tree. Over and over it repeated its courtship display. Each time it walked slowly along the top of the wall, its head feathers raised into a little crest, its wings beating so rapidly they formed a blur. It appeared almost on the point of taking off. The parading walk, the vibrating of the wings were interspersed with darting flights into the top of the apple tree close to the resting female. After half a dozen repetitions, the displays abruptly ended. The female bolted away toward the North

Woods. For a moment the parading male failed to realize she had left. Then it shot into the air and labored off in frantic pursuit. The wall and the tree were left deserted. But the little avian drama had not ended. It had been merely transferred to the setting of another stage.

MAY 18. Nellie and I, this morning, have the impression of occupying ringside seats at a rodent battle of the century. Again two chipmunks are fighting for possession of the small territory of a flat stone capping one of the walls. It is a rich territory for here Nellie is in the habit of sprinkling sunflower seeds.

The two little animals stand on their hind legs, their eyes narrowed to slits, batting and clawing each other. They break apart and leap together again. They fall into biting embraces, whirl and roll so swiftly their striped bodies appear merging together. The battle continues without a sound—action, and more action—for one long and furious round.

It ends with one contestant fleeing, high-tailed, along the wall, the other, for the time being, master of its fruitful little foothold on the earth. And all the while our chimney swifts race unconcerned about the house and our tree swallows sweep back and forth above the combatants in smooth flight, with easy glides between their wing beats, giving the impression they are waltzing through the air.

This chipmunk excitement has hardly died down when friends of ours, Bill and Vinnie Stocking, drive down the lane. They have come from the annual antique show at the parish house of the high-steepled white church in the village. One of the attractions of the show has been a dress and bonnet Vinnie is wearing. They were made in 1845, thirty-nine years after our Trail Wood house was built. We all examine the bonnet with special interest for it is decorated with an intricate pattern of horsehairs woven into the cloth. Pale tan, they still shine in the sun, although they were originally collected from the tail of some animal that fed in some pasture in the region more than a century and a quarter ago.

MAY 19. The heat mounts as this day advances. It climbs to eighty-five degrees, then to ninety degrees, ninety-one degrees, to a peak of ninety-three degrees by early afternoon. A gray squirrel comes, hop and pause, hunting for buried nuts under our largest hickory tree. Then the hops cease and the pause lengthens itself into minutes. The little animal stretches out on the grass. It is flattened like a bearskin rug—a squirrel overwhelmed by spring fever. We—like the squirrel, the birds, the chipmunks, the woodchucks—lie low, anesthetized by lethargy in the sudden heat.

Late in the afternoon we watch blue-black clouds tumbling upward in the northwestern sky. They mushroom higher, draw nearer, loom over us. The faint, irregular cannonading of distant thunder grows louder. It unfolds across the sky. First comes a spitting of rain, then a deluge, then the full fury of the storm. Almost continual is the glare of the lightning. Its jagged paths cut across the dark backdrop of the storm clouds, here, there, to the north, the west, and the east. The rain, hurled by the furious gusts of one of the most violent storms of the year, at times obscures the scene around us.

Then comes the hail. All down the lane, across the fields, along the brook, on the roofs of house and sheds, flattening vegetation, ripping through leaves, pounding among branches, the gleaming fusillade thumps and rattles and crashes. Hard and shining—pearls of ice—the hailstones strike and bound. They collect in white carpets in all the depressions. Any bird caught in the air without protection, subjected to the battering of these ice bullets, would have been beaten down stunned or dead or with a broken wing. Yet the violent tumult is hardly over when we see red-winged blackbirds alighting in the yard, arriving from the cover where they found protection from the tempest. We watch them searching for food over the saturated, hail-strewn ground.

Nellie and I walk down the lane to view the damage left in the wake of the storm. All across the lowland areas beside the brook, the great leaves of the hellebore and the skunk cabbage have been punctured and torn. But we notice a difference between the two. Ragged holes have been ripped in the skunk cabbage while the hellebore leaves have been shredded lengthwise. Numerous prominent veins run down the latter leaves. These strengthening members separate the leaf tissue into elongated compartments, unlike the fewer, more angling veins of the wide skunk cabbage.

In consequence, the hail has shredded one and perforated the other.

M A Y 2 0 .   Across the bridge over Stepping Stone Brook, up the curving path past my writing cabin, around the huge prostrate juniper outspread beyond the cabin—like an immense green pancake, fifty feet across—I ascend to the summit of Lichen Ridge. In spite of yesterday's rain and hail, the frosty-hued carpets of the *Cladonia* lichen are still dry and brittle, crunching underfoot. I see but few of the *Andrenid* bees. The main activity of these early spring tunnelers is over. Most of their little mounds of excavated earth have been flattened by the storm.

Leaving the ridge and pushing my way through a narrow gap in a high sheep wall, I wind on a meandering course up the slope of Juniper Hill. Thirty years ago, this area was open meadow. After the sheep left, cows pastured at the edge of the woods and deer sometimes browsed among them. At the top of the slope I stand for a while wandering with my eyes through all the endlessly varied shadings of green—May green, spring green—spread out below me. They reach across the meadows and run along the branches of the trees. From the pale, almost ethereal green of the slender aspens about my cabin, through the apple green of lush pondside plants to the emerald green of the meadows, I go ranging through all the subtle diversity of tints of this one color that have been derived from the palette of spring.

By way of Shagbark Hickory Trail and Whippoorwill Spring, I circle home. All through the afternoon woods I am never beyond the voices of the towhees calling back and forth. Beside the moss-rimmed pool that catches the falling water of the spring, I find heart-shaped tracks, large and small, imprinted in the moist soil. Two deer, a doe and a fawn, have paused here and lowered their heads to drink from the pool. As I bend over these tracks, I am jerked erect by a loud rushing sound almost beside me. No more than a couple of paces away up the steep side of the little glen, a ruffed grouse has launched itself into the air. Camouflaged and crouching low, it had escaped my notice completely.

After the bird has hurled itself away among the trees, I dip up a handful of cold, clear water from the little pool. Drinking in this primitive way, I always experience a sense of special satisfaction. I am back to fundamentals, touching bedrock. I ask for nothing; I need nothing. I am depending on no impedimenta or paraphernalia. I am traveling light. As always in this simple rite of drinking from my cupped hand, I seem enjoying for a time the ultimate in natural simplicity.

It is not long after I leave the glen behind and begin following deeply shaded Veery Lane toward home, that I become aware of the barred owls. Two have begun hooting before night has come. All the way up the slope and as I stand at my back door, I hear them calling back and forth among the trees near Ground Pine Crossing.

M A Y 2 1 .   Although summer, officially, is a month away, I see a summer scene around me. I walk under trees green with enveloping foliage. I stride through grass already lush. I stop beneath a wild black cherry tree in the pasture and, looking up, see the cream white of panicled blooms. Kingbirds have come back to their wild apple nesting tree on the slope of Lichen Ridge. And when I linger on the path along the southern lowland edge of the pond we call Azalea Shore, I wind among the pinxter flowers, the pink azaleas, in full bloom. In the main they are flowers for the sense of sight rather than for the sense of smell. Their attractiveness lies in their color rather than in a sweet perfume. So slight is their fragrance that for many they are virtually odorless. A few weeks later, along this same path, the flowers of the white swamp azalea will open in their annual sequence. Although they may be less appealing to the eye, they will envelop the passerby with a pervading fragrance that has in it a touch of the spicy scent of carnations.

M A Y 2 2 .   Leaning once again on the pasture bars, as I did at the end of that early April day, I listen to the tapping of a downy woodpecker and the lowing of a distant cow. I breathe in the perfumed air of the awakened earth, watch the shadows slowly lengthen across the fields, note the subtle shifts in the coloring of the landscape as the sunset ebbs. Such an evening as this we have looked forward to for a long, long time—all through the months of winter.

Across the pasture I see the gash in the side of the tree where the pileated woodpecker hunted for ants. Beyond it, sinking into darkness, the North Woods extends as far as the Old Colonial Road. And beyond that, wilder and more remote, the Far North Woods stretches on for more than half a mile to Griffin Road. All these acres belong to us. We hold the deed. They are legally ours. Yet what do we own? Not the right to abuse or destroy or "skin the land and move away." In one of his penetrating sentences, Aldo Leopold pointed out: "We abuse land because we view it as a commodity belonging to us." Land is a trust as well as a possession. The land has its rights and so have all the wild creatures it supports.

What our land means most of all to us, I believe, is contact. The great gift of possession is the privilege of contact. Here we have real contact, enduring contact, continuity of year-after-year contact with one small part of the earth's surface. Here, surrounded by our Trail Wood acres, we feel a mutual relationship with the land. We merge with it. It owns us as much as we own it. We are transient dwellers here, just as the fox and the field sparrow and the spring azure butterfly are transient dwellers in these same surroundings. Only we have what fox and sparrow and butterfly can never have. Because of ownership, we can protect and preserve as well as enjoy the land.

M A Y 2 3 . It is jumping rabbits that provide our entertainment at the end of this May day. Returning from the waterfall after sunset, Nellie and I halt at the small gate beyond the wild plum tangle. Ahead of us two cottontail rabbits are frolicking on the lawn. They race over the grass. They leap high into the air. They veer and swivel, cut into curves, and loop into spinning circles. They leapfrog in great bounds over each other's backs. They bolt headlong into the plum tangle and explode out again.

Once we see one of the animals dash directly toward a flicker searching for ants on the ground. The bird shoots up two or three feet in the air in a sudden airborne leap while the cottontail races beneath it. Then it drops back exactly where it was before and resumes its probing of the ground for insects. A little later a rapid change in direction sets a rabbit careening toward a worm-hunting robin. It, too, bounds into the air. But it flutters only a few feet to one side and lands again. Like the woodcock I saw earlier in the spring showing slight concern when a cottontail fed close by on its singing grounds, the flicker and the robin show only prudent caution, not panic, in the presence of these unaggressive vegetarians.

This leaping, spinning, bounding, zigzagging of the cottontails represents more than just a display of the exuberance of spring. It has its own special significance. For the earthbound quadrupeds it is the equivalent of the streaking plunges of the courtship flights of the hawks at the time of their mating. Around and around the yard, in and out of the plum tangle, the smaller male is pursuing the larger female. What we are watching is the courtship performance of the cottontails—that ecstatic prelude to baby rabbits in a fur-lined nest in a depression in the ground and then to the small, perfect miniature cottontails we will see hopping about, round-eyed, in the sunset of some early summer day.

MAY 24. Walking in the deepening darkness this evening we glimpse the first fireflies of the year—two sparks of light brightening, then fading, then brightening again. We watch them go drifting away low above the night-clad slope of Firefly Meadow. Their wandering, airborne little lanterns mark the commencement of a new era of the spring.

On this evening a soft, faint breeze flows idly out of the south. It is filled with the fragrances of May collected mile after mile from flower and stream, from new leaves and crushed grass, from all the host of growing things that have unfolded, expanded, and added their particular scents to the air at dusk. Far away, somewhere off toward the hill pastures of another farm to the north, a whippoorwill begins its calling.

Here at Trail Wood there is always a tug-of-war between two months—May and October. Which brings the finest hours of the year? On some mid-October day when we stand in the sunshine, breathing in the crisp air of fall, surrounded by the glory of the autumn foliage, I am sure we will vote for the tenth month of the year. But tonight, watching these first fireflies, listening to the lone whippoorwill in the darkness, savoring all the late-May scents carried on the breeze, remembering the bird song and the wild flowers of the day, we have no doubts. *These* are the best hours. *These* are the best days. The minutes of the very weeks we are living in, the hours and the days of this fifth month, are merging together into what—it seems to us now—surely must be the finest time of all.

MAY 25. I have just glanced down the hill toward the pond. What I see is the most incredible sight I have encountered at Trail Wood. Near Summerhouse Rock, Nellie, fully clothed, is wading out into the deeper water of the pond. Already she is a dozen feet from shore. There the water is well above her hips. For a moment I stare incredulously. Then I shout. Nellie looks up but, bending forward, continues her advance. I sprint down the slope. As I run I observe that she has a stick in one hand and is reaching out as far as she can, pawing the surface of the water. When I am pounding along the path at the pond's edge, I catch sight of a black object floating on the surface just beyond reach of the stick. It is a dead grackle.

A few minutes later as we come back up the hill to the house, Nellie sloshing along in water-filled shoes and dripping dress, she relates the events that preceded what I have observed. In a leisurely circuit of the pond, when she was drawing near the flat, tilted surface of Summerhouse Rock, a swirl of fighting grackles swept past her and out over the water. In the melee, one of the black birds was borne down, driven lower and lower. Nellie saw it strike the surface with a splash, struggle with flailing wings but struggle in vain, unable to lift itself into the air again. At the time the floating bird was only five or six feet from the edge of the pond. Snatching up the first stick she could find, Nellie tried to reach out and pull it toward her. But a stiff breeze carried the bird away. Second by second the gap widened as the helpless grackle drifted farther into the pond. As it continued its struggles Nellie followed until the bottom pitched steeply downward. But always the stick fell short of the ensnared bird. Its efforts grew weaker, then ceased entirely, and it floated lifeless on the water.

On several other occasions, we have seen smaller and lighter birds rise on the wing again after being forced down into the pond. Once the harrying of a kingbird drove a young redwing into the water close to Cattail Corner. As soon as its tormentor ceased its attack, it lifted into the air and flew to the cattails. But a grackle is a comparatively large and heavy bird. Once it had touched the surface of the water it had been trapped, helpless, unable to rise again.

MAY 26. The wild cherry tree is near the end of its blooming. Here, downwind, I stand once more, as I did earlier this month in the lee of the apple tree, in a storm of streaming petals. This time the petals are tiny and shining white, turning and glinting, hardly larger than real snowflakes. The steady drift continues all the time I remain examining the deep red of the large flowers of the Furnessville peony of my Insect Garden. The buddleia bushes that I originally planted to attract butterflies have all disappeared, killed by our harsher winters. But amid the weed tangles, cut by paths winding down the slope, the peony has flourished. I have known this same plant for more than half a century, first at my grandfather's Lone Oak Farm near Furnessville in the Indiana dune country, then transplanted to Long Island, then here, where it has been rooted for nearly two decades. Already this enduring plant must be well past the three-quarters-of-a-century mark. We sometimes call it our Memorial Day peony, for usually by the latter days of May it is in full bloom, the rich red of its many flowers standing out amid the dark green of its thriving foliage.

MAY 27. Years ago, I talked with someone who had been a companion of John Muir's on walks in Yosemite Valley. His pace, I was told, was unhurried. Muir stopped often to enjoy a view or examine a tree. For ten or fifteen minutes at a time he would sit down beside some favorite wild flower along the way. It might take him ten hours to walk ten miles.

On this late-May morning, along the varied paths I follow, my pace is even slower. Everywhere, in this time of new and tender leaves and plant growths, those variously formed and tinted swellings we call galls are enlarging and taking shape. When I turn over the elongated leaves of the wild goose or horse plum, *Prunus americana*, I find as many as half a hundred rodlike growths projecting out and clustering together on the underside. They are pale green. When I come to the wild black cherry, I discover reddish-hued, club-shaped galls massed on the upper side instead of the underside of the leaves. Green globular galls are swelling on the stems of the new goldenrod. And like tiny pancakes of yellow-green with a central dot of red, flat and roughly circular, other galls are splattered over the underside of maple leaves. Still others project in small reddish or greenish cones from the upper side of the witch hazel foliage. And anchored to oak leaves, the smooth-skinned globes of the oak apples are already beginning their growth.

Entirely different are the developing galls on the meadow-sweet. They extend along the underside of the central vein of the leaf. Bright green elongated, inflated bags, they suggest pea pods. These legumelike galls are remarkably strong. I have been surprised by their failure to burst like little balloons when I have exerted hard pressure between a thumb and forefinger. Their sturdy walls are thin and translucent. When you hold such a gall up to the light, you can see the silhouettes of several slender larvae feeding within.

All galls produced by insects—and their varied forms number in the thousands—harbor larvae inside. Each starts in the same way. The abnormal plant growth is induced by an irritant introduced into the tissues when the eggs are laid or added as a byproduct of the life processes of the larva. Wide is the variety of the egg layers—flies, wasps, moths, aphides, beetles, sawflies, lace bugs, and gnats among them. The spherical oak apple is produced by a flylike insect scarcely a quarter of an inch long; the pine-cone willow gall by a minute gnat; the cockscomb elm gall by an aphid; the spindle-shaped gall of the goldenrod by a mottled brown and gray moth about three-quarters of an inch in length.

In all cases, the egg layer is insuring a snug home and a well-stocked larder for its progeny. The larva or larvae that develop within the gall feed on the tissues around them, beginning at the center and working outward. In some instances there are galls within galls, tiny galls inside larger galls, the product of different insects. Thus one gall may represent a community of developing larvae. Moreover, besides the multitude of galls produced by insects, there are others induced by mites, spiders, and bacteria. The bacterial root galls of the legumes represent the principal agents of nature in making atmospheric nitrogen available for the use of plants.

Through the ages galls have been used in many ways. In China, a thousand years ago, one aphid was artificially reared and the growths it produced were employed for medical purposes and as a source of dye and tannin. In Roman days galls were resorted to in prophesy. The famous Aleppo gall, or gallnut, produced by a wasplike insect on the leaves of oaks in western Asia and eastern Europe, was widely used for centuries to provide an astringent, a tonic, and an antidote for certain poisons. The Greeks valued it as a source of dye for their wool. And at one time it was a specified ingredient in formulas for the ink used by both the United States Treasury and the Bank of England.

M A Y 2 8 .   The first froghopper of the year. The first bubble nest of foam shining out among the goldenrods. The first small white milestone in the course of the season.

Each spring we look forward to the initial foam nest of the frog hopper as earlier we look forward to the first spring azure butterfly. On what species of plant, we always wonder, will we encounter the earliest of these innumerable spittlebugs, these immature leafhoppers within their sheltering coverlet of bubbles. There, protected from sun and enemies, they sink their sharp little beaks into the plant tissues to drain away the sap which they utilize both for food and the production of their masses of enduring bubbles.

One spring we saw the first froghopper foam clinging to a nettle; another spring on a Saint-John's-wort. This year the plant is a goldenrod. As the season advances we will encounter similar small masses amid such diversified sources of sap as fleabane and everlasting, thistles and tansy, asters and daisies, grasses and the foliage of bushes and trees.

On this May morning, as so often in the past, I stoop and pluck a grass stem. With it, I carefully part the foam to glimpse the little occupant within. At such a moment I am always conscious of a special kind of charm in these midgets with their smooth, plump, moist, and shining little bodies, their dark, round eyes, their highly individual manner of life. After my grass-stem inspection, I draw the coverlet of bubbles back in place. This mass—about the size of the nail on my little finger—provides evidence that warm weather has come to stay. The forward and backward movements, the advances and retreats of the weeks of April and early May are giving way to the more equable weather of the latter days of our New England spring.

MAY 29. With a picnic lunch packed in a market basket, we set out about nine o'clock this morning. A leisurely half hour later, we deposit the basket on the heavy chestnut plank that is supported by twin piles of flat rocks to form a bench beside the cascade where Hampton Brook cuts across the Old Woods Road. Unencumbered, we begin our search for the largest of our orchids.

At one time or another, at various places at Trail Wood, we have encountered eight species of orchids. One, the helleborine, *Epipactis helleborine*, we found only once and never saw again. Early each June we visit our single stand of the large twayblade, *Liparis lilifolia*, where the Old Colonial Road starts its curving climb up Old Cabin Hill. In July we look for the clustered greenish-yellow flowers of the ragged fringed orchis, *Habenaria lacera*, in the lowest part of Woodcock Pasture and the deep pink of the calopogon, *Calopogon pulchellus*, in the far corner of Firefly Meadow. When August comes, we hunt along the Old Woods Road for the spotted coralroot, *Corallorhiza trifida*, with its flowers supported on leafless stems above twisted roots that resemble a small mass of coral. Then, too, the downy rattlesnake plantain, *Goodyera pubescens*, the most numerous of our orchids, scattered all through the woods, bears clustered white flowers on a woolly stem. In this month, but only on rare occasions, we have found the small purple fringed orchis, *Habenaria psycodes*. Then, when early September arrives, the path around the pond carries us through scattered groups of nodding ladies tresses, *Spiranthes cernua*, with their white flowers ascending the upper stem in a double spiral. Last year we counted more than 100 along the path.

But on this May morning it is the earliest, as well as the largest and most showy of our orchids that we are hunting—the pendent, pink, elongated floral bubbles of the moccasin flowers, the pink lady's slippers, *Cypripedium acaule*.

There is, to the west of the Old Woods Road, a gentle tilt to the land that carries it down among a scattering of oaks and maples to the edge of a serpentine strip of wetland where a slow seepage of water drains away to Hampton Brook. This gravelly woodland, with its acid soil, is our wild orchid garden in the latter days of May. And here, on this morning in the sunshine, we wander from bloom to bloom.

Nellie catches sight of the first flower close beside an oak tree. It dangles in the air fully a foot above the two deep-green, shiny leaves outspread at the base of its stem. The flower seems inflated, buoyant, almost airborne as it swings in the breeze. Each stem supports a single bloom which may be two inches or more in length. Its delicate pink is veined with darker markings.

Soon our eyes, becoming accustomed to the flowers, see them scattered close at hand and far away across the slope before us. The count rises to ten, a dozen, fifteen, twenty. Known by many names—Noah's ark, Indian moccasin, squirrel's shoes, two-lips, old goose, and nerveroot—the pink lady's slipper is familiar from Newfoundland to the mountains of North Carolina and Tennessee. The yellow lady's slipper, *Cypripedium calceolus*, the whippoorwill's shoes, also grows in our region but we have never found it at Trail Wood.

We wander from bloom to bloom, each giving the impression of having been blown from tinted glass. Within these solitary flowers, hidden stigmata receive pollen from the backs of visiting wild bees—often from those of the little *Andrenid* bees which have matured in the dark chambers of their burrows in the earth.

By the time the noon hour has arrived, we have found and enjoyed more than thirty of the moccasin flowers. Well content, we sit in the spring woods on the chestnut plank beside the cascade and eat our lunch. Then, with the basket empty, we start for home.

MAY 30. Like Alice on her way to Wonderland, I, on my way to the brook below my Insect Garden, have met a preoccupied rabbit. Alice's rabbit was white. Mine has only a tail and underparts that are white. Alice's rabbit wore a waistcoat and carried a watch. Mine is a common cottontail.

But the preoccupation of Alice's white rabbit seems a characteristic of the brownish animal I meet slowly hopping toward me up the garden path. It gives the impression of being wrapped in thought, absorbed in other things, paying little heed to its surroundings. As though unaware of my existence, it advances toward me in short, deliberate jumps. It is only a few feet away when its reverie is shattered and it returns to awareness with a jerk. For an instant its startled gaze surveys my motionless form. Then its long leaping hind legs go into action. Its body shoots in an arc through the air and, as quickly as Alice's rabbit popped down the hole descending to Wonderland, my cottontail disappears into the weeds beside the path.

Not far from where its powder-puff tail flashes out of sight, May apples are blooming creamy white under the dark umbrellas of their leaves. I run a forefinger along the smooth texture of the curving petals and, leaning down, enjoy the golden beauty of the circlet of stamens held within each floral cup. Then I push my way through a spicebush tangle and emerge into the small open space beside the brook where it tumbles over my loose dam of rocks. I sit for a time on the rustic bench watching the race of the water, then open the book I have carried with me. Its orange-hued covers are decorated with stylized dragonflies. Printed in gold are the title and name of the author: *The Life of the Fly,* J. H. Fabre. Within, placed at a page I have read innumerable times, is a bookmark, a yellowed postcard printed more than sixty years ago by the publisher, Dodd, Mead & Company. It invites the reader to sign the card, attach a one-cent stamp, and send it in to receive "without obligation of any kind" an illustrated booklet about "Fabre's life and wonderful work."

And so for another time I lose myself in the adventures of this pioneer French entomologist who carried on his classic researches in a pebbly, thistly bit of wasteland near the village of Sérignan, in Provence. The man becomes as interesting as the insects he writes about. It was his work, recorded in his voluminous *Souvenirs Entomologiques,* that opened the eyes of generations of readers to the unsuspected wonders that lie in the capacities and habits of the small and often overlooked creatures that fly past us or walk about us on six legs. He led them into a world still filled with mysteries.

I put the book down remembering two small encounters of the past—one at the pool below the waterfall, the other beside

Whippoorwill Cove. A golden-colored mayfly bobbed in the air above the waterfall pool, dipping to deposit its eggs. Suddenly it fluttered upward. It ascended higher and higher. Then I glimpsed some small dipterous insect following close behind. What was this pursuer? What was the story behind the event I saw? In the calm of evening at the edge of Whippoorwill Cove a gnatlike insect with fuzzy body and plumed antennae drifted on the still water. As I peered closely I saw, whirling around it, a dark smaller insect which I could never see clearly because of its tiny size and the speed of its movements. It spun about the floating gnat, darted at it, batted it, returned again and again to its swift attack. Each time it struck I could see the gnat bumped and jarred on the water.

I suspect that in both instances the attacking and pursuing insects were small winged parasites but I am still mystified, unable to understand completely what I have seen. At Trail Wood, as at Sérignan, as around the whole circumference of the globe, wherever anyone stops and observes the complex activity of the insects, he finds himself in a realm of riddles, of small events he is unable to explain.

MAY 31. An hour has gone by. All during this time we have been engrossed in a charming spectacle. Four young chipmunks, so bright, so perfect, so newly minted, have appeared at the mouth of the burrow in which they were born. They blossom from the opening, each facing a different direction, each half in and half out of the entrance of the tunnel. For the first time in their lives they are seeing the wide green world around them. For the first time they are smelling the multitudinous scents of spring instead of the damp, earthy odor of their darkened burrow. For the first time they feel the warmth of sunshine on their fur.

Chipmunks are well grown before they emerge from underground. We watch this quartet responding to new sights and sounds. They lift themselves higher to scan their surroundings, then disappear in an instant at some startling movement or unfamiliar sound. For a long time they seem bound to the hole, anchored in place as though attached to a rubber band. Then one timidly pulls itself out on the grass, advances a few inches, then shoots back to safety again. When one is frightened, all are frightened. Denoting the waxing and waning of their courage, their heads rise and fall like the heads of baby birds in a nest.

As time goes on, one after the other they embark on daring journeys, first the length of their bodies, then half a foot out into an unfamiliar world. During this hour when we are watching the four young chipmunks, we are observing them at the time when they are making their first tentative steps toward breaking home ties.

The sight of their advances and retreats recalls to Nellie a recent dream of hers. She remembers she was making a farewell tour of our trails before starting on some long journey. Saying goodbye to a chipmunk in her dream, she lifted it in her hands and when she put it to her ear she could hear it purring. Then she woke up. Another time she awoke from an even more improbable dream, associated with another animal. It concerned being back on Long Island and meeting our black and white cat sitting on a street corner. When she asked it what it was doing there, it told her it was waiting for the light to change so it could cross the street and catch a bus for Florida. One of the great missed opportunities of my life, probably, has been a failure to keep a complete record of the natural history of Nellie's dreams.

J U N E 1 . All the time I have been working this morning, slashing brush, chopping down small trees, cutting a new trail between Ground Pine Crossing and Whippoorwill Spring, the woods around me has rung with the voices of scarlet tanager and towhee, catbird and wood thrush and veery. I often lean my ax against a tree or stand with my bush knife or clippers idle in my hands to listen.

This is pleasant, leisurely work filled with diversions all the way. Each time I cut through some shaded tangle floored with moss, letting the sunshine in, I look around me, wondering about the little alterations that will follow. Every change in the out-of-doors represents an advantage to one plant, a disadvantage to another. At one place I spend some time beside an oak sapling only about three feet high. It is adorned like a Christmas tree with half a dozen galls like round masses of wool, beautifully decorated with spots of reddish brown. Within, protected and surrounded by ample food, dwell the larvae of the small cynipid wasp that is the gall maker. Oak trees alone are associated with more than 300 kinds of galls. I examine closely one of the woolly globes clinging to the sapling. When I handle it I find it breaks apart in sections like a ripe raspberry.

At the end of this morning's work, I walk on to Whippoorwill Spring in its shaded little glen to dip up cupped handfuls of its cold flowing water. Several times when I have come this way on previous occasions, I have carried a small thermometer in my pocket. Each time that I have thrust it under the falling water it has recorded the same temperature—forty-five degrees Fahrenheit. That, apparently, is the "normal temperature" of this woodland spring.

My wristwatch tells me it is close to noon when I leave the woods and strike out across the pasture toward the house. The sunshine streams down from overhead. Two hawks spiral upward without a wing beat above the open fields. I watch their weightless shadows skimming over the ground. They curve around me, go wheeling away before me, race across the meadow grass as the soaring birds turn continually between me and this higher sun of June.

J U N E 2 . Shouldering up into the sunshine on the slope west of the pond, a gray glacial boulder splotched with the pale green of *parmelia* lichens absorbs and retains the warmth of the sun's rays far into these June evenings. As we have done for a number of years on this second day of June—my birthday—Nellie and I have come this way to pick the small sweet red fruit of the wild strawberries. In the microclimate of extended heat surrounding this half-buried boulder, they ripen first of all.

We sit on the warm earth beside the rock enjoying a handful of the wild fruit. We feel a great content. We listen to nearby sounds and faraway sounds. On the quiet air of evening, the soft warbling of a bluebird carries from the hickory trees. Below us the pond lies placid, outspread, mirroring tinged light from the sky. Behind us, as dusk seeps through the woods, echo the evening songs of the wood thrush and the veery.

What finer time of year could one ask for in which to be born than on a day in early June? Here on this old farm, which seems the perfect setting for two so interested in the out-of-doors as

Nellie and I, another birthday finds me as absorbed in the continuing story of the living world, as deeply appreciative of our great good fortune in dwelling in its midst as ever before. Far away is the noise, the crush, the fume-laden atmosphere of all the great cities of the world.

An airliner drones overhead. We look up. High in the sky it shines in the rays of the now hidden sun. It is winging its way from a big city to a bigger city. In the mood of this moment, viewing it from the angle of our special outlook, it appears to us to be going from bad to worse over the best.

J U N E 3 . Down the northern slope of Old Cabin Hill, along the ridge that parallels the abandoned railroad right-of-way that runs along the western boundary of Trail Wood, among the massive gray blocks of the glacial boulderfield, we are never beyond the calling of the ovenbirds. The voices of these ground-nesting warblers are part of all the hours of this June day. We leave one behind only to hear the ringing "Teacher! Teacher! Teacher!" of another ahead of us.

The common name of this warbler—the ovenbird—is derived from the nest it builds in some dense clump of vegetation, a nest with an overarching entrance suggesting the Dutch oven of the Trail Wood fireplace. As a rule, Nellie and I do not hunt for bird nests. We leave them undisturbed. Only once in the past, in all our woodland walks, have we accidentally stumbled upon one of these oven-shaped nests. But on this June morning we chance upon another.

While Nellie pauses to examine a fern beside one of the glacial rocks, I push ahead through a low and wiry stand of huckleberry bushes. Almost at my feet a darting little form appears and scuttles away before me. At first I think it is a chipmunk or mouse. Then I catch a clearer glimpse. For an instant, among sparser vegetation, I can focus sharply. I recognize the olive-hued back of an ovenbird. It utters no sound. Its wings are slightly raised. They flutter as it runs. It tempts pursuit; it seems so easy to overtake and catch. But I resist temptation.

Standing still, I look carefully about me. Hardly two feet from where I stand, I make out the little bower nest of the bird in a tussock of grass under a huckleberry bush. Within the dim interior I glimpse a cluster of four small white eggs speckled with brownish markings. Nellie and I examine with care, from a distance and through our field glasses, the nest and its eggs. Then we obliterate our trail as much as possible lest it lead some predator to the spot.

On another June day, exactly a century and a quarter ago as this is written, Henry Thoreau witnessed the same performance I have seen. His entry in his journal, dated June 7, 1853, provided the first account ever recorded of this "rodent run" of the nesting ovenbird. "The ovenbird," he wrote, "runs from her covered nest so close to the ground under the lowest twigs and leaves, even the loose leaves on the ground, like a mouse, that I cannot get a fair view of her. She does not fly at all. Is it to attract me, or partly to protect herself?"

This performance of the ovenbird is one of innumerable distraction displays—such as the killdeer's broken-wing diversionary performance—that birds employ to draw away danger from the vicinity of their nests. More than 200 species of birds listed in the American Ornithologists' Union's checklist of North American birds utilize some form of feigning to divert attention to themselves and away from their eggs and young in times of peril.

A wood thrush, on one occasion, was observed holding the interest of a blacksnake heading for its nest by engaging in a broken-wing display on a branch in a tree. On the surface of a Maine lake a loon put on a similar performance. Some small birds flutter like butterflies in front of a predator. "False feeding" is also employed. Kirtland's warblers and woodpeckers and killdeer all have been seen carrying food to places where their nests were not when they were under observation.

Thoreau ended his account of the "rodent run" of the ovenbird with the question: "Is it to attract me, or partly to protect herself?" Behaviorists now suggest that birds do not feign injuries or put on other displays purposefully to draw enemies away from the nest. Rather, they say, the action is the result of an inner conflict. The birds are torn between a desire to stay and a pull to flee. But whatever the source of the action, its consequences are obvious. The sight of the bird in apparent distress catches the eye. It holds the attention. It impells the intruder to follow the performer. And so it is led farther and farther from the location of the nest.

J U N E 4 .   I fall in the pond before breakfast! Along Azalea Shore, a broken limb blocks the way. I tug at one of the dead branches, trying to drag the obstruction aside. It snaps without warning. Frightening the fishes and sending two mallards quacking and skittering down the pond, I pitch backward with a great splash into the shallows. Now it is my turn to come squelching and dripping up the slope to change my clothes—with a new mishap to explain to Nellie.

After breakfast, I clear away the fallen limb without further misadventure. Then, on my way to a morning's work in my writing cabin, I pause along the way to watch the polliwogs. Black swarms of little tadpoles sweep outward along the muddy bottom of the shallows at my approach. Some have hatched from green frog eggs, some from toad eggs. Ballooned out into far larger size, the second-year tadpoles of the bullfrogs rush away propelled by the whipping of their slender tails. At Driftwood Cove, I come upon a score or more of the smaller swimmers resting, like black notes on a sheet of music, on two parallel twigs submerged among decaying leaves. Every floating mass of algae or decomposing vegetation provides a haven for the alarmed tadpoles.

It is among one such mass, entangled in a dense growth of waterweeds, that action shifts my attention from the polliwogs. A large bullfrog rips into sight swimming for its life. It attains the shore and I see one side of its body appears raw and covered with a kind of whitish slime. Then I notice the blunt head of a large snapping turtle thrusting this way and that among the swaying vegetation. Like a dog that has lost the scent, it is hunting for the frog, a frog that has made a miraculous escape and has attained the safety of the land.

J U N E 5 .   Seven feet of spiderlings. This is the first thing I see as I begin following the paths that wind through the weed tangles of my Insect Garden. Pouring from the brownish pear-shaped egg case of the large yellow and black web-building spider, *Argiope aurantia*, a multitude of tiny arachnids follow a long, tightly stretched thread of silk that extends almost horizontally to an anchorage in a plant top. Massed together down its length, as on an elevated highway, the spiderlings follow the thread toward their destination—the weed jungles through which they will disperse. There may be 1,000 eggs in one of these overwintering egg cases. But so great is the mortality of spiders that at the end of summer only a single adult pair of all the many spiderlings may remain.

Another dozen steps along this same path brings me to a dry, last-year's Saint-John's-wort, a dead and weathered plant at the edge of green new growth. And here I find a *polistes* paper-making wasp arrived before me. It is gouging out mouthfuls of the plant fibers and—with jaws that bite inward from the sides—chewing them into pulp paper for use in building its nest. I stand here trying to recall another time when I have seen one of these wasps getting its raw material from such a source. Customarily I see the insects collecting bits of softened wood from old fenceposts and, on several occasions, from the bars on which I leaned at the edge of the pasture.

The third thing I encounter on my circuit proves to be an enigmatic discovery. Scattered over the seed heads of a clump of curly dock are clusters of small tannish-yellow balls, each about three-eighths of an inch in diameter. They suggest airy bubbles. For the walls of each are formed of a kind of network or lattice construction. Through the minute openings, I can see the pupa cases of their makers curled up comma-wise in the interior, one inhabiting each bubble cocoon.

As I stand here wondering what I am seeing, what insect will emerge from the interior of these small openwork globes, my attention is caught by a movement in the midst of the cocoons. A flat, grayish predacious bug moves from one to the other to where four of the little spheres are clustered close together. I see it thrust its slender sucking proboscis through a tiny opening in the lacelike exterior of one of the cocoons. For a time it stands motionless, draining away the life fluids from the transforming creature within. Then it withdraws its proboscis and moves on to the next cocoon. But the pupae of the curly dock, each with its own insubstantial shelter, are numerous. Some will survive. But the puzzle of their identity remains. Will I ever know what shape they will assume when they leave their transformation chambers and emerge into the world as adults?

J U N E 6 . Each dawn before breakfast I walk a tenth of a mile down the lane to the road for the morning paper. During these spring days, each walk yields its own special moments of interest. In the sunrise on this morning I stop to watch a rabbit farther down the lane. It halts, moves on, then halts again. It is sampling the vegetable smorgasbord of plants beside the wheel tracks. It tries a bud here, a leaf there. It nibbles on a dandelion. It makes a couple of hops and settles down to feeding on new fare. For a cottontail rabbit in the spring, all outdoors is a banquet hall. It appears to be working its way through a seventy-course meal.

I advance slowly. It moves on down the lane ahead of me. Without showing alarm, it keeps its distance. Then I quicken my pace. It watches me for a moment, then bolts into the weeds, seeking a rabbit's safety in flight.

More than once in this record of our outdoor days I have had occasion to comment on the inoffensive, live-and-let-live character of the cottontail and on the acceptance by other wild creatures of its peaceful intentions. But there is another side to the rabbit's nature, as was demonstrated a year or so ago on another farm half a mile from here. The pacific cottontail, so easily put to flight, becomes a doughty defensive battler when face to face with a great emergency.

At the edge of the yard on this farm, a mother rabbit had made her nest. One day, as the owner of the farm looked that way, he saw the cottontail catch sight of a large blacksnake weaving through the grass toward the fur-lined depression containing her young. On the instant, she bounded to meet the advancing serpent. When she was almost upon it, she leaped into the air, passing low over its body and lashing out with her hind feet. The serpent writhed and struck blindly. Again and again the cottontail attacked, raking the serpent's body with the claws of its hind feet. Long red scratches appeared on the shining length of the blacksnake. Whenever it swung toward the nest, the cottontail—usually so timid, usually so peaceful—renewed her hit-and-run attacks. They ended only when the bleeding reptile crawled away—leaving the baby cottontails unharmed within their nest.

J U N E 7 .   Sunshine filters down through the new leaves. It warms the mossy rocks and the woodland mold. We have been sitting here for some time where Hampton Brook winds through the Far North Woods among skunk cabbage and hellebore, under overleaning trees, along quiet stretches where the transparent water slides above a sandy bed. Idly we watch the water striders drift or skate across its surface. This is the good time of year. These are the Elysian days, the days we dreamed of in January and February and early March. Around us stretch the calm woods and the tranquil hills. The sun shines. The brook flows. The June air is redolent with earth perfumes. We seem in a woodland paradise where all is peace.

Yet even here the calm is transitory. No lasting peace is anywhere on earth where life exists. Each living thing has its foes; each creature lives a mortal life. Each nestling bird, each branch in a tree competes with its kind for food. For all, danger in many forms lies waiting, temporarily sleeping perhaps, but never gone entirely. The beauty we see in this time of bird song and flowers is the beauty of *form* and *color* and *sound*. The spirit that extends through all nature is one of never-ending competition, of parasite and predator, of a shifting balance of power, of harmony achieved through discord. Everything alive, plant and animal, spends its days in a world set at odds, surrounded by perils, competing, each in its own way, in a realm of strife.

Like the general deploying troops before a battle, thinking not in terms of individuals but in terms of divisions and armies, nature is concerned with classes and genera. It is up to the individual to survive as best it can. Nature is not a mother concerned for the welfare of each member of a family. Nature is not friendly or even well intentioned toward the individual. Nature is neutral. But if disaster to the individual is not averted, neither is it planned. If there is no compassion, there is also no malice. There is only the working of natural laws.

An apple falls from a tree and lands in the grass. It crushes an ant and misses another ant. The apple is directed on its downward course only by natural laws. It is pulled by gravity, perhaps deflected slightly by the wind. The dead ant was not the target of its fall. No intention was involved in the course it took. One ant is killed and one escapes. But there has been no setting aside of natural laws to bring about the destruction of one and the deliverance of the other. All the ants in the world, all the hawks and rabbits and men and wild flowers, exist amid the same natural laws. They are surrounded by the impartiality of nature.

We are on earth under certain conditions. We adjust to those conditions to survive. To live is to be in peril. It is up to the individual to be alert, to avoid injury and death, to survive as long as it can. Nature looks on without concern if you step on a rotten branch, fall out of a tree, and break your neck; if you walk out on thin ice, break through, and drown; if you eat poison mushrooms and die—just as nature displays no interest in the mouse caught by the weasel, the rabbit surprised by the fox, or the bird that falls into the talons of the hawk. It is up to us—to all living things—to man and mouse and rabbit and bird to be on guard.

Within the framework of conditions as they are, each living thing comes into being and spends its certain time on earth. It lives among hazards. It exists surrounded by violent death. Wherever we walk in nature, one form of life is consuming another form of life and being consumed in turn. Carnivorous parasites, hidden from sight, are devouring the tissues of their living hosts. Predators are lying in wait. The red fang and the bloody claw are as much a part of nature as the drifting cloud, the rainbow, and bird song. To ignore this fact is to distort the truth. To see only the beautiful, only the pleasant is to walk through the out-of-doors wearing blinders. But the same is true for those who find only the violence, the bloodshed, the suffering in their contacts with nature. There are both the terror and the pain and the brightness and the beauty in nature. Both must be seen and recognized. Neither can be ignored. Nature is not gentle but there are warm and gentle breezes. For even the most timid and hunted of creatures, there are moments of peace and natural pleasure.

Time is made up of fragments of time; the hour is a mosaic of moments. What we observe of tranquillity or tension in any given glance during any given moment never tells the whole story. If the peace we see, as we sit beside this woodland brook on this day in June, is a temporary peace, the terror we witness is also a passing thing. Both are part of the everchanging whole. The harmony of such a sunny hour as this, this seeming truce around us, is not an illusion. It is a moment of reality as much as the moment of strife.

J U N E 8 .   Remembering my thoughts of yesterday as I walk through the woods today, I think again of the ant and the falling apple. Its death results from the functioning of natural laws. This is clear. This we see. But why *that* ant? Why *that* individual instead of the other? This is no concern of nature's. But it is the all-important concern of the individual. The laws of nature function and nothing interferes either to aid or destroy us. Yet one is aided and one is destroyed. We call this luck or chance or fate.

I recall a small item I once read in a newspaper. In a lonely part of a southern state a man became lost in the woods. Late at night he came to an isolated tavern beside a road. All was dark. He pounded on the door. Only a few nights before the tavern had been robbed. The owner put his head from an upper window and shouted for the man to get away from his door. The hammering continued. He shouted again telling the stranger to get away from his door or he would shoot. Getting no reply, he shot. The man at the door was killed. Then it was discovered he was a deaf mute.

At a ferry slip on Staten Island in New York Bay a long line of cars once rolled down an incline toward the ferry. At one point the mechanism holding the boat in place slipped, allowing the craft to drift outward. A gap suddenly yawned in front of one of the cars containing an elderly couple. Their car plunged into fifty feet of water and both were drowned. The failure of the mechanism resulted, understandably, from mechanical causes or human error. This explains why *a* car fell into the bay. But it does not explain why it was *that* car, in a long line of cars, why it was *that* couple among all the people involved.

Everything that happens happens according to natural laws. But over and over again we ask: Why does it happen to one and not to another? There is nothing planned, nothing premeditated. It is a matter of time and space and movement. It is being at a certain place at a certain time. It is happening to stand where two lines meet. But why are we there? Why me?

The only comforting thing I have ever heard is a saying of the ancient Greeks: "Fate can control everything; but fate does not control everything." In other words, at any time fate can step in and spoil the little sum we are adding up. But it does not always step in. We have a certain latitude between natural laws, which are inflexible, and fate, which is unpredictable. Within that latitude, we can influence, by our own efforts, what befalls us and what we become.

J U N E 9.   Two members of the iris family, one deep violet-blue, the other a lighter, clearer shade, bring beauty to opposite ends of the pond this morning. All across the dam, I am in the time of the blue-eyed grass. The path I follow at the water's edge is bordered by a scattering of the unassuming little flowers, half an inch across, each six-petaled, each golden-centered, each rising on a slender, grasslike stem. There is something friendly and companionable about this small blue and gold bloom of spring and early summer. This is a reaction many people have felt. It is reflected in one of the simple common names bestowed on this flower: the blue-eyed Mary.

At the opposite, the western, end of the pond, all down the curving declivity where the seepage of intermittent Woodcock Brook provides abundant moisture, the outstanding feature of the landscape is a wide, descending cascade of blue, the lighter violet-blue of the fleur-de-lis, the blue flag, *Iris versicolor*. When I stand in the midst of this river of flowers a little later on, I find the larger blooms are as much as five inches across, a diameter ten times that of the demure blue-eyed grass. Each blue flag rides at the top of a slender stem rising to a height of two feet or more amid the green swords of the elongated leaves.

All these graceful blooms around me appear poised, balanced, as though they had alighted on the tips of the stems. As I am thinking of this, something flutters down from the air, something that provides a special accent to the floral beauty about me. A black and yellow tiger swallowtail butterfly drifts past, turns, sweeps down with slender legs extended and comes to rest clinging to one of the blue flowers. There for a time it hangs, its wings outspread, basking in the sun.

As I sometimes do, as I complete my circuit of the pond I turn over an occasional flat rock to see the hidden life beneath. Here it is a brownish wolf spider, there two black field crickets, again a colony of tiny ants. It is in this way that I discover living beauty of a new and striking kind. What I see is a cricket. But it is no ordinary cricket. Its body is a rich red; its wings a creamy white. So far as I can remember I have never seen one of these insects so brilliantly marked. But somewhere in the back of my mind, the sight—as the saying goes—rings a little bell. I recall a professional entomologist once describing such a combination of colors. The memory of his explanation of the phenomenon throws light on this strikingly beautiful insect surprise I have uncovered. In some instances, he told me, after they attain their wings in their final molt, black field crickets have brighter tintings for a day or more before they assume the polished patent-leather blackness of their normal hue.

J U N E 1 0 . Probably I will never do it again. Certainly I have never done it before. I have just pulled a chipmunk's tail. It is hard to say which of us—the chipmunk or I—is more surprised. It has happened this way.

Leaving the pasture, as we come home, Nellie and I draw near to the apple tree beside the terrace wall overlooking the slope to Hampton Brook. From a lower limb we have suspended half of a coconut shell to hold sunflower seeds for chickadees and nuthatches during the winter. This year we have continued feeding on into June—a fact that has not escaped the eye of one of our chipmunks.

As I draw close, I notice that its head is invisible, thrust down inside the shell. But its tail is dangling down on the outside. On an impulse I creep silently toward it and, never expecting to succeed in my intention, give the hanging tail a little tug. The chipmunk blasts out of the shell, scattering the sunflower seeds. Its wild leap into space carries it to the grass below. There it bolts away and vanishes in a crevice in the wall. In all the long history of Trail Wood, it no doubt will be the only chipmunk ever to have its tail pulled by a man.

In these June days, the activity of other chipmunks is all around us. We watch one run toward the house, leap upward in a kind of flying wingover, and snatch a small hairy caterpillar from the clapboards. From time to time, these rodents add meat of various kinds to their mainly vegetarian diet. In the slowly fading light at the end of another day we were surprised by the unexpected actions of a chipmunk on the terrace. We saw it dig a little hole in the ground and then, like a gray squirrel burying a nut, hide a discarded prune seed. After the hole was filled, it patted down the earth carefully with its forepaws before it dashed away.

The family of young chipmunks that we watched breaking home ties so short a time ago is now scattered. The same animals we saw timidly making their initial contact with the world aboveground now streak this way and that, filled with life, advancing in sudden explosions of energy. On the flat stone at the end of the wall near the apple tree, where some of the remaining birdseed is sprinkled, one of the parents is feeding amid cowbirds and bluejays. Two of the young chipmunks arrive to join in the feast. There is a sudden whirl like pinwheels under the flutter of the alarmed birds. It ends with the young animals in flight and the old one sitting in the middle of the rock resuming its eating of grain. Chipmunk parents are neither affectionate nor indulgent once the young have left the burrow where they were born. As soon as they are well launched aboveground, the new chipmunks are on their own. Home ties are broken and the family falls apart.

JUNE 11. Nellie has just come up the long lane from the mailbox bringing more than mail. She arrives with news of something special for me to see, something to record on color film. Down the lane we go, my camera slung over my shoulder. We cross the bridge, pass the mulberry tree, leave Wet Weather Brook behind. We have almost reached Kenyon Road when we stop beside the wall that separates the lane from the triangular patch of wetland we call Pussy Willow Corner. There Nellie turns me loose to discover what she has seen. When I find it, it is worth the search—a vision of double beauty, the beauty of the plant and animal worlds combined.

Against the background of the lichened wall, a large jack-in-the-pulpit lifts its green spathe, striped with delicate vertical darker lines, to a height of more than a foot above the loam. Clinging to it, its wings outspread, its long ribbon tails unrolled, its twin antennae thrusting up like miniature fern fronds, a graceful luna moth has completed its transformation. The ethereal green of its coloring seems to glow in the morning light. Perfect, unblemished, it hangs without a tremor, its wing hardening in the air, while I record my pictures. And so it will hang during all the hours of the day until the sun sets and evening comes. Then, when dimmer light fills the lane, it will lift buoyantly into the air. On untried wings, with ribbon tails outstretched behind it, it will ride away into the dusk, its active aerial life begun.

The finding of the luna is the great event of the morning. But toward evening another quiet adventure, coming in the calm of sunset, adds to our enjoyment of the day. We are sitting on the plank bench beside the brook below my Insect Garden watching those slow-flying insects of the watercourse, the alderflies, go bobbing past on their light-and-dark-patched wings. These insects not infrequently are mistaken for damselflies. They both appear weak and uncertain on the wing.

Upstream, above the slow current of the pool impounded by my rock dam, little honey-colored mayflies rise and fall as they lay their eggs in the water. From head to tail, they are hardly more than half an inch long. Continually they drift in our direction. I sweep my hand through the air and catch, for a moment, one of the winged dancers. When I release it and it flutters away, I notice an orange-hued sphere, so minute it is almost invisible, clinging to the tip of my forefinger. Something new, something never before experienced, has happened to me. The mayfly has deposited one of its innumerable eggs on my hand. I have, for a moment, taken the place of the brook.

JUNE 12. Standing beside the mound of flat rocks that forms the hired man's monument on this pasture vantage point beyond the brook, I listen to the singing of our two red-capped sparrows of the open fields. One is a chipping sparrow, the other a field sparrow. The chipping sparrow rides on a mullein head. The field sparrow clings to a low bush. On and on goes the dry, one-pitch song of the chippy while the clear-toned trill of the field sparrow ends and begins again.

In the course of the years at Trail Wood, we have become acquainted with many individuals among these small and common sparrows. Of them all, two have made the deepest impression. The field sparrow we saw in a moment of rare beauty; the chipping sparrow in a time of stress and danger.

It was in another spring dawn, when the level rays of the sun shot across a dew-laden pasture, that I came upon the field sparrow. Looking toward the sun, I glimpsed the little bird, with its pinkish bill and reddish cap, perched in a grass clump, surrounded by a glitter of dewdrops. But they were no ordinary globes of moisture shining with a white diamond light. From my viewpoint, all the droplets spangling the slender leaves acted as prisms. They blazed with colored light, with intensely brilliant blues and reds and greens. For that moment, the modest little sparrow seemed the centerpiece of some priceless ornament. It appeared set in the heart of shining, varicolored jewels.

It was under one of our apple trees, near the end of a particularly hot afternoon in early summer, that I encountered the chipping sparrow in its time of crisis. It materialized suddenly in the grass almost at my feet, a young bird from a nest on the lowest limb of the tree, a nestling still unable to fly. Leaving its home too soon, it had fluttered to the ground. It looked about, gave a little hop, and gaped wide its yellow-lined mouth, begging for food.

There followed an absorbing and dramatic sequence. A dozen feet away, a tangle of thick grass and weeds hid the base of a stone wall. It provided the only protection nearby. I stepped back beside the tree trunk and almost at once one of the parent birds alighted with food in its bill. The spot where it descended into the grass was in front of the young bird but a foot or so nearer the wall. The fledgling hopped toward it to be fed. Then the parent took wing and, a few minutes later, returned with more food, landing again a little farther toward the goal. Once more the vulnerable little bird hopped in its direction. In this way, as I stood watching, it was moved in a series of short advances until it had attained the sanctuary of the dense weed cover.

Always associated in my mind with the chipping sparrow and its parched, stacatto song, repeated so often in the heat of early summer days, is an eloquent paragraph in William T. Davis's evergreen book of nature rambles, *Days Afield on Staten Island*. It runs as follows:

"A man who concerns himself principally with the artificial, and who thinks that the world is for stirring business alone, misses entirely that divine halo that rests about much in nature. To him all things are certain. He can have a particular tree cut down or an ox killed at command, and he is ever busy spinning a web of affairs. You see him hurrying across the street with rapid strides, for hasn't the Valley Railroad declared a dividend? Such things must be, but they are not the safest springs of pleasure. We must not put by entirely the chippy singing in the apple tree, or the white clouds, for Nature declares a dividend every hour— and dew drops always pay par to the summer leaves."

JUNE 13. A green bowl accompanies me down the slope to the edge of the pond this morning. I am off to gather wild food for our breakfast.

At Cattail Corner I halt beside a stand of the green sword leaves and upright stalks. At this season of the year each stalk supports two closely packed masses of flowers. The lower, resembling an elongated greenish-brown sausage, comprises the female flowers. The upper, now yellow with the pollen that will descend in a fertilizing shower over the pistillate blooms below, is formed of the male flowers. The familiar brown cattail heads of fall and winter, with their thousands of densely packed seeds with silken filaments attached, result from the fertilization that is taking place on these June days.

Leaning out, I tilt a stalk to one side, hold the bowl beneath, and shake the yellow pollen into it. Moving from head to head, I make a harvest of the floral dust. Before I stop, more than a cupful fills the bottom of the bowl. Mixed with batter, it will contribute to a special delicacy of this time of year—cattail-pollen pancakes. Served with maple syrup, they possess a delicate haunting flavor vaguely suggesting corn fritters.

Whenever we awoke last night, we heard the hoarse foghorn bellow of the male bullfrogs. Who would think of describing it as a siren song or a call of enticement? Yet each of the males, yellow-throated now in the breeding season, is in truth singing a raucous lovesong at the water's edge. It is the male that determines where the eggs will be laid. Through their stentorian, unmelodious voices—whose sound has been variously described in such words as "Be drowned, better go round" and "Jug-o-rum, more rum"—the males are attracting the females to them.

Now as I stand for a moment breathing in the cool, moist air of the June dawn, I glimpse the result of the batrachian din. Among the aquatic plants a dozen feet away, I see a thin, outspread, gelatinous film floating near the surface of the water. This layer of jelly, nearly two feet across, may well contain 10,000 or more minute bullfrog eggs.

I set down the bowl in the midst of a patch of clover and walk along the path for a closer view. Beyond, the shine of another mass of eggs catches my eye; beyond that a third. Following the water's edge, I count six such masses. Thus in a single night, in this one pond, half a dozen batrachians have released more than 50,000 bullfrog eggs—possibly more than 100,000, for as many as 20,000 have been counted in one mass of the floating film.

As I am turning back, I pause to watch four or five of the second-year tadpoles clustered at the edge of a stand of waterweed. There is a flurry of movement. A streamlined lizard-formed shape shoots from among the aquatic vegetation. I recognize it as the rapacious larva of the *Dytiscus* diving beetle. The adult beetles, also predacious, catch prey as large as small fish. On this occasion, the swiftly launched attack of the water tiger— with its sickle-shaped mandibles ready both for gripping its prey and sucking the juices from its body—fails. The tadpoles scatter and escape. The aquatic predator retires once more into the waterweed tangle. Hanging head downward, breathing through the tip of its tail thrust above the water, it remains motionless, awaiting the approach of other prey.

So time slips by as I loiter along the path on my return to the green bowl and its pollen harvest. Which accounts for the fact that this morning we breakfast late on our cattail-pollen pancakes.

J U N E 1 4 . In the gusty, unstable wind of this hot June day, we watch our two red-shouldered hawks pitch and veer, tilting violently in the invisible turbulence of the air. They ride the element as though mounted on bucking bronchos. But they do it effortlessly. They counter the forces of the wind with the speed and skill of expert fencers. This is the realm for which they were designed. They seem exulting in their powers, using to the utmost every sense, every inborn ability. To me, standing earth-bound, how free they seem, how unfettered, how wild and how alive.

It is wildness that elevates the senses. It is tameness that downgrades them. Most of the human beings of the world use their senses fitfully, incompletely. Rarely do we meet an individual as fully alive as a fox that has just heard a mouse in a grass clump.

Watching the freedom of the hawks in the ever-changing wind, I recall another hawk—a bird removed from the element for which it was destined when it hatched from the egg. Nellie and I had pulled into a roadside attraction near the Suwannee River in Florida, when we were coming *North with the Spring*. Among the so-called attractions were several cages covered with chicken wire. One, about five feet long, three feet wide, and two feet high held a captive marsh hawk. Its world had suddenly shrunk, had been compressed from the whole sky to the limits of this one small boxlike cage. Over and over, the bird fluttered up against the wire above it. Before we drove away, sickened by the sight, we saw how the captive hawk had rubbed all the feathers from the top of its head in its hopeless efforts to escape.

The unpleasant impression made upon us by caged creatures is augmented by the realization of how they have been robbed of almost all opportunity to enjoy the natural play and pleasure of their senses. The wilder the creature, the more extensive the area over which it roamed, the more far-reaching its use of land or sky, the more the unpleasantness manifests itself.

I will never erase the image of one coyote caged in a Salt Lake City zoo. Nellie and I were traveling west through the autumn when we spent half an hour or so among the creatures on exhibition. Most were lying quietly. They appeared to have given up, to have become reconciled to their captivity. But among all wild creatures, as among men, there are a few who never give up, who never become reconciled, who keep on fighting to the end.

Such was a leopard I once saw in a cage where big cats were being trained for circus performances. Whenever anyone approached, it hurled itself at the steel bars, crashing head-on in its fury. It never became tractable. It never made peace with its loss of freedom. I heard later that it literally committed suicide, endlessly plunging headfirst against the steel bars of its cage until it died of its injuries.

And such was the spirit of the coyote we encountered at the Salt Lake City zoo. Imprisoned in a small cage beneath two leopards, perhaps terrified and made desperate by their nearness, never forgetting its former freedom, it chewed and tore with its white perfect teeth at the heavy wire front of its cage. It left off this only to begin a digging motion with its forepaws along a smooth pipe that had been placed at a descending angle in the cage. Perhaps this was the only place where its paws would slip, giving it the feeling—the futile yet comforting feeling—of digging free. The wood all around the pipe was grooved and scarred by its frantic efforts. Without resting, it continued its digging motions in a kind of frenzy. Then it took up biting the wire again. We turned away sorry that for even a moment we had patronized the front of its cage. An animal desperate for freedom, shut away from all that means life to it, provides no fit spectacle for any human being called civilized.

J U N E 1 5 . The hot midday breeze flows around me as I lean, as I so often do on these sunny days, on the weather-polished rail of the bridge over Stepping Stone Brook. The moving air is laden with a spicy fragrance. Drifting across dense stands of sweet fern along the little overflow stream, it arrives redolent with the rich scent of the heated leaves.

For a time I remain here, basking in the sunshine, enjoying the wild scents, watching an amber-hued dragonfly dipping its tail in the water as it lays its eggs in the pond a dozen feet out from Driftwood Cove. From a dead limb thirty feet up in a pondside maple a kingbird pitches down in a long plunge and snatches the insect from the air. I hear the snap of its bill and see the glitter of sunshine on the vibrating wings of the dragonfly disappear from above the surface of the pond.

Its life is over. But in its final minutes it has more than balanced nature's books. Like the toad I saw last month extruding its strings of eggs even as the water snake swallowed it, the dragonfly was insuring the perpetuation of its species. The end of its life has followed, not preceded, that act.

Three or four days ago, I watched while a butterfly met a similar fate—but with a difference. I was coming down the slope toward the old apple tree at the beginning of Veery Lane when a spicebush swallowtail, immaculate, apparently just emerged from the chrysalis, drifted by in the sunshine. I watched it rise and clear the tree. There was a flash of yellow; once more the sharp snap of mandibles. A great-crested flycatcher rose to a higher treetop, the large insect fluttering in its bill.

In an infinitesimal way, the life and death of the butterfly, unlike the life and death of the dragonfly, formed a subtraction rather than an addition to the ledger of nature. It had had no time to provide for the perpetuation of its species. It had died before it could contribute another of its kind to take its place.

J U N E   1 6 .   The mystery of the strange little globes, the lacelike cocoons on the curly dock near the sundial of my Insect Garden, is solved at last. A week and a half have passed since I first noticed them beside this winding path. This morning I come to them again. At once I observe a change. Activity envelops them. There is constant movement among the dozen or more brownish spheres at the tip of the highest stem of one of the dock plants. Ragged holes are being eaten through the network of crisscrossing lines of solid material forming the sides of the globes. From these openings small grayish beetles are pushing their way to the freedom of adult life. Each is preceded out of its hole by a long, down-curving snout. So the identity of the mysterious makers of the cocoons is revealed. They are tiny snout beetles or weevils.

Vast and varied are the members of the suborder to which they belong, the *Rhynchophora*. There are pea weevils, bean weevils, rice weevils, coffee weevils, granary weevils. Some of these insects live in the hips of roses. Others lay their eggs in the staminate flowers of coniferous trees. One whole group specializes in fungi. The larvae of some live in the smut on corn and wheat. There are leaf-rolling weevils and potato-root borers. The worms of wormy cherries are the larvae of a fruit weevil. Most famous of all these agricultural pests is the cotton boll weevil of the South. The little insects I see, clad in their somber gray, pushing their way through the doorways they have eaten in their cocoons belong to the genus *Hypera*.

I see the small emerging insects creeping down the stem of the dock and disappearing among the leaves. They are using their legs as expertly at the beginning of their adult lives as they will at the end. I watch them go, wondering what hazards they will run, what adventures they will meet. The chances are I will never know. In fact the chances are that—as was the case with the orb weaver's spiderlings—as I watch them disappearing into the surrounding jungle of the weed tangles, I am seeing them vanish from my life completely. Infinitely remote is the possibility that I will ever again encounter any of these inconspicuous creatures, diminutive and gray-clad, among the rank vegetation of this unkept natural garden of mine, where the insects are welcome.

J U N E   1 7 .   When I made this path circling the pond close to the water's edge, I made it for wider use than I anticipated. It is not only our path, it is also the path of raccoons and foxes and skunks and crows and grackles. It is a duckling path as well as a man path. When I look across Whippoorwill Cove, I see a half a dozen baby mallards strung out single file behind a duck that is leading the procession. They all are plodding along, out of step, toward that favorite mallard resting place, the eight-by-eight-and-a-half-foot flat surface where Summerhouse Rock juts out into the water.

In these June days feathers are strewn around the rock. For this is the time of year when the drakes are molting, going into the eclipse plumage of their flightless period when their brilliant colors are lost and they resemble the females. Out toward the center of the pond, half a dozen of the light body feathers of the molting ducks run along the surface in the breeze. Each trails a tiny wake behind it. Sometimes in a freshening breeze, several take off like living things, ride through the air, and then alight again. Above them, plowing back and forth in their feeding, pass and repass a trio of barn swallows. More than once, as I watch, I see one swerve, dive, snatch up a drifting feather, zoom up, drop the feather, swoop, and pick it from the air as it drifts downward.

And all the time, the flying swallows call back and forth as though in amiable conversation. Few birds appear more sociable than these swallows. I have spent hours in their company when they have nested in our middle shed. The male accompanies the female while she gathers mud and builds the nest. Oftentimes as the female broods the eggs, the male rests close by on the same beam that holds the cuplike structure. While I have worked at odd jobs in the shed, I have heard the soft twittering of the two engaged in a kind of continuing dialogue. The companionable swallows!

JUNE 18. Fireflies drifting over the weed tangles, a lighted window in the house behind us, the stars above, and our two flashlight beams probing before us as we advance—these are the only sources of light we see. All the rest of this familiar scene about us is eclipsed by the night.

After the long June twilight has ended and before the rising of the moon, Nellie and I are on our way down the slope that leads to Hampton Brook to see what we can spotlight along the edges of the stream.

These twin beams of illumination wander down the path; they race through the interlacing stands of weeds and orchard grass on either side. A flick of a wrist and my beam plunges away before us. Two brilliant spots, side by side, flare up in the darkness. The eyes of a nocturnal rabbit have caught and reflected the light. It is probably the same unalarmed cottontail that I encountered in the dawn along the lane earlier this month. Enclosed in the darkness, it is even more calm. It permits us to come closer. It hops only a yard or two, squats to watch us advance, then moves on a few leisurely jumps before it pauses again. Only when we have followed it almost to the brook, when it finds itself pinned between us and the stream, does it swerve abruptly and go bounding away, the rising and falling white of its tail disappearing along the path toward the waterfall.

With our flashlights switched off, Nellie and I sit for a time on the bench beside the little cascade. Walled in by the dark, we listen to the water music of the stream rushing over and among the stones. This is the song of the brook. Rocks are the instruments on which the flowing water plays.

Then with flashlights snapped on again, we work slowly upstream. The sound of the cascade recedes behind us. It is replaced by the low murmurs, the small gurgles of the more smoothly running water. At times the chorus of the gray tree frogs—a kind of slumberous, rocking-chair repetition of sound— swells in the night, then fades away. My beam plunges into a small pool among the rocks where dark little dace are swimming. It adds a glistening coat of light to a green frog settled on a pad of moss at the stream's edge. It catches half a dozen water striders motionless on the surface film of backwater below a decaying log. Nellie swings her light up the opposite embankment of this little valley that the flow of the brook has carved out of the land. It comes to rest on something white and shining among the weeds— the foam nest of a froghopper.

So we move through the night. With beams that wander and probe and pause, we continue this exploring in a world that is compressed into two small round- or oval-illuminated areas—the little stages on which shapes and movements occupy our sudden spotlight.

J U N E 1 9 .   A stick floats at the edge of the pond between Twin Rocks and the rustic bridge. Like Friday's footprints in the sand of Robinson Crusoe's island, it reveals a presence unsuspected. No more than an inch thick and hardly three feet long, it attracts my eye this morning as I pause with a basket filled with books and manuscripts on my way to my writing cabin among the aspens.

The stick is white. Its naked wood shines out in the darker water. When I fish it out, I see it is clipped off as though with the sharp blade of an ax. And all along its length toothmarks reveal where the smooth, greenish bark of an aspen or "popple" sapling has been gnawed away. Along Azalea Shore I come upon three other stranded barkless sticks.

Just down the slope from the house, while we slept last night, a wandering beaver has explored our pond under the cover of darkness.

Although the animal came and went unseen, I can tell the age of our nocturnal visitor. It is in its second spring. When young beavers are two years old, they are crowded out of the community and set out in search of a new home. Wherever one of these roamers goes it carries with it the skills of its race. And those ingrained, unlearned skills are considerable.

Years ago a lake was planned in the Hampton area that now

forms the James L. Goodwin State Forest. There was considerable debate over where the dam should be located. Before a
decision was reached, beavers moved into the lowland stretch the
lake now occupies. They constructed a dam of their own. As it
turned out, these wild engineers had chosen the best site among
those being considered. It was there that the large, permanent,
man-made barrier of earth and concrete was erected.

The thought of a beaver felling an aspen sapling and feeding
on its bark within sight of our bedroom windows adds a thrill of
wildness to this day. I can picture the lone wanderer coming in
the darkness down Hampton Brook from the beaver pond in the
Far North Woods. I can picture it skirting around our waterfall,
moving against the gradual flow across a swampy lowland, reaching Stepping Stone Brook, and ascending its short run to reach
the pond. In all likelihood this may be its only visit. Our pond, so
close to the house, I conclude—somewhat hastily—could not be
to a beaver's liking.

With a tinging of regret, I notice that its one visit seems satisfactory and the fact that probably it will not return to stay seems
fortunate. I have visions of beavers taking over the pond, dam-
ming up Stepping Stone Brook, raising the level of the water,
flooding out our footpath, felling aspens around my cabin. I continue on over the bridge and up the slope reflecting on how
wildness and our possessions are so often at odds.

We plant a vegetable garden and almost at once the little rabbit we have watched with delight becomes our enemy. We set
out a patch of sweet corn and the raccoon, whose life we have
shared with scraps of food in fall and spring, takes on the character of a burglar in the night. We cultivate raspberries or blueberries and the catbirds and robins become our competitors. Our
possessions alienate us from the life of the out-of-doors. Who can
own anything and still be completely a part of nature? Are not
those who have the truest, the most unbiased outlook on nature
those who have only a rambler's lease, those who, like Thoreau,
wander over land they do not own?

J U N E 2 0 .   Under the deliberate ebbing of the sunset, this
last of the spring days comes to its tranquil end. The air is so still
in this transition hour, we catch the faint, faraway jingling of bob-

olinks from a meadow beyond Kenyon Road. We hear a wood thrush singing in the dusky glen below Whippoorwill Spring.

Looking toward home from the higher ground of Nighthawk Hill, we watch the aerial merry-go-round of half a dozen chimney swifts gyrating about the house. The cheerful crackling sound of their incessant calling carries across the field. When we look through our binoculars, we see songbirds active all across the newly cut lawn—bluebirds, robins, song sparrows, and brown thrashers gathering food for nestlings and flying away with bills crammed with what appear to be small green worms.

We find ourselves breathing slowly and deeply, savoring the sweet June air. In this hour of change, this hour of transition from spring to summer, we have the impression of being set in the midst of tranquil skies, tranquil earth, tranquil everything. It is such an ending of a day as Margaret Craven describes in *I Heard the Owl Call My Name* when the leading character in her story walks at a river's edge clinging to a lovely day, repeating to himself: "Don't go—not yet—not yet—"

But, as the author says, "the day slipped away as fast as any other." So in this time of fading tinted light, we, too, wish to cling to this day and to this spring. But the light fades and the spring goes. Before we awake in the morning another season will be here.

THE
WALKS
OF
SUMMER

JUNE 21. In the long living room, the wall clock that for more than eighty years has been striking the hours chimes with its double note ten times. Nellie and I step outside into the soft darkness of the first of the summer nights.

Before dawn this morning, in a flick of time, the Summer Solstice came and slipped by. Season slid into season; spring ended and summer began. Now, in the warm and humid night, the lowland fields spread around us ablaze with tiny aerial lights. We are wandering in the midst of one of the great firefly displays of the year.

All across the Starfield, all over Woodcock Pasture, all down the slope of Firefly Meadow, beside the pond and among the trees along the brook, a winking, glowing glitter spangles the darkness. In this June night, who can guess how many billion fireflies are on the wing above the dark fields of eastern Connecticut? We walk out into the midst of the drifting lights. For a long time we stand halfway down the slope to the pond. The luminous insects trail past us. They arc over us, slant before us, alight on us. They glow about us, wink, drift, hang in the air, descend like falling sparks, and like sparks expire, but unlike sparks glow again.

The deep "trummmp" of a bullfrog carries up the slope from

the pond. Beyond the faint shine of the water, Azalea Shore rises in an ebony wall of darker darkness. And all along its length, the living lights weave shifting patterns of illumination, dots and lines that are reflected in their turn in the black mirror of the still water. Flashing on, winking off, flashing on again, the insects go drifting by. I reach out and scoop one from the air. It shines in the hollow of my hand. The glowing patch on my palm comes and goes. Then the light streaks away in a curving ascent. The firefly is on the wing, joining all those disembodied lights dancing above the darkened meadow.

Farther down the slope, where we glimpse vaguely white daisies in the dark, we stand beside a firefly imprisoned in a spider web. As long as we watch it, its light shines with a steady, unwinking glow. Has it been bitten by the spider?

Time passes swiftly here in the heart of the firefly display. June, with us, is the month when the luminous insects reach their peak of abundance. And the hours between ten and twelve, on nights moist and warm, is the time when the greatest number of the insects are in the air. The flashing on and off of their lights is more apparent than the dimmer glow of the earthbound females, wingless and creeping among the grass stems. We come upon several, their small, greenish-tinged lights set, like a lamp in a window, to catch the eye of the flying males. Signaling back and forth by means of light, the luminous beetles find their mates in the darkness.

Walking in the silence of the night in the midst of a thousand fireflies is an ethereal, almost spiritual experience. It is akin to losing yourself in the sweep of an aurora in the autumn sky. At times the whole hillside seems to shimmer with the myriad lights, as though waves of luminescence were sweeping across the darkness. In certain places they accumulate into concentrations like constellations of stars.

The little moving lamps shine on every side, above us and below. Looking up, we see them high over our heads, fireflies and stars intermingling. Looking down, we see them weaving just above the grass tops. They spangle the dark trees, drift and wink along the heavy bulk of the stone walls. They ascend above the tips of the ebony silhouettes of the ashes and maples along the brook to trail their moving lights against the unmoving pinpoints of the stars. When we look back up the slope, we see our white cottage glowing dimly under the dark trees surrounded by fireflies. In this year of abnormal firefly abundance, this is one of our magic times, one of our own Arabian Nights at Trail Wood.

First we hear the calling of a gray tree frog, then the mechanical clatter of an approaching helicopter. With its red and green navigation lights winking on and off, it passes overhead. What, we wonder, do all these fields below look like from the air? We envy the occupants of the passing sky machine gazing down on the outspread landscape where all the lowland fields are shot through and through with the intermeshing lights of these myriad insects weaving their intricate patterns in the night.

On and on the display continues. And always there is something new. One drifting insect flies into my face. We see another swooping downward, its light like a falling star. In other years we have walked among these insect lights when ground mist lay in veils along the hillside, in the moonlight, when heat lightning played beyond the dark skyline to the west.

And always at such a time, as tonight, we have responded primarily to the unearthly beauty of the scene around us, to this fairyland of insect lights. We have forgotten all about luciferin and luciferace and adenosine triphosphate—all about the complicated chemistry of what we see, all about the miracle of cold light, never duplicated in the laboratory, its riddle still unsolved. Instead we have given ourselves up to the magic of the soft night with its swarming lights drifting by in silence. This is our memory when, back enclosed by the walls of a house once more, we hear the old clock striking again. This time it adds two more chimes to its total.

J U N E 2 2 .   Late this afternoon, after an hour among the wild flowers of the meadow, Nellie comes home with an adventure to relate.

At the far side of the Starfield there is a dense stand of silvery-green narrow-leaved mountain mint, *Pycnanthemum tenuifolium*. A foot or more in height, the slim tapered leaves intermingling in a cloudlike maze, the plants, at this time of year, bear at their tops rounded masses of buds. In July, these buds will expand into white or greenish-white flowers with tiny lower petals reddish tinged. Although it gives off only a slight perfume, when mountain mint is in bloom it swarms with nectar gatherers. The air around the stand hums with the coming and going of bees. Wasps and flies of varied colors, glinting in the sun, alight and move about from floret to floret. Small butterflies flutter in and out among the flower heads. No other spot in this north meadow is the scene of greater insect activity.

After she had checked the flower heads to see how far the buds had developed, Nellie turned back along the path across the Starfield. Just before she reached the point where the trail to Ground Pine Crossing branches away to the west, she caught sight of a moving line of black and white descending the slope along the path toward her. A mother skunk was leading a procession of five baby skunks in her direction. The animals, approaching single file, appeared unaware of her presence. Usually a mother skunk makes such forays afield under cover of darkness. In this instance the family was abroad before sunset.

Nellie stepped aside off the beaten path ten or a dozen feet and stood motionless among the weeds. The skunk parade ignored her entirely. Advancing deliberately, the five little skunks strung out, trailing behind the adult, the family group went by. Like a spectator watching a procession marching past, Nellie saw the line of moving black and white animals descend the slope and, following the path, pass the mountain mint and disappear among the trees.

J U N E 2 3 .   Detours begin our morning walks on these June days. On our way to the woods or the pond or the fields we pause to enjoy our old-fashioned roses. They are now at the height of their blooming—damask roses, moss roses, musk roses, pink roses, yellow roses, red roses.

Pink are the flowers of the autumn damask, *Rosa damacena bifera*. It dates back to Roman times. Spanish missionaries who brought it to the New World called it The Rose of Castile. Scarlet-crimson are the petals of that other fragrant damask rose, the Kazanlik rose, *Rosa damacena trigintipetala*. Since before 1850 it has been grown in the Kazanlik Valley of Bulgaria, where it is still cultivated for extracting attar of roses. Buttercup yellow is the earliest of our roses to flower, Harison's yellow, *Rosa harisonii*, now past its main blooming. From 1830 on, it was a favorite of the American pioneers. It followed their westward movement. In California, the descendants of slips planted so long ago, now gone wild, mark the sites of ghost towns.

The present beauty of these old roses, their fragrance, and their long associations and histories blend together in these days of their luxurious blooming. Here is the Bourbon rose, Zepherine Drouhin, a semiclimber with its clear pink, its delicate scent, and its thornless stems; the moss rose, Salet, which blooms repeatedly, its flowers giving off a rich and musky perfume; the Portland rose, Jacques Cartier, fully double, pale pink, and intensely fragrant; General Jacqueminot, a hybrid perpetual, flowering intermittently with smooth and velvety petals, crimson-hued; and the common moss rose, *Rosa centifolia muscosa*, whose pink and scented flowers have been a feature of European gardens since before 1727.

And among these roses of yesterday there is one whose history we do not know—a bush beside a wall. It is covered with double blooms, fragrant, and beautifully pink. The individual flowers measure as much as four inches across. The petals themselves are deliciously perfumed. I first saw it as a boy, growing in profusion in my grandmother's garden in the dune country of Indiana. We call it Grandma Way's rose.

After we leave the fragrance and varicolored blooms of all these cultivated bushes behind and strike out across the fields on this morning, we meet another rose—a rose more ancient than the rose of Rome. With the coming of summer days, that lover of dry soils and rocky pastures, *Rosa carolina*, the wild pasture rose, is opening its five-petaled blooms, each petal purely and delicately pink. In the sunshine of summers long before the first Europeans landed on these shores, this wild rose bloomed as it blooms today. When I first meet it, each year, in the earliest days of its new season, its simple beauty impresses me anew. The pasture rose remains one of my own special favorites among all the wild flowers of the summer fields.

J U N E 2 4 .   Its name is *Conocephalum*. It grows across the brook from the bench below my Insect Garden. Without true roots, without true leaves, a plant more primitive than the ferns, it clings close to the ground, spreading in a dense carpet that suggests seaweed left stranded at low tide. The surface of each broad, flat thallus is broken up into polygonal areas. It has the appearance of being plated. It suggests the skin of a lizard. This feature is the source of the plant's more familiar common name: liverwort. A microscopic examination of a cross section of an animal's liver reveals similar plate or scalelike areas. "Wort" means plant, so liverwort means, literally, "liver plant." In more primitive times, this resemblance led to the wide employment of liverworts as an imagined cure for liver complaints.

On this afternoon, I cross the rushing water of the little cascade, stepping from stone to stone. When I bend down, pick and crush a fragment of one of the plants and hold it to my nose, I catch a rich, spicy aroma. The *Conocephalum* has still another name: The Fragrant Liverwort.

Always in moist soil, always near the banks of Hampton Brook, we have found this primitive plant in a number of places. Nellie first came across it in the lowland woods above the waterfall. Later, one year, I discovered a solid carpet extending over nearly a dozen square feet, just downstream from the bridge. The following spring, when the snow melted, abnormally high water scoured out all that area and carried the carpet away. I remember how richly the air was filled with its scent when I walked there before I had identified the plant.

A. J. Grout, in one of the later editions of his classic guidebook, *Mosses with a Hand Lens,* writes of this ancient plant: "When crushed the thallus of *Conocephalum* gives out a pleasant spicy odor with which as a boy I became familiar on trout-fishing trips. The source of the odor I never learned until after the second edition of this book was printed."

That little aside takes only forty-two words. Yet how pleasant it is to encounter in a guidebook on such primitive plants as mosses and liverworts. The juices of the living world seem to flow back into the desiccated specimens of the herbarium. It makes the reader long to see and hold and smell the fragrant liverwort himself.

Such older textbooks did much to stimulate our interest. The modern streamlined, more impersonal guidebooks are no doubt more efficient for those desiring first of all to know what they are seeing. They save time for the ones already interested. But something is lost—the warmth, the charm, the human asides of the former, more discursive guides. They, too, aided in identification. But they also contributed something more, something that seems to me important. They both identified the thing and aroused an interest in the thing identified.

JUNE 25. Winding across the green cushion of dew-covered moss and curving up and over the side of a low gray rock runs a shining silver path. It is the mucous ribbon trail a slug has unrolled in the night. All around it drops of moisture clustered on the moss glitter in the sunshine, catch and hold the light. This is the time of day in which to see at their best the small wild gardens of Trail Wood—the moss gardens, fungus gardens, fern gardens, lichen gardens, little secluded wild-flower gardens—each with its own appeal, its own particular beauty and interest. We search for them, blunder upon them, find them roofed over with fallen leaves.

It is while we are on such a search this morning along the upper slope of Juniper Hill that we come upon something else, something entirely different shining out amid dark green moss. It is round and metallic. It is also wet with dew. The lost has been found! What we have discovered is the pocket magnifying glass we dropped more than a week ago. The dews of many nights and the soaking downpour of two rains have harmed it not at all. I wipe it dry and put it in my pocket. The amplifier of tiny scenes, the doorway into little secrets, this small hand lens, so often used, is ours to use again.

JUNE 26. Without turning my head as I go past, I look out of the side of my eye. I note how the robin is perched on the edge of its nest in the midst of the lilac bushes. It is frozen in place, food in its bill. Apparently convinced I have not observed it, it remains unmoving until I have passed. If it had shifted position, if it had flown away, it would have revealed the location of its nest.

This is the event of a few days ago. This morning, when I come near, secretiveness is gone. Excitement surrounds the lilac tangle. The young are almost ready to leave the nest and the two parent birds are alert for predators. I watch one of the robins pursue a marauding grackle from tree to tree in the direction of the water-fall. Birds know birds. They recognize the harmless and the dangerous ones. Grackles are notorious as nest robbers and killers of young birds. Sometimes I see both robins making a concerted rush when a grackle comes too close. The three birds—one dark and long tailed laboring ahead and two robins in close pursuit—disappear among the trees. But most of the time, the robin gives up after a short chase and I see the pursued grackle make a wide detour and come skulking back again.

Of all the birds, with their manifold shapes and colors and habits, the grackles are for me the hardest to like. Their harsh calls grate on my ear. Their pale eyes repel me. Their flight is heavy and labored. Compared to the smooth, graceful buoyancy of the swallow, it is like the jolting progress of a laden cart. Their dark plumage—except when glinting and iridescent in brilliant sunshine—adds little to their appeal. Their predatory habits and their destruction of the young of other species are repulsive. They are without grace of movement, warmth of eye, charm of voice, or appeal of habits. Yet in recent years these dark birds have increased steadily while more attractive species have declined.

Beauty of form and voice and eye are not the primary considerations in survival. The industrious feeding and the omnivorous appetites of the grackles undoubtedly have played a major role in their abundance. Because as part of this feeding they consume insects injurious to crops, their increase can be considered a plus for agriculture.

J U N E   2 7 .   For a quarter of an hour, I have been sitting on the stone bridge spanning Hampton Brook. Heat presses down. Birds are silent in the midday lull. The stream murmurs and gurgles, washing over the rocks below me.

What has kept me here is something occurring on a leaf. Within easy reach of my hand, where an elderberry bush leans out over the stream, one of the leaves is thatched with a mass of a hundred or more slanting, brownish, elongated eggs. Each ends in a small tubercle—that sign that the eggs were laid by one of those primitive nerve-winged aquatic insects, the *Neuroptera,* of the watercourses.

As I watch through the same pocket lens I found on the dew-spangled moss of Juniper Hill two days ago, I see tiny lizardlike creatures emerging from the cylinders of many of the eggs. First to appear are the massive heads with their twin caliperlike projecting jaws. One by one the larvae wriggle free, let go their hold on the leaf, pitch downward through the air, and disappear in the rush of water below. What I am observing is the beginning of life for the orl- or alderfly. On the hottest of the summer days, I see this insect, in its adult form, bobbing along in slow, fluttering flight, always above the water path of the brook. It is most often at midday that the females deposit their eggs, usually selecting a leaf that overhangs swift-flowing water. Now, in this subsequent midday hour, the eggs are beginning to hatch.

Between that hatching and the final transformation into the adult, the lives of these aquatic larvae are spent swimming and creeping about, clinging to the underside of stones or buried in mud. Oxygen is obtained through gills projecting from segments of their bodies and from their long tails fringed with similar filaments. The submerged creatures feed on smaller aquatic larvae such as those of the caddis flies and the smaller mayflies, while they in turn hide from their own greatest enemy, the voracious hellgrammites, or conniption bugs, those larvae of their larger relatives, the dobsonflies. After about a year underwater, those that have escaped these enemies crawl up the stream bank, bury themselves in the earth above the water level, and pupate. The life cycle is completed by the winged adults that emerge and rise in their slow and awkward navigation of the air. It is these adults that are imitated in the artificial dry flies that anglers call "alders."

While I am watching the events on my leaf beside the bridge, Nellie is having an adventure of her own.

Descending the trail past my cabin on her way back from Juniper Hill, she is approaching the bridge over Stepping Stone Brook when she sees something small ahead of her fluttering at the edge of the path. She swings her field glasses in its direction. The "something" is a late-nesting black and white warbler. Standing there, she sees for the first time one of these small ground nesters engaged in a broken-wing act. Somewhere close by is a little cup of grass, rootlets, and bark fibers lined with fern down, a nest such as we once discovered in vegetation near the edge of the woods at the top of Juniper Hill. Nellie quickly moves away over the bridge in order to leave the little warbler and its hidden brood undisturbed.

Meeting under the hickory trees as we both came home, we compare our little adventures of the day.

J U N E 2 8 .   The robin's nest in the lilac bush is empty. This morning we see one of the fledglings, speckled of breast, following its parent across the grass, begging for food. Earthworm after earthworm disappears, crammed into its gaping mouth by the older bird. But these are emergency rations, temporary meals. They soon will end and the fledgling will be on its own. But during this period it is watching food being gathered, observing how to provide for itself.

Last spring, in Pleasant Valley, a dozen miles from here, a friend of ours, Ellen Gillard, was watching a young robin being fed on the ground when she saw this time of free meals come to an end. The parent bird ran over the grass with a large earthworm in its bill. But instead of thrusting it into the open mouth, it dropped it on the ground, pecked at it a few times, then flew away. The fledgling wandered about. Then it returned to the worm and began pecking at it. In the end it succeeded in breaking it up into several pieces, each of which it gulped down. Its babyhood was over. It had fed itself. It was on its way to being a self-sustaining bird.

I recall a similar instance in the life of a young barn swallow. Not long after it had left the nest, I saw it perched on a telephone wire, waiting to be fed. Time after time the parent swallow swooped low, fluttered, and in a quick transfer of food thrust insects into its wide-gaping mouth. Then came a change. The swoop and the flutter were repeated but the food was not transferred. Instead the parent swallow dropped it through the air close to the perching fledgling. The first time nothing happened. But the second time the young bird plunged after the falling insect and snapped it from the air.

How much are young birds taught about food-getting by their parents? It is difficult to say. Instinct plays a dominant role in the lives of these wild creatures. But the experiences and observations during the first days out of the nest evidently play a part in forming their feeding habits. This concerns not only the gathering of food but the food that is selected.

Summer scene at Trail Wood. A wild cherry tree towers in the foreground and hickories rise beyond the sheds and the house.

A small ribbon snake suns itself among the flat rocks of an
old stone wall that runs along the top of Firefly Meadow.

As graceful as it is deadly, the destroying angel mushroom, *Amanita phalloides*, rises from the damp soil of the woodland.

The Insect Garden, with its central sundial and its trails winding through weed tangles, slopes down to Hampton Brook.

A small crab spider, clinging to a spicebush swallowtail
butterfly, drains away the vital juices of its paralyzed prey.

At the mouth of its knothole-cavern home in an apple tree,
a gray tree frog sleeps through the heat of the summer day.

A baby cottontail rabbit is surrounded by acres of food in a
lush summertime world filled with green and growing things.

From the stump of a dead tree, the large, strikingly marked elm mushroom, *Pleurotus ulmarius*, appears at summer's end.

One other recollection in connection with a young robin illustrates how food at first rejected comes to be accepted, thus expanding the range of recognized sources of nourishment. After feeding a fledgling three or four earthworms, on this occasion, the parent arrived with a beetle in its bill. This it crammed into the open mouth. The speckled robin promptly spit it out. The older bird picked it up and the same sequence was repeated. This recurred over and over again. I kept count. For twenty-seven times the beetle was offered to the young robin and for twenty-seven times it was rejected. Sometimes the fledgling spit it out a foot or more. Then, on the twenty-eighth try, it gulped down the less-appetizing food. For several seconds the parent bird looked around. It seemed to suspect that the fledgling had given the beetle a quick flip to one side. Satisfied at last, it flew away in search of other food. In this lengthy process the young bird apparently had learned to add variety to its diet.

J U N E   2 9 .   All across this sunny hillside I come upon hawkweeds nipped off by deer. All along this dusty path I see the small and heart-shaped hoofmarks of a fawn. Halting here, where the warm air is redolent with the fragrance of sweet ferns in the sun, I picture the little deer browsing in the dusk, halting where I halt, stretching out its slender neck, snipping off the stems of a hawkweed here, a hawkweed there. Visualizing this, I remember another fawn that on a summer's day three or four miles to the south of here experienced a rare adventure in the midst of a mowing field.

Haying time arrived. The clatter of mowing machines wheeling around the outer edges of the tract commenced. Several times the farmer saw a fawn that had been hiding in the long grass change its position, moving farther toward the center of the field. Most of that day, biting away their swaths, the cutting bars of the machines continued leveling the hay. The radius of their circles continually decreased.

Around and around the hidden fawn they clattered. Closer and closer they came. Smaller and smaller grew the area of standing grass. Late in the afternoon, just before the last of the hay came down, with no hiding place left, the fawn wandered away across the shorn field until it reached the edge of the neighboring woods. For a long time it stood there in the shade of an oak tree, looking back at a field so inexplicably altered.

On this same farm, during a subsequent summer, another event of unexpected natural-history interest resulted from the method used to call heifers from the pasture at feeding time. One of the hired men would pound on the bottom of an empty pail. The young animals, recognizing the sound, would come running toward the spot. On one occasion, each time the hammering on the pail began, an echo seemed to rebound from the edge of the woods where the fawn had disappeared. But the echo continued on after the pounding stopped. A pileated woodpecker, apparently mistaking the sound for the drumming of a rival, was answering each tattoo on the bottom of the pail by pounding with its bill on a hollow limb.

J U N E 30. Why do baby birds have short tails? We are looking down at a small cardinal crouching in the grass near the stone steps leading to the entry door. It has fluttered down from a nest hidden among the massed pillar roses blooming beside the house. When it becomes alarmed and launches itself on whirring little wings for a downward course to the safety of a forsythia bush, it seems to have no tail at all. It has the appearance of a larger bumblebee.

After it is gone, Nellie and I discuss all the explanations of why baby birds have short tails. The shorter the tails of fledglings, the less they will get in the way in the cramped confines of a nest. Tails are unneeded until flight begins, so they can develop last of all. And so on. Our discussion ends in laughter. For our serious consideration of something rather obvious has recalled a charming passage in J. H. Fabre's *The Life of the Fly*. In it he recounts his first scientific experiment, made when he was a child of five or six:

"There I stand one day, a pensive urchin, with my hands behind my back and my face turned to the sun. The dazzling splendor fascinates me. I am the moth attracted by the light of the lamp. With what am I enjoying the glorious radiance: with my mouth or my eyes? That is the question put by my budding scientific curiosity. Reader, do not smile: the future observer is already practicing and experimenting. I open my mouth wide and close

my eyes: the glory disappears. I open my eyes and shut my
mouth: the glory reappears. I repeat the performance with the
same result. The question is solved: I have learnt by deduction
that I see the sun with my eyes. Oh, what a discovery! That evening, I told the whole house all about it. Grandmother smiled
fondly at my simplicity: the others laughed at it. 'Tis the way of
the world."

J U L Y 1 .   When July comes, summer shifts into high gear.
Days of hazy, shimmering heat arrive; hayloads roll in from the
mowing fields; gray young starlings and new red-winged blackbirds gather in flocks and settle on the open pastures.

In July we are in a month that looks forward and backward—
forward to autumn, backward to spring. Watching the new birds
flocking together, seeing in them one of the earliest signs of fall,
we experience each year the same feeling Thoreau recorded in his
journal: "How early in the year it begins to be late!"

But at the same time that we sense this rushing onward toward the autumn, we are conscious that everywhere around us
there remain evidences of the spring. We are still in the time of
little things: little rabbits, newly fledged birds, sprouts of trees
struggling upward, young woodchucks feeding in the meadows,
tadpoles and minnows swarming in the shallows at the pond edge.
We are still in the time of the new, the time of the rejuvenation
of the world.

J U L Y 2 .   In the cool of the evening, after the fires of the
sun have been banked for the day, I stand at the edge of a
meadow watching the beginning of a life among the grasses. It is
the newest of the new. It is a baby cottontail so tiny I could hold
it in one of my hands. When it hops about in sudden starts and
stops it suggests a frog jumping. When it lands, it virtually disappears in the grass. When it nibbles at shoots and stems, I see
its upthrust, almost translucent ears waving slightly like long
leaves stirring in a breeze. It is perfect, without a blemish, a new
product of the year. Its eyes are large and round and filled with a
look of innocence and wonder. This perhaps is its first journey
alone into the world. What will be its fate in life? On this day who
knows—perhaps not even fate.

In the quiet of this after-sunset hour, the baby rabbit moves about in a world of peace—no guns, no dogs, no hawks are here. It is sad knowledge to comprehend that if rabbits everywhere lived in such peace—with peril removed, with enemies gone— the result would be a world overrun by starving rabbits inhabiting a desert. This we can understand completely and still we can wish this one small individual rabbit well.

J U L Y 3 .   Five young cowbirds feed in the sunset. I see them ahead of me on one of the meadow paths, a cluster of grayish birds, part of the new installment hatched this spring. And where have they been hatched? One, perhaps, in the nest of a yellowthroat, one in the nest of a song sparrow, one in the nest of a blue-winged warbler, one in the nest of a towhee, one even in the nest of a brown thrasher. So they have started life—fed and cared for by so many varied foster parents, each an odd bird in the nest, each becoming fledged away from its kind. All of these five eastern brown-headed cowbirds have started life in a different nest and each alone in the nest. But now they are alone no longer. How soon, in the early summer, their kind bands together!

In the life of these birds heredity determines all. Infinitely diverse are the species duped by the cowbirds, the imposed-upon nest builders that brood the secretly implanted eggs and feed and care for the parasitic nestlings. Mainly they belong to four families: the finches, the warblers, the flycatchers, and the vireos. Sixty-two species and subspecies of finches and forty-four species and subspecies of warblers are victimized by the cowbirds in North America. The activity of these parasites is recognized as the chief limiting factor in the increase of many of the smaller sparrows and finches. And in the jack pine region of Michigan it is the cowbird that forms the greatest natural threat to the survival of the rare Kirtland's warbler.

But in no matter what surroundings the Cowbird hatches, no matter what the character or habits of its foster parents, these birds—coming into life under such diverse circumstances—remain almost unaffected by their early experiences. The young cowbird from the song sparrow nest has no characteristics of the song sparrow, the young cowbird from the towhee nest no characteristics of the towhee. As I look at these five young birds that so soon have banded together with others of their kind, I cannot guess which was fed by the towhee, which by the song sparrow. In the story of the cowbird, how small is the part written by environment, how great the part written by heredity.

JULY 4. A black and yellow bumblebee blunders amid the tangles of the meadow grass. It bumps against the stems, flounders among the slender leaves. Slowly it works its way upward, at last breaks free.

Watching this small event in the rising heat on this morning of the Fourth of July, I have the sensation of having stood here before, of having experienced this same thing long ago, but with a different insect playing the leading role. Then I realize that what remains so tenaciously in my memory took place before I was born. It occurred on another continent. It was seen by the eyes of another man who set it down in the pages of a book. What I am recalling are the first sentences of "The July Grass," written by Richard Jefferies generations ago in England: "A July fly went sideways over the long grass. His wings made a burr about him like a net, beating so fast they wrapped him round with a cloud." So our memories of the out-of-doors and our memories of our reading coalesce.

In this manner, this Fourth of July of the insects begins.

In former times, when this day brought freedom from the confinement of an office, I used to set aside the major portion of it for exploring among the small inhabitants of the garden-to-attract-insects that I had set out between the trees of an ancient orchard on a Long Island hillside. How precious seemed those "insect Fourth of Julys." Now, in the wild profusion of this old farm, all of Trail Wood forms a natural insect garden extending away over fields and woods for a hundred acres and more.

I wander here, I wander there as the hours go by. I part the weeds and grasses with my hands. I probe with my eyes. I look under stones. I turn over leaves, trace the cabalistic scrawls of leaf miners, enjoy the multiform shapes and colorings of the insect galls, pause to follow the wavering flight of a swallowtail butterfly in the sunshine.

Sunshine and cloud shadows alternate on this day. When the

sun pours down its warmth, I find dragonflies clad in metallic colors basking on the old stone walls. Along the brook, black-winged damselflies go bobbing by, now in the shade, now in the sunshine. When I examine with my small hand lens the galls on the underside of the hickory leaves, where the larvae have matured and transformed into emerging adults, I note the curious appearance of the exit holes. Instead of being round, they are ragged and irregular as though an explosion had blasted them open.

Once I come upon a cluster of insect eggs with the top of each egg ringed with spikes like the circlet around the head of the Statue of Liberty. And from many of these eggs, I discover, small red predacious stinkbugs are appearing. Red, too, cherry red, are the massed bodies of a colony of aphids on the stem of a goldenrod. A few are winged. Many are small and new. Even as I watch them through my pocket lens, I see one more addition made to the colony. It emerges, headfirst, from its parent alive and complete, a product of the virgin birth that is characteristic of the females of these tiny drinkers of sap.

On this July day, I notice how the roving bumblebees seem to favor, above any other wild flower we have, the small blue blooms of the motherwort, or lion's ear, a member of the mint family, and how the small banded wild bees have alighted on all the black-eyed Susans, whirling around and around in their swift gathering of the pollen.

Such is the news of the fields, these are the things I see as I walk up and down our acres on this July the Fourth.

J U L Y 5 .   The noisiest of all our Trail Wood birds in these early days of July is the pair of great-crested flycatchers nesting in a hollow apple tree along Azalea Shore. The loud, carrying "wheep!" and the "chattering laughter" of their calls come across the pond at all hours of the day. They have a husky quality as though the birds were hoarse from calling.

For several seasons these flycatchers have built their nest in this same decaying tree. Most often such birds choose as their nesting sites hollow apple trees, usually close to woodlands. But on occasions pairs have constructed nests and raised broods in such unlikely cavities as in a stove pipe, a lard bucket, and an abandoned wooden pump.

With both parents contributing to the task, it often takes as long as two weeks to build the nest within the cavity. To form the foundation and bring the nest itself within a foot or so of the opening, the bottom of the hole is filled with a mass of such oddments as seed pods, old leaves, pine needles, pieces of paper, woodchuck fur, rags, string, bark fibers, chicken feathers, onion skins, rootlets, and—as has so often been noted—snakeskins, the latter frequently left dangling from the entrance of the hole. At the top of this mass, the female hollows out the nest, lining it with finer and softer material.

This morning I have been sitting in the shade of a clump of bird cherry trees on the hillside overlooking the pond, watching through my field glasses the comings and goings of the parent birds. In these hours when the young are almost ready to leave the nest, when a maximum of nourishment is required, all during the daylight hours the two birds continually search for food. I see one come flying in with a large dragonfly fluttering in its bill. About ninety-four percent of the food consumed by these flycatchers consists of small animals, almost entirely insects. The destination of all this food converging on the nesting hole is the five open mouths of the five fledglings.

In following the events across the pond this morning, I notice an interesting thing. During their search for food, each flycatcher has its own separate hunting territory. One invariably flies east along the southern edge of the water and over the dam into Firefly Meadow. The other, just as invariably, crosses the pond and forages for insect food in the lowland of Woodcock Pasture. Another thing I observe: the two birds always turn the same way— one right, one left—when they emerge from the nesting hole. On their return they sweep around the same side of the tree in carrying food to the fledglings. Their habits are fixed, their procedure established. As long as I watch them, they arrive and depart around their respective sides of the tree trunk. And each follows almost the identical path through the air to and from the fields where it searches for food.

J U L Y 6 . With the same sense of puzzlement and mystery I have known so many times before, I stand gazing up at this same wild grapevine beside this same woodland trail in the South

Woods. From the ground it ascends almost vertically to attach itself to the lower branch of a tree thrusting out a dozen feet above
my head. Each time I come this way, I have the sensation of observing nature performing the Indian rope trick.

How did the vine ascend through the insubstantial air? How
did it reach its present anchorage well out from the trunk of the
tree? It does not zigzag upward from branch to branch. There are
no other limbs between its support and the ground. It is rooted
below and attached above several feet from the trunk of the tree.
There is no sign it ascended along the trunk. Yet here it is, drawing its almost vertical line through the air to reach the outthrust
branch above it.

Experts have stood beside me here. They have explained on
more than one occasion how the vine, at some earlier time, must
have climbed on lower branches now dead and fallen away. Or
how it must have ascended on bushes or smaller saplings that
have since disappeared and so attained the limb to which it is attached. Or how in some windstorm of the past it must have been
dislodged from other limbs and flung in its present position.
Something of the kind, no doubt, must have happened; I can see
no other explanation. Yet always, as I stand here, I find myself accepting what I am told on faith. And always, in the back of my
brain, rebellious, recalcitrant cells seem jeering: "If you'll believe
that, you will believe anything." They prefer the mystery.

So when I come this way once more and gaze up at the dark
ascending line drawn from the mossy woodland floor on which I
stand to the branch above me, when I see this single strand
mounting through the air and anchoring itself to the limb overhead, I am still only partially convinced. I experience again the
old emotion of standing in the presence of some elongated living
sphinx of the woodland, its riddle still a puzzle to the mind.

J U L Y 7 . It is four o'clock in the afternoon. For the past
two days, whenever I have followed this path leading to the brook
and the waterfall, I have seen a face peering from the round entrance hole of a bluebird box. The second brood is almost ready to
leave the nest.

Now, sitting at a distance, watching through my field glasses,
I see the fledgling at the opening gradually push itself farther and

farther out. In the afternoon sunshine I notice how it is framed in by the silvery gray of the weathered wood. It turns its head. It peers down at the ground. It looks up at the sky. It surveys the wide new world spreading away around it. As the minutes pass, its shoulders and speckled breast follow its head, and for a long time it hangs with much of its body outside the opening. Then, without warning, when the minute hand of my wristwatch points almost exactly to four thirty, it pushes itself outward, trusts to its untried wings, and flies buoyantly away to alight in an abrupt descent in the old apple tree nearest the bridge.

Until it lands, the two parent birds follow it with soft warbling calls. Then they return to the vicinity of the box. A second head has appeared at the opening. Once more I see the same sequence of events. But on this occasion the timespan is contracted. My watch shows it is a little before four forty-five when this second fledgling leaves the nest.

A third head fills the opening. This time I work nearer, camera in hand. Immediately the parent bluebirds are circling around me with continual cries of alarm. Their calling attracts other birds. The nearby tangle of the wild plums is filled with the excited clamor of avian voices. Hopping from branch to branch, joining in the hubbub, are two orioles, a song sparrow, a chipping sparrow, a white-breasted nuthatch, two catbirds—one with a green worm in its bill—and an indigo bunting. As I have noted many times, the launching of a brood, especially when danger threatens, is a period of concern for more than just the parent birds. A wave of excitement runs through the other birds of the area.

By a little after five o'clock, the fourth and last of the brood has broken home ties. I see the adult bluebirds carrying food to the quartet of fledglings now scattered on different perches. For each of these young birds, its habitat has suddenly expanded from the confines of a crowded box to the vastness of the out-of-doors. From now on during the succeeding summer days we will see four new forms, with spotted breasts, perching on the wires or flying from tree to tree—forms graceful in the air, forms that in the course of time will be clad in the plumage of the adult, in one of the most lovely manifestations of the color blue, in bluebird blue.

J U L Y 8 .   I look up at the sky. I look down at the pond. Neither there nor here can I become more than a fleeting sojourner. Aeons of evolution have decreed that only on the thin crust of the globe, only on the solid surface of the earth with its superimposed film of oxygen-bearing air, can I make a permanent home. But how varied, how appealing, how beautiful are our surroundings on this outer skin of the planet.

So far, the probings of unmanned spaceships to neighboring planets where comparable life seemed most likely have found only dead surfaces on lifeless globes wheeling in space. Surrounded as we are by infinitely varied life—the life of the plants, the life of the animals—we seem the fortunate ones, set with awareness amid conditions unique in all our galaxy. So, all the more, this fragile natural world in which we spend our days should be appreciated and protected and cherished. It should be recognized for what it is, something irreplaceable.

J U L Y 9 .   All this still, sultry afternoon, cumulus clouds boil upward along the western horizon. We watch their mountains of vapor tumbling, changing form, mushrooming higher into the sky. When we come home across the fields from Ground Pine Crossing about four o'clock, an electric sense of change fills the air. The sky to the west and north grows more glowering. Black mixes with its blue.

We sit on the back steps watching the storm roll toward us. Beyond the horizon little noiseless flicks of lightning flare and die. That peculiar illumination that precedes the breaking of a summer storm spreads across the fields. The faint muttering of thunder grows louder. At times the faraway lightning flickers in rapid succession, the light fluttering in the sky. Then the flashes grow longer, brighter, draw closer. The distant thunder becomes nearer thunder. Bringing a quick chill to the air, a wind charges out of the west. The first large drops of rain shatter on the twigs of the lilac bush and spatter the ground below.

Looking out from inside the house, we watch the deluge sweeping the fields, the gusts pounding the trees, the leaves ripped away and hurled scudding downwind. Reverberating continually, the thunder crashes around us. It seems to multiply and unfold across the sky. Overhead we hear the snap of electrical charges meeting our lightning rods. More than once the flash and crash come almost simultaneously, indicating a close strike by the thunderbolt. Half a mile away, beside Hampton Brook in the Far North Woods, one of these bolts, as we discover later, smites and shatters a fifty-foot swamp maple. Blasting the trunk apart, it hurls a giant sliver, fifteen feet long, fifty feet away.

Although, unlike the thunderstorm of May, this one brings no hail, it runs its course as one of the most violent electrical disturbances of the year. The lightning's glare, the deafening afterclap shaking the house—on all sides of us the violence and tumult rage on. For more than half an hour, with intensity unabated, the storm continues. It flattens the grass in the meadow. It snaps small branches from the trees. It knocks out our electric current.

But at last the thunder moves off toward the east, rolling away among the hills. The blackness ebbs from the blue-black clouds. The wind eases off and the rain ends. The western sky grows brighter. We walk out into our dripping, battered surroundings to survey the damage. We have experienced on this afternoon a time of super-Fourth-of-July fireworks after the Fourth has gone.

J U L Y 1 0 .   All this day in the wake of the storm our barn swallows have been diving on everything in fur. Their fledglings, overflowing the nest in the center shed, are almost ready to take to the air. As the hours pass, the concern of the two parent birds mounts.

Before breakfast I am attracted by a twittering commotion and see them, one after the other, plunging down, strafing a gray squirrel that is nosing through the terrace grass searching for food. The little animal meets the onslaught with the same stratagem it uses when a hawk makes a pass at it in a tree. Then it flips behind a limb, putting the branch between it and the strike of the predator. Here it takes shelter behind a horizontal aluminum trough that carries away water from the downspout at one corner of the house. Each time the swallows tilt downward into their twittering dives, it disappears, pressing itself close to the side of the trough. Then its head pops up again while the birds circle for a fresh assault.

Midday comes and the swallows glimpse a neighbor's cat prowling down a meadow path beyond the sheds. They scud to a new assault. In a rapid-fire succession of plunges and veering climbs, they streak past within inches of its head. The cat flattens its ears and hugs the ground each time they dive. Eventually it turns back and the swallows resume their scything of the air for insect food.

Completely different was the response of another cat of which I once heard, a black part-Persian living on an upstate farm in New York. Each summer when the swallows nesting in the barn commenced harrying it when their young were ready to fly, it responded by rolling over on its back with its paws uppermost. At the plunge of the swallows, the cat would wait until they were almost on it and then sweep its clawed forefeet through the air. Immediately the attackers became wary, checking their plunge or veering to one side before they came within reach.

The last running attack of the swallow's day comes with the evening arrival of a foraging fox. Its appearance from the plum tangle is greeted with cries and dives repeated again and again. Each time a bird shoots down, almost clipping an ear, I see the fox duck its head. Twice it jumps to one side. So the fox and the diving birds move across the yard until the animal—following its single-minded course in search of food—reaches the protection of the apple tree.

J U L Y 1 1 .   Of all the small creatures I see on my walk today, two make the deepest impression on my mind. One is a moth, the other a fly. And both are in trouble.

It is among the globelike clusters of the flowers of a milkweed that I discover the moth, a day-flying *Ctenucha virginica*. It is dining on nectar. But it is dining dangerously. Each of the clustered flowers is equipped with five insect traps—tapering slits into which the feet of visitors may slip and become imprisoned. Larger, stronger insects, such as bumblebees, can jerk their legs free. But when they do they bring with them, attached like snapped-on clothespins, twin paddles of pollen which perforce they transport to other blooms. Less robust insects, unable to extricate their legs, remain trapped until they die.

When I approach, I see the *Ctenucha* fluttering violently. It is shackled, unable to rise, one leg imprisoned in a milkweed trap. Then it is in the air, free and on the wing. Where it had been, a movement, something waving about, attracts my eye. It is one of the insect's legs. Unable to pull it free from the milkweed it had—with no apparent discomfort—pulled it free from its body.

Less fortunate is the fly, one of the *Asilidae,* those swift insect hawks, the robber flies, that employ long, hooked legs to snatch their prey from leaves or from the air. In a sudden dart after quarry, it has ensnared itself in a spider web. One of its forelegs has become anchored over its left wing. The thin, almost invisible threads of silk hold like cables. No more than one-fourth the size of the fly, the spider, gray with banded legs, edges warily nearer. It reaches out with its forelegs and touches the fly, then backs hastily away, then returns again. Always avoiding the robber fly's long sucking proboscis with its stilettolike tip, the spider intermittently strokes and taps its dangerous victim. Ten minutes go by. The struggles of the imprisoned fly grow weaker and cease. When I leave, the spider, clinging well behind the menace of the proboscis, is beginning to drain away the vital fluids of its larger prey.

J U L Y 1 2 .   In the long, slow ebbing of light at the end of this July day, two foxes come to the apple tree below the terrace. The first arrives some time before the second. We watch it moving in smooth-flowing motion, sniffing around the base of the tree, searching for fragments of food. Then we see it suddenly stop, lift its head, remain motionless, pointing its sharp nose toward the north. It appears to grow more uneasy. Two or three times it changes position slightly.

Then, trotting out of the Starfield and across the yard from the north, comes the second fox.

It approaches without hesitation. As it draws near, the first fox engages in a curious performance. It turns aside. Its tail enlarges, grows fluffed and arched, curving like a quarter moon. Its uneasiness increases. When the newcomer is only a few feet away, it performs a slightly ritualized maneuver such as we have never seen before. It runs in a slow half-circle around the other fox, its tail still arched and fluffed. Then it abruptly flops over on its side and lies still in the grass.

The other fox ignores it; goes about the business of hunting for food. For a minute or more the prone animal lies motionless. Then it jumps to its feet. Its tail has returned to normal size. It joins its companion in nosing about in the grass beneath the apple tree.

What is the meaning of what we have seen—the significance of the actions of the fox? At first I wonder if we have observed some kind of mating performance. But I discard that idea immediately. It is entirely the wrong time of year. Apparently the ritual of the fox in making its circling movement and falling on its side has been a sign of submission. The second arrival is the top dog fox, the dominant one. Like the action of the wolf that is being bested in a fight and falls on its back, its legs in the air and its jugular vein unprotected as an acknowledgment of defeat, the sudden supine position of the first fox acknowledged the authority of the second, stronger fox when it arrived. Through such ritualized sign language various animals avoid needless bloodshed and injury.

J U L Y 1 3 .   In the dawn of this Saturday in July, nearly seventy miles north and east of Trail Wood, I hurry along a path beside Fairhaven Bay. This is the day that for so many years has been a high point of the month, the day of the annual Thoreau Society meetings in Concord. Below a red cottage, originally built as a hunting lodge by Thoreau's jailer, Sam Staples, and later the home of the Concord historian and Thoreauvian Ruth Wheeler, I find my friend, Walter Harding, waiting beside a green canoe. Here, for at least a dozen years, we have kept a daybreak rendezvous.

The world authority on Thoreau's life and author or editor of a score of related books, Walter Harding has been secretary-treasurer of the Society since its founding more than thirty-five years ago. Each year, before his cares of the day begin, he and I paddle along the dawn-quiet waterways that contributed so many pages to Thoreau's journal.

In the stillness we hear the slight hissing sound as we push out from shore through the massed green of floating duckweed to begin paddling with long, regular strokes down the length of the bay, that widened portion of the Sudbury River. Across the water's silky sheen we see a faint silvering of mist veiling the fields Thoreau crossed on that long-ago day of sunshine and rain when he fished at Baker Farm. When we reach the upper end of the bay and enter the narrower serpentine of the stream, we float for a time with paddles still to watch a female wood duck, followed by a string of sixteen ducklings, weave in and out among the buttonbushes.

The sun has cleared the trees and a breeze blows down the river by the time we reach and pass under the arch of Lee's Bridge. Here we find ourselves in the midst of sudden, ethereal beauty. The rays of the rising sun are meeting the ripples on the river. The light, cast upward, runs in waving lines of brilliance over the curve of the masonry above us. And at one point, as we look up, we see the dark shadow of our canoe stealing through the shining maze of all these moving ripple waves of light.

At times, coming back, we drift, watching the bullfrogs with only their eyes and noses thrust above the surface among the lily pads. Once we pass a floating water chestnut, an aquatic plant formerly troublesome along this river. The hard, sharp, pointed nutlike seeds, about the turn of the century, were used as heads for souvenir hatpins sold in Concord. They were called "Thoreau Hatpins."

The bay is opening up before us again when we come upon a string of shining bubbles floating on the dark surface of the stream. They remind me of an entry in Thoreau's journal. In it he recounts how he looked over the side of his boat and saw himself reflected in the curved mirror of a floating bubble. Our paddling ceases while Walter and I look down trying to catch a glimpse of ourselves in the bubbles. The main result is that we almost tip over the canoe.

So, as we come paddling back down the length of the bay, we contemplate what immortality we might have attained among Thoreauvians, what romantic legends might have been built up around us, if we both had perished following Thoreau's example. Then it occurs to us that unless one of us escaped to tell the story no one would ever know the cause of the tragedy. Our end would have been merely like that of the people run over by the train at the railroad station in the first chapter of *Walden:* "—and it will be called, and will be, 'A melancholy accident.'"

At last our canoe slides through the duckweed, entering like a ferry at its slip the place where our passage has left a narrow channel two hours before. As we pull the canoe up on the shore, we promise ourselves another outing in tomorrow's dawn—this time on the paths of the Great Meadows, downstream beside the Concord River.

J U L Y   1 4 .   Where Thoreau, on a July day in 1851, looking from Ball's Hill, saw light reflecting from grass blades in the Great Meadows beyond the Concord River so that it appeared "a sea of grass hoary with light, the counterpart of the frost in spring," Walter Harding and I, on this July day, walk amid wide and shallow ponds. Instead of expanses of lowland mowing grass, fields of water spread around us. Where Thoreau watched platoons of mowers "advancing with regular sweeps across the broad meadow and ever and anon standing to whet their scythes," we see wild ducks tacking back and forth on the wind-ruffled surface of the ponds. The Great Meadows today is a federal sanctuary administered by the U.S. Fish and Wildlife Service. By use of impounded water, its character has been largely changed.

Stopping often, sweeping our field glass-magnified gaze over the shallow water, we advance along the central dikelike road that cuts across the former meadows to the river's edge. This Sunday's dawn is filled with the activity of wildlife. We halt to watch a muskrat, plastered with mud, feeding at the edge of a stand of cattails. We see a small painted turtle, stuccoed with the bright green of duckweed, pushing its way through a floating carpet of this miniature water plant.

Red-winged blackbirds—surely the most plentiful species of songbird in North America—are all around us. They rise from the road edges in swirling clouds. They pour from trees along the river. They stream through the air above us. They alight on the huge leaves of the American lotus. A resident flock of Canada geese, airborne after a labored takeoff, turns, catches the sun, and wheels away, beating off to some private feeding ground in a straggling V that grows smaller and smaller until even the far-carrying voices of the waterfowl move beyond our hearing.

But we are never out of hearing of the smaller water birds—the quacking of black ducks and mallards and, at times, the strumming calling of coot. Monarch butterflies and tiger swallowtails drift ahead of us along the roadway.

It is when we are coming back along the dike-top road that we have the special encounter of this daybreak walk. After watching for some time several ducks feeding in a secluded bay, we glance ahead along the road and become aware of a young Virginia rail advancing toward us. Turning to one side, then to the other, it picks up bits of food. Without showing any sign of recognition, it draws steadily nearer, heading directly between us. It is hardly more than seven feet away when one of the redwings gives its loud, quick note of alarm. The young rail, without a glance in our direction, swerves and darts into the protection of the roadside weeds. On this July morning the redwings are watchmen of the Great Meadows.

Looking back for one final survey before we leave, Walter and I observe a kingfisher, with strong, evenly spaced downstrokes of its wings, cross the road, following the straight line of its flight. Just so, on an April day when high water from the river had flooded the Great Meadows, Thoreau had seen a kingfisher pass by him, hurrying, he noted, "as if on urgent business."

J U L Y 15 . Home again at Trail Wood. Toward evening I try an experiment.

In a book I have been reading recently, *The Wild Realm: Animals in East Africa*, the noted anthropologist, Louis S. B. Leakey, describes the stratagem used by cheetahs to get within striking distance of gazelles and other fleet-footed grazing prey. On the open plains there is no hiding place. Stalking predators are always in sight. If they make a direct, head-on approach, the alarmed quarry is in flight before they can get near. The ruse they resort to is what Leakey terms "the diagonal open approach."

By taking a slanting course, perhaps first this way, then that, they gradually move closer. The prey sees them clearly all the time. But it sees them only in profile. "I have concluded," Leakey writes, "that many animals do not regard a carnivore as a threat when they see it in profile." It seems merely passing by. But its diagonal line of advance brings it steadily nearer. When it has reached its striking distance, it suddenly whirls and charges.

In the after-sunset light this evening I have a chance to put the stratagem of the cheetah to a test. I see two cottontails nibbling grass a hundred feet apart. I play the part of the cheetah, they the part of the gazelles. Walking slowly past the first rabbit on a diagonal course, watching it from the side of my eye, I head directly toward the second animal. Here, so far from Africa, the results provide a dramatic corroboration of the effectiveness of the cheetah's ruse. The first cottontail, as long as I pass by in profile, even though I draw steadily closer in my angling advance, pays little attention to me. But the second rabbit—although it is nearly twice as far off—becomes alarmed and bolts away as soon as it becomes aware of my direct, head-on approach.

J U L Y 16. Along varied trails on this day I note the changes of mid-July. The flowers of the wild columbine are gone; those of the Canada lily are going; the flat masses crowning the yarrows are turning a tarnished white; hop clover has gone to seed; and where the globe clusters of the milkweeds have passed their prime, dry and fallen blooms litter the green troughs of the leaves below. Queen Anne's lace and black-eyed Susan and bouncing Bet are at the height of their blooming. The narrow-leaved mountain mint is coming on. But the catnip's flowering time is on the wane and the Philadelphia fleabane, so prominent in the fields of early June, has been replaced by those smaller, less conspicuous relatives, the daisy fleabane, or lace buttons, *Erigeron annuus,* and the paler, narrower-leaved lesser daisy flea-bane, *Erigeron strigosus.* Bluejays call again after the silence of their secretive nesting time. The first band-saw song of a cicada cuts through the heated air. Baby woodchucks—now grown larger—give their high, shrill whistle with a long trill or twitter at the end. Beside the pond, the fragrant swamp azalea, with its sticky, or "clammy," white tubular flowers, has reached and passed the height of its blooming. Beside the bridge, pale green clusters of forming fruit have replaced the white masses of the el-derberry blooms, while beside the Starfield path other clusters of shining, red, and ripened fruit, filled with tartness, decorate the low chokecherry trees. Blossoms now, in mid-July, have become seeds and fruits, eggs warblers and wrens and robins. So the sum-mer presses on.

J U L Y 17. The flicker and I are both doing the same thing. We are both stretched out in the grass. We are both, on this cloudless midmorning in July, taking sun baths—I with my shirt stripped off, the woodpecker with its wings outspread. Here we lie in the sunshine hardly a hundred feet apart.

Already the temperature has climbed nearly to ninety degrees. The humidity is high. We are beginning one of the "squirrel-walking days" of summer—a day when in heavy, oppressive heat we see all the gray squirrels walking instead of jumping when they move about. The air seems half water, clinging like a hot, wet garment, pressing close. Every movement along the way has been made like a deep-sea diver, encased in rubber and brass and lead, trying to walk on land. When I come to this open grassy spot I throw my shirt aside and lie down for one of my summer toastings in the sun.

After I have been here for some time, I become aware of the flicker. Accepting my unmoving body as something inanimate, it has swooped down to a landing in the grass. For several minutes it hunts about for ants. Then it ceases this activity, extends its wings, and remains quiescent, bathing in the sunshine. I watch it—again without turning my head, reflecting on how many interesting things I see these days out of the sides of my eyes.

Minutes pass. I feel drowsy and content and filled with well-being. I wonder what the flicker knows and feels as it is similarly engaged. The scent of the soft carpet of grass on which I lie, the smell of the dry summer earth fill my nostrils. I close my eyes and let my ears report the activity of the summer day around me.

I hear the husky "wheep" of a great-crested flycatcher, made small by distance. I catch the high, buzzy chittering of young swifts, nearly ready to leave their twig nest within our fireplace chimney. But dominant over these sounds is the clear repeated calling of a quail somewhere beside the brook downstream from the bridge. Its bright "Bob White" carries far through the still, humid air. It is the voice of health, of buoyancy, of exuberance, of energy. It lifts my spirits. It seems proclaiming its conquest over the heat languor that has enveloped all the rest of nature.

Then there is an abrupt airy rushing sound. I open my eyes and see my companion, the flicker, mounting through the air. Its time of rest and bathing in the sun are over. There are ants to find! It is up and away about a woodpecker's business. I sit up and pull on my shirt. Our dual sunbathing is over. Moving through the heat at a creeping pace, with no swift wings to carry me, I return to work awaiting me indoors—to my own equivalent of hunting ants.

J U L Y   1 8 .   The fine drift of mizzling rain begins in the afternoon. It is less a downpour than a heavier mist, a gentle, indolent descent of moisture without wind. Plants relax. Lichens on trees and rocks, almost minute by minute, take on a richer green. We wander along the stone walls and out over the open fields, breathing in the cooler, moister air laden with the awakened perfumes of the summer earth.

Where it clings to a goldenrod, we stop beside a resting monarch butterfly. Its wings are folded together, pointing straight down. The fine rain has accumulated in little drops on the scales of its wings—wings that fail to become wet, that appear to be shedding water like the oiled plumage of a duck.

As we stand beside the butterfly, we become aware of a low cracking, crunching sound. It comes from the direction of the old apple tree nearest the bridge below my Insect Garden. We look that way. The lowest limb on the tree's southern side is shaking as though some heavy animal is leaping about among the twigs and branches. The splintering sound swells and we see the whole lower limb split from the trunk and come crashing to the ground.

This venerable tree must be close to a century old. In its long life, how many gales it has endured, how many blizzards and ice storms in winter! Its twisted limbs have survived, without breaking, the tail ends of several hurricanes. But for decades the slow work of insects, the inroads of fungus, the aging of its fibers have gradually sapped the strength of this lower limb. Now, on this day in its old age, it is no hurricane, no windstorm, no ice load that forms the stress that causes it to reach its breaking point. It is merely the accumulated moisture of this drifting rain, the slight extra load of droplets of water on the leaves and twigs and bark, the droplets of a gentle, windless summer shower. So small a weight, apparently, has tipped the scales and has been sufficient to bring it down at last.

J U L Y   1 9 .   Seeing everything through a fine mesh of bronze wire, Nellie and I are ending our walks today by eating a leisurely picnic supper in the little screened-in summerhouse that looks down on the western edge of the pond. Here we sometimes watch the full moon rise above the black lacework of the treetops beyond the dam and stretch the yellow path of a "moonglade" down the length of the dark pond. Here we often see wood ducks come speeding in at dusk to land, with their wild, squealing cries, in Whippoorwill Cove. Here, in the fading light of spring evenings, we listen to the songs of veeries and wood thrushes and scarlet tanagers and rose-breasted grosbeaks.

Now, with the sunset light fading from the sky, we eat our sandwiches and fruit and drink our iced tea in that quieter time of slow transition when the summer day becomes the summer night. The thin humming of mosquitoes carries from beyond the screen. Several times a small explosive smack reveals where some impetuous robber fly or dragonfly has slammed against the wire in pursuit of prey.

It is a little past eight o'clock when the mouse alarm clock goes off. Outside the screen, on a crosspiece of the framework just under the roof, a white-footed mouse has made its nest. Each evening, at almost exactly the same time, it descends to the ground to begin its nightly foraging among the grass and weeds that border the woods. First we hear the sound of its running feet on the wire. It suggests a gust of wind striking the screen. Then we glimpse the white of its underbelly as with starts and stops of varying length it makes its headfirst rushes down the southern side of the summerhouse. When it nears the bottom we see it arc away in a long leap into the grass. Sometime between now and morning, sometime when we are asleep, it will ascend the screen again to the nest where, curled up and hidden, it will sleep until another evening comes.

Between eight and eight thirty, during these long evenings in July, the late chorus of the birds continues on around us. Voices of towhees and wood thrushes, veeries and catbirds still come from among the trees. After a time, something starts up the green frogs' twanging-banjo chorus. Then the air is filled with the low-pitched bellowing of the bullfrogs. I always listen to them hoping to hear what I have never heard. A friend of ours, Dr. John Buck of the National Institute of Health at Bethesda, Maryland—an authority on synchronization in the activity of animals, particularly in the flashing of fireflies—once asked me to keep track and see if

I ever heard bullfrogs calling in unison. I never have. But I must admit that the sound sleep of country nights has interfered with my research.

It is when Nellie is packing up the thermos bottle and the paper plates that we hear a snarl and squall from the wooded hillside behind us. Somewhere between the pond and Wild Apple Glade, a tortoiseshell cat—its fur a combination of yellow and black and tan and white—has its home. It is as wild as any creature born in the woods, as alert and self-reliant. So far it has evaded foxes and dogs and owls. Sometimes we see it sitting in a clearing on the hillside looking out over the pond like a householder sitting on the porch of his house. We have named it, from the color of its fur, Marmalade. The snarl and the squall are succeeded by silence. Has the fox at last cornered this cat gone wild? Or has it escaped? Will we see it again?

JULY 20.   On this evening walk of ours, at the end of another torrid day, we come upon the deep red of wood lilies along the skirts of Juniper Hill; we find a firefly asleep, its lamp still unlit, clinging to the furry underside of a mullein leaf; we smell the honeyed fragrance of the sweetest flower of our north meadow, the purple bloom of a pasture thistle; we watch a small butterfly in the sunset weaving in and out, up and down among the forests of the mowing grass. With such small and pleasant things we round out our day. Small they are. Pleasant they are. But unimportant they are not. They are the enduring things. They are part of all that steadfast, unconquered, timeless, simple progression in nature that Thomas Hardy pointed out will "go onward the same though dynasties pass."

JULY 21.   The sun comes out; the sun goes behind a cloud; the sun comes out again. I stand below the bridge, the water of the brook swirling among the stones beside me. My camera is sharply focused on the fledglings in a catbird's nest built in a swamp rose bush at the edge of a tangle of alder, spicebush, and wild grape vines. The parent birds arrive with their bills crammed with food; they leave with their bills empty. But the ap-

pearance of the sun and the arrival of the parents—the two ingre-
dients of the picture I am seeking—seem never to coincide.

For a quarter of an hour, I stand almost unmoving, waiting for
the right split second in which to take the picture. My body is
still but my eyes roam about me. They see things they otherwise
would have missed. Something new becomes apparent during almost every moment I remain there leaning against the stone masonry of the bridge.

Close to my feet, half hidden in the grass of the brookside, I
discover a green frog. Anchored a few feet away to a twig projecting out from between stones of the bridge is the nymphal skin of a
dragonfly. Lodged on rocks around which water foams are bright
yellow fragments of the wings of a tiger swallowtail butterfly. It,
no doubt, has been caught by some bird, perhaps a kingbird,
which stripped off the inedible wings and let them flutter down
before its meal began. I look up from these colorful remnants of
the luckless insect just in time to see another tiger swallowtail,
buoyant and graceful, come sweeping over the bridge to pass
above my head and flutter away upstream along the brook.

Thus the time goes by. But it is not time wasted or regretted.
When at last I snap the shutter and return with my camera up the
lane, I have a picture plus—a picture plus the memory of little
things I have seen and enjoyed while I waited.

J U L Y 2 2 .   The heat of the day, the dark of the moon, the
setting of the sun, the thawing of the ice, wind and snow, heat
mirages and silent nights—unending are the changes of nature.
And now, in July, the running of the sap, the unfolding of the
leaf, the blooming of the flower reach their climax in the ripening
of the fruit.

These days I nibble my way along our trails. In openings and
beside gray stone walls, in Monument Pasture and the Far North
Woods and along the Big Grapevine Trail, I gather handfuls of
red raspberries, ready to fall from their stems at a touch. Blackcaps are ripe along the lane. Here I pluck a sassafras leaf to chew
its juicy stem and there a black birch twig full of wintergreen
flavor. Wherever I go on these walks in July I sample wild fruits
and wild flavors. During the weeks that lie ahead, this enjoyment

of the bounty of the woods and fields will be expanded by the variety and sweetness of dewberries and blackberries, wild cherries and huckleberries as each ripens in the sunshine of the summer days.

JULY 23. My sampling of the growing harvest of wild fruits takes place, this morning, along the ridgetop beyond the glacial boulderfield in the Far North Woods. Here, while a wood pewee drawls its short "pee-a-wee" song in the rising heat of the day, I find what I came to see—two small flowers of the higher, drier woods.

The first is a close relative of the ladies tresses orchid, *Goodyera pubescens*, the downy rattlesnake plantain. Ladies tresses bloom in the open fields but the downy rattlesnake plantain is rooted only in woodlands, in dry and shaded places. Its eye-arresting rosette of leaves lies almost flat on the woodland floor. Plantain-shaped, dark green, each leaf is decorated with a network of almost white interconnecting veining. The effect suggests the scales of a serpent and is the source of the "rattlesnake" in the plant's common name. Among the mountains of North Carolina it thrives at elevations as high as 4,000 feet. From the center of several of the rosettes rises a single woolly stem supporting at its summit a cylindrical mass or spike of greenish-white buds and beginning flowers. I extract my pocket magnifying glass and feel again the elation of seeing the tiny individual blooms expand into orchid flowers. So I ascend from bloom to bloom up the spike and on into the buds of the still unopened flowers nearer the top.

Only a dozen times this morning do I come upon these floral spikes. But in many places I stop beside another plant, the second object of my search. It, too, rises in a single stem. About it, sharply serrated, mottled with white along the veins, its tapering, deep-green leaves grow in whorls. Two, sometimes more, remarkable flowers dangle from the top of the stem—flowers that are now nearing the end of their blooming. The waxy petals are white or pink-tinged. These, also, are woodland blooms that are appreciated fully only when viewed through a magnifying glass. Seen thus, each flower expands into the form of an inverted king's crown. These blossoms I am enjoying are those of the striped wintergreen, *Chimapila maculata.* For some reason unknown, coming down from far-earlier times, this pleasant woodland plant has been burdened with such unpleasant common names as ratsbane, wild arsenic, and dragon's tongue.

JULY 24. Like a summer cloud trailing its shadow across the meadows, a diaphanous mass of green algae drifts above the mud of the shallows beside the dam. For two or three minutes, I have been standing motionless watching a rather pretty red-banded water snake, about two feet long, sliding in and out among the aquatic vegetation a dozen feet away. Suddenly, as it comes nearer, it becomes conscious of my presence. Its body stretches up. Its head appears above the surface, turned toward me with its dark forked tongue flicking in and out. I stare at it and it stares at me.

This continues for perhaps thirty seconds. Then, apparently dissatisfied with what it sees, the banded serpent writhes away, cutting through the cloud of drifting algae. And now something unexpected happens. A new actor appears on the scene. What has been a commonplace occurrence becomes something exceptional. A large bluegill sunfish that has been swimming lazily in the warm water, shoots ahead in close pursuit of the water snake. It follows its wriggling course through the cloud of algae, swimming above it, keeping pace with its advance. All across the sumberged green cloud it accompanies the invisible reptile. When the water snake emerges on the other side the fish loses interest and turns away.

I stand there for a time thinking about what I have seen. Obviously the sunfish was not interested in the snake itself, only in what the snake was producing. Its pursuit of the serpent lasted only as long as it was pushing its way through the algae cloud. The interest of the fish apparently was like the interest of the tree swallows that accompany me, swooping close and circling around, when I walk through the meadow grass in summer. As the swallows are intent on snapping up small insects frightened from the grass tangles, so the bluegill was intent on the bits of food floating upward from the disturbed algae in the wake of the swimming snake.

J U L Y 2 5 .   We both lean close to the goldenrod leaf. Cling-ing to it is a winged, brownish insect. It is about the size and color and shape of a *Polistes* paper-making wasp. Thin yellowish bands around the abdomen heighten the resemblance. Even its differently shaped wings, because of transparent sections, appear to have the same form as those of the mimicked wasp. These fea-tures we notice at a glance. But a closer inspection leads us to its remarkable forelegs. They are enlarged and spined like the fore-legs of a praying mantis.

Here, beyond the western end of the pond, we are seeing that rarely encountered insect, the false mantis or false rearhorse, the little *Mantispa*. Only once before in my life—in the woods near Fairhaven Bay, at Concord, Massachusetts—have I encountered it. Like the mantis, which is fully three times its size, the *Man-tispa* has the same greatly elongated prothorax as well as the same spined forelegs that can be snapped shut over its victims. It has the same triangular, pointed face. Yet the two insects are not even members of the same order. The praying mantis belongs to the order of the crickets and grasshoppers, the *Orthoptera;* the *Mantispa* to the order of the lacewing flies and the ant lions, the *Neuroptera.* Evolution, curiously enough, has produced in two entirely different orders individuals with similar forms and similar ways of obtaining their food.

Seldom observed and always widely scattered, the insect we are seeing was first described only a century and a half ago. Even after that it remained a creature of mystery. It was not until 1869, when a Viennese scientist name Brauer published the results of careful observations, that its remarkable life story became known. His studies were made with a European species but it is assumed that the life of the American *Mantispa* has a parallel history.

The insect we watch on the goldenrod is feeding on small, soft-bodied, sap-sucking aphides. Always, Brauer found, the adults live by stalking or lying in wait and capturing prey with their forelegs. In this same month of the year, the month of July, he observed the females attaching to leaves numerous rose-red eggs, each perched at the top of a stalk like the elongated eggs of the familiar lacewing fly. Their appearance is similar to that of the fruiting bodies of certain mosses. From these eggs hatch agile, lizard-shaped larvae much like the immature lacewing flies, those devourers of aphides, the aphis lions.

It was by accident that Brauer stumbled upon the secret of the subsequent life of the immature *Mantispids*. After overwintering without food, they emerge in spring and search for the wolf spiders that live under stones and carry their egg sacs with them when they move about. Forcing their way into these sacs they live as parasites surrounded by food, gorging themselves on the young spiders then pupating to emerge as a "second larva" of somewhat different form and then continuing the feast. After a second pupation, they appear from the sac as adults and their habits change. Instead of sedentary parasites, they become active predators. Since Brauer's original discovery, an entomologist in Brazil has reported that a South American species lives as a parasitic larva not in the egg sacs of spiders but in the nests of wasps.

We watch for a long time this strange, almost lonely representative of its species. We have encountered on this day one of the rarest inhabitants of Trail Wood. In all our walks along all our varied paths, we probably will never come upon its kind again.

J U L Y 2 6 .   Evening has come. And so have the little foxes.

For several days now, in the twilights, a family of gray foxes, an adult and three cubs, has appeared suddenly at the edge of the Wild Plum Tangle at the borders of the northeastern corner of the yard. They are attracted by the bits of suet and the scraps from meals we scatter under the apple tree below the terrace.

Tonight Nellie and I watch the three small foxes playing like kittens on the lawn. They chase each other. They make flying leaps. They race in circles. They are bowled over and jump to their feet again. Then abruptly the frolic ceases and they all begin pouncing on small crickets in the grass.

The foxes know we are here. They know we see them. But they pay little attention to us as long as we remain unmoving. Once when I altar my position just as one of the cubs is looking my way, an odd performance takes place. It sits on its haunches, staring intently in my direction. Then it moves its head quickly to one side, then to the other. It raises it and lowers it. This it repeats four or five times. Like the mourning dove that shifts its head from side to side when looking at something that puzzles it, the little fox is changing its viewpoint to observe me more closely. A minute or two go by. Then, as I continue to remain without further movement, it loses interest in me and returns to its cricket hunting.

Always when we have an opportunity to watch a litter of young wild animals, we are fascinated by the differences in attitudes and character exhibited by the different individuals. One of the three cubs is the most trusting, the boldest. Another is considerably more wary. And the third is the shyest, the most timid of all. It sits beside the plums, ready for a quick retreat into the tangle. It watches its more daring companions ranging over the yard in a search for food that carries them as far as the terrace and the apple tree. But it remains where it is until one of the foragers discovers a good-sized piece of suet. As it comes trotting back across the grass with this prize in its jaws, the timid cub comes to life. It gives chase, trying to hijack the food in transit. The two disappear, racing away down the path toward the brook. A little later, returning from its unsuccessful sally as a freebooter, the shy fox—as shy as ever—takes up the same position again, sitting facing out over the yard, its back protected by the maze of the wild plums behind it.

J U L Y   2 7 .   Gradually the contents of the pail that has accompanied me on my walk this morning rises nearer the top. Wherever the ground has been disturbed, I stop and look for the fleshy stems and leaves of the purslane.

As one by one I pick the plants and add them to my pail, I notice how easily the stems break off. Purslane, or pussley, *Portulaca oleracea*, is a plant with unexpected capacities. For example, it has the ability to reroot itself wherever even a fragment of a stem falls on moist ground. A wild relative of that low-growing domesticated plant with flowers of many colors found so often blooming in country gardens—the portulaca—it sprawls over the ground. Its succulent stems, tinged with terra-cotta pink, bear small, thick, dark-green leaves and radiate out from a central root system. Instead of the eye-catching reds and pinks and oranges and yellows of the garden portulacas, the wild purslane has only tiny, inconspicuous blooms a quarter of an inch in diameter and always the same color—yellow. Not only are they minute but they are open for but a few hours during the morning and this only on days when the sun shines brightly. Yet in spite of their tiny size and restricted blooming time, they are found by a sufficient number of small bees and butterflies to insure their pollenation.

Fields and wasteland provide the favored home of the purslane. Thoreau found it growing in the disturbed soil of his bean field at Walden Pond. He not only hoed it up; he ate it. During earlier times, the plant was widely used in America as a pot herb and sometimes added to salads and eaten raw. Captain John Smith, in *A Description of New England,* published in 1616, refers to the use of "purslin" in "brothes and sallets." In such far-flung lands as China, France, India, and Mexico, purslane has been sold regularly in the food markets.

When my pail is full I return to the house with my morning's wild harvest. Nellie washes and boils the plants. Then we—as Henry Thoreau had done—make a "satisfactory dinner" of succulent purslane. We add salt and butter and find the taste mild, not rank or weedlike. The younger plants, and especially the small rounded leaves that break from the stem with the slightest strain, prove best of all.

J U L Y 2 8 .   The calm air of this evening appears filled with tree swallows. They turn continually over the pond, swoop down, splash on the surface. Already, so early in the year, the birds are flocking. It is hard to realize that the contents of eggs that hatched, it seems, only yesterday will but a few weeks hence be whirling over the saw grass of the Everglades, part of the great movement of migration.

I stand on the dam near the spot where I saw the sunfish following the water snake through the cloud of algae. I watch the birds in their comings and goings. I listen to their bright cries on the wing. Then, after a time, my attention is attracted from the active to the passive, to something small and submerged, to something that on occasion is also associated with drifting clouds of algae.

The whole bottom of the shallows along the dam, I notice, is sprinkled—as with a dense scattering of seeds—with tiny, flat, rounded objects. Again I am encountering something new beside the pond. Kneeling on the grass, I reach out and scoop up several of the seedlike objects in my hand. Each is only about a quarter of an inch across. Washing away the mud, I carry them up the slope to the house. When Nellie and I examine them through a fourteen-power magnifying glass, they expand into the form of tiny freshwater clams. I trace them down in a guidebook. Their scientific designation is *Pisidium;* their appropriate common name is the pill clam.

Why this sudden population explosion of pill clams beside the dam? Why on this particular summer? Whatever the answers to these mysteries, these minute newcomers to the pond are creatures of unusual habits and special interest. Not only do pill clams burrow in the mud, but on occasion they pull themselves up the stems of waterweeds. At other times, large numbers of these tiny freshwater mollusks ride about, transported within filmy clouds of drifting algae.

JULY 29. Green imprinted on silver, a line of delicate pawprints lead across the dew-laden grass in the dawn. Some small creature, probably a meadow mouse, has left its trail among the droplets that glitter in the low rays of the rising sun. I look behind me. My own larger tracks record my wandering progress through the shining beauty of this moisture-laden dawn.

Glancing toward the sun, I see miniature rainbows shimmering across the flat webs spread out on the grass by spiders. Fine droplets of dew cover the silken threads from end to end. Acting as prisms, they merge into a maze of tiny glinting spots of brilliant red and blue and green.

Farther off, in the direction of the sun, color and action combine. I follow the movements of a robin taking a bath in the dew-drenched grass. It flutters its wings, ruffles its feathers, fills the air with flying drops of moisture. Like the little field sparrow I saw in the June dawn set amid the colored prisms of the morning dew, it is enclosed by a glittering cloud shot through with little flashes of rainbow colors.

Everywhere in the physical world—from the vastness of the aurora and the sky at sunset to loveliness in miniature, to the hues of the tiniest petals, the irreplaceable patterns of the transient snowflakes, delicate pawprints left in morning dew—we are surrounded by beauty in myriad forms. In his old age, the Indian poet Rabindranath Tagore put his feeling about all this endless beauty of the world into this sentence: "When I go from hence, let this be my parting word: what I have seen is unsurpassable."

J U L Y 3 0 . With three sandwiches, two tomatoes, two oranges, and three plums, Nellie and I head for Juniper Hill. These evening hours of ebbing light, after the heat of the day, provide a perfect time for a July picnic.

The spot we choose on the upper slope is carpeted with cinquefoil and set between spreading clumps of juniper that resemble immense green bowls partly filled with the brown of dry leaves carried on the winds of last fall. Near the base of one clump, a flat mushroom, half a foot across, resembles a pancake browned in the skillet and ready for serving. It is one of those striking *Boletus* mushrooms that from July through September each year appear on this hillside.

Here at Trail Wood we live in a land of echoes, of far-carrying sounds. Something in the topography of the Little River valley before us, of Hampton Ridge behind us, of stone walls and solid woodlands bordering open fields is conducive to projecting sounds. In the vicinity of several little bays and curving places along the walls of trees that edge our meadows echoes abound.

Sitting here, eating at leisure our picnic supper, we amuse ourselves by noting all the different sounds projected to our ears. There are the close-at-hand sounds—the wood thrush that sings, the towhee that calls no more than thirty feet away. There are the intermediate sounds—the trumping of bullfrogs around the edges of our pond, the high, emery-wheel burring of an evening cicada, the screaming of kingbirds darting about a wild apple tree. There are the distant sounds—the remote bawling of a cow, the muffled barking of a dog, the far-off laboring of the engine of a truck. They arise mainly from the wide river valley now filled with a faint shining haze at the end of this dry and dusty day.

Our sandwiches, our tomatoes, our oranges, our three plums all are gone. But we sit on. We sit on finding our entertainment in our sense of hearing. And as long as we remain, the diverse sounds of the July evening—some loud, some faint, some new, some familiar, some near, some far away—continue on, diminishing gradually as the long dusk settles down.

JULY 31. The yowl of a cat. The sound of an animal tearing through the weeds and underbrush. A flash of yellow and black and tan and white. And we—on our way down to the summerhouse after sunset—see Marmalade, the cat gone wild, explode from the woods at the western end of the pond. It is running for its life. Only a few yards behind, a gray fox is gaining ground at every leap.

The cat reaches the base of a red maple tree rooted beside the cascade down which Whippoorwill Brook foams into the pond. We hear the scrabbling sound of claws on bark and see its varicolored form shoot up the trunk like an ascending rocket. It seems to us to be going as fast straight up as it had been over the ground below. It is only when it reaches a branch twenty feet in the air that it halts its frantic scramble upward. There it crouches, looking down at the fox circling around and around the base of the tree. Once more this cat of the woods has escaped its mortal enemy. In spite of the silent flight of prowling owls in the night and the swift running of foxes, its charmed life still continues.

Usually it is the red fox rather than the gray that corners and kills cats. But occasionally a hungry gray in its hunting will attack such an animal when it finds it wandering in the woods. For some time we see the outwitted predator remain beneath the tree where the cat has found sanctuary. It moves away, looks up, comes back, unwilling to leave the spot.

At last we see it leap from boulder to boulder across the brook and then trot along the pond-edge path beside Whippoorwill Cove. It stops at intervals to thrust its pointed nose into grass clumps where meadow mice may lurk. When it disappears in the deeper dusk of the undergrowth along Azalea Shore, we see it no more. But we can follow its invisible progress as it moves down the edge of the pond by the loud, smacking alarm note of a brown thrasher keeping pace with its advance.

AUGUST 1. Strewn through the woods west of Juniper Hill, boulders lie scattered like gray cattle sleeping among the ferns. So they have remained for ten thousand years, since they were deposited here during the retreat of the glaciers. On the detailed maps prepared by the U.S. Geological Survey, this portion of our farm, like the similar tract on the ridgetop north of Old Cabin Hill, is represented by a cluster of crosses and bears the designation: "Boulderfield."

There is a strange, aloof, ancient air about this resting place of Ice Age rocks. Most often when we approach it we find it wrapped in silence. It seems remote from the present. It strikes us as a spot where events of some special nature interest should take place. But—for some reason we never can explain—this has not been so.

When first we saw it, our impression was that this must be a serpent field as well as a boulder field. It has a snaky look. It seems the perfect habitat for reptiles. Yet in all our years at Trail Wood we have never seen one serpent of any kind here, not even a striped garter snake. Our second impression was that among the rocks of this shaded woods we were likely to come upon some new species of fern to add to our Trail Wood list. But we never have.

Looking back over a span of nearly twenty years, as I sit here on the soft cushion of moss covering one of the gray boulders, I cannot recall a single outstanding adventure or new discovery or notable memory connected with this silent, almost eerie place. The chief interest of this south boulderfield lies in the nature and history of itself. In the back of our minds over the years the area has come to symbolize vaguely The Place Where Nothing Happens.

Yet I always approach it with the feeling that new adventures, unseen, must lie all around me. I always expect to glimpse something exceptional in some quick glance from the side of my eye. Here, as in that secluded dune-country woods to which I was taken once as a small child and which I could never find again— that lonely tract of which I wrote in *The Lost Woods*—there is a sense of action stilled by my presence, of standing in a charmed circle where all life pauses and checks its activity until I pass on.

A U G U S T 2 .   Do ladybird beetles ever drink milk?

A few minutes ago, I would have had to answer that question with an "I don't know." Now I can be exact and positive. The answer is—at least in one instance here beside the pond at Trail Wood—"Yes, they do."

We have been sitting for some time in our screened-in summerhouse overlooking the water, once more eating a picnic supper at the end of a day of heat. Somewhere in the course of this leisurely meal, I spill a small drop of milk. The first I am aware of it is when I notice a black-and-red ladybird beetle, or spotted ladybug, which must have entered with me, riding on my clothes. I see it reach the white droplet and halt beside it. I look more closely. The drop grows smaller, shrinks steadily. The beetle is drinking the cow's milk. How many times, I wonder, has *that* happened?

Do bats ever hold their wings motionless and glide through the air?

From this same viewpoint beside the pond, on another evening in August, I learned the answer to that query also. Nellie and I sat watching the erratic zigzagging of a little brown bat as it hunted for insects over the water. Once, as it passed down the length of the pond, the flying mammal adopted a surprising change in the movement of its wings. Its normal fluttering progress, the veering flight that is characteristic and makes it appear unstable in the air, suddenly ceased. Following an almost straight line, it advanced with rather slow and deliberate wing beats. It seemed buoyant in the air, like a short-eared owl hunting low over a sea meadow. Then, before it resumed its normal flight, with the twistings and midair halts that are part of its aerial feeding, we saw it hold its wings out rigid and, for a moment or two, without moving them, glide on a downward course.

A U G U S T 3 .   In general, the more alert we are in the out-of-doors, the more hours we spend abroad, the more we know and understand, the more of interest we will see. But for even the most alert observer, there are times when we seem wandering amid deserted halls. In pursuit of that worthy goal—to see accurately and interpret what we see correctly—we encounter tides of fortune. Luck brings us to a certain spot at the right time; or the event that would have absorbed our attention takes place only minutes before we arrive or minutes after we are gone. For us all, there are hours with comparatively little to reward our wandering. But other hours compensate, those hours when, all around us, we encounter wild forms of life in the midst of their intuitive, age-old activity.

In recalling this, as I walk in the Far North Woods, I remember a plaintive chapter in a book written by a woman bird watcher who spent the better part of one day sitting still in the woods waiting for things to happen. She had taken at face value Thoreau's prediction that if you sit in one place in the woods sooner or later every inhabitant will pass by. At the end of her unproductive hours, she noted that her mistake had been in not giving due importance to one phrase in what she had read. That phrase was: "sooner or *later*."

AUGUST 4. There are, in old, rocky New England pastures, certain hillside patches that are indicators of drought. The soil is thinner, more arid; the vegetation more sparse. The effect of drying heat and rainless days becomes apparent first of all in such locations.

I have just stopped beside one area of the kind. With its low cinquefoil and stunted goldenrod, its running dewberry vines and, here and there, its clusters of red-tipped "British soldiers" lichen, it has about it a sense of primitive remoteness. Of little use to agriculture, it seems more allied to the wilderness than to farming and grazing fields. Such areas remain small footholds of wildness even though they are included in some owner's deed.

I stand looking about me while the midmorning heat soars. The querulous cawing of young crows carries from the woods that enclose the Ground Pine Crossing Trail. Their voices have a parched, disgruntled sound. Throughout the day we will hear them raised in complaint. Young crows are not birds that suffer in silence.

When I glance down at my feet, my eye is caught by a hundred tiny movements. Small black insects are walking or hopping over the ground among the stunted grasses and scattered herbs. I squat down, peering closely. What I see is a multitude of immature field crickets. Some appear hardly larger than flies. So the orchestra of the late-summer and early-fall fields has its small beginnings.

For a while longer I stand here. Just as I am about to turn

away, my ears catch a slight sliding or scraping sound behind me. I wheel around and discover that a spotted turtle is plodding in slow motion toward me. It is only five or six feet away. To it my motionless form, no doubt, seemed as inoffensive as a tree trunk. But when I turn it stops. I take a step closer. It pulls in its head. Its feet and tail follow and it closes its polka-dot shell. But before its head is tucked away, I note its eyes are orange-hued. The spotted turtle I have encountered is a female. For among these creatures it is the females that have orange eyes, the males that have dark brown eyes.

I pick up the turtle and examine its smooth, dark shell decorated with small round markings of yellow. The number of these spots varies greatly with different individuals. They tend to increase with age. In some rare instances they may reach a total of a hundred or more. I am interested to find a spotted turtle on this dry pasture slope some distance from the pond. Throughout its range from Maine to Florida it is closely associated with pools and ponds and bogs and ditches. But my turtle is not lost. During tests in animal behavior laboratories, such creatures have demonstrated a keen location sense and a special ability in finding their way through mazes. When I put the turtle down, I point it, quite unnecessarily, downhill in the direction of the pond before I start for home. There, in a reference book in my library, I am glad to discover this turtle acquaintance of mine may well live for many years. The life span of spotted turtles in some instances has extended across more than four decades.

AUGUST 5.   In pink and purple, chartreuse and cherry red, orange and buttercup yellow, lavender and tan, Nellie's daylily garden, blooming in a hundred varieties, runs in a wide descending ribbon of color close to the stone wall beside the lane. Each of the varicolored blooms has expanded for its one day of life. The names of these flowers range from Vagabond King, Winning Ways, and Prairie Moonlight to Grandfather Time, Ice Carnival, Silver King, Bold Rankin, and Soft Whisper.

It is beside the last name that I halt in this midmorning at the beginning of my walk. Its petals are creamy yellow with a tinting of pink at their ruffled edges. But it is something else that has caught my eye. The broad petals of the flower form the stage on which a small tragic drama is being enacted. One of the graceful spicebush swallowtail butterflies, with its dark wings relieved by patterns of blue-green, pale straw yellow, orange, and yellow-green, has alighted and thrust its head deep into the throat of the lily in a search for nectar. Usually so quick to take flight at my approach, it remains unmoving, its wings raised and held together. It seems unaware of my presence. And unaware it is.

For now I catch sight of something else I had missed—a smooth, globular, pea-sized body, yellow like the yellow of the lily. Waiting for prey, a crab spider has been lurking within the flower on which the swallowtail has landed. I can reconstruct the sequence of events that has occurred: the alighting of the swallowtail, its concentration on obtaining nectar, the waiting spider, camouflaged by its color, attaching itself to the soft abdomen of the butterfly, injecting its immobilizing anesthetic, then beginning the draining away of the life fluids of its prey in the long feasting that will end only when hunger is satiated.

The most striking feature of this encounter is the disparity in size of predator and prey. The baglike body of the crab spider is only a few times larger than the head of the butterfly. Usually it is flies, honeybees, occasionally little skipper butterflies—smaller insects—that are its prey. Never before have I seen so large a victim overcome by one of these lurking predators. Yet its venom has been sufficient to subdue and paralyze its larger victim in a prelude to its death.

The spicebush butterfly never moves again. I return a dozen times during the day. On their floral stage the two actors remain with positions unchanged. When evening comes and the flower's transient life is over, its petals begin to close. In this gradual action that marks its end, I see the lifeless butterfly slowly swallowed up. When for a last time before going to bed I come back in the darkness, my flashlight illuminates a dead butterfly entombed within a dead flower. Only the ends of its wings peep from the elongated tube of the enclosing petals.

A U G U S T 6.   Along Azalea Shore this morning big frogs are eating little frogs. Among the long stems of the pondweeds, large-mouthed bass are weaving in and out, sliding through the water, making sudden rushes to gulp down minnows, often the young of their own kind.

All this, as I slowly make my way around the rim of the water, appears far away—this endless stream of life devouring life. It seems remote from our world, occurring in another unfathomable realm of far-different values and viewpoints. These countless individual tragedies leave us virtually untouched. Who ever can really put himself in the place of a tadpole or a minnow? Moreover, here by the pond, I see so many beginning frogs, so many minnows in the shallows. These are cold-blooded creatures in a cold-blooded world. This is nature's method of controlling overpopulation. Our emotions remain almost unaffected by death on such wholesale terms.

But when, a little later, I stand beside a post where a hawk has stripped away the feathers of a young grouse, letting the plumage drift to the ground before consuming its victim, I respond in a somewhat different way. Here the prey is a larger, warm-blooded creature whose individual life seems comparable to our own. The tragedy is no longer general, wholesale. It has become individual, specific. Here we can identify with the victim in a way impossible among the swarming minnows and wriggling tadpoles. At such a time, while we understand and even agree with the cold logic of nature's disregard for such an individual life as this, we know a sense of regret for its passing, regret for *our* passing, also.

But in all the countless deaths that come on a summer day, in what we call the "deadly competition" of nature, once more a clear distinction is needed. Technically, competition that produces death is "deadly." But when we say this, what we say carries the overtones of *intention*. To bring death is not the intention of the predator. It is incidental to its design. The hawk that plunges toward a flying bird, the weasel that noses its way toward a rabbit's nest, the blacksnake that weaves upward among the branches of a tree to reach the nestling birds—none is intent on producing death. Its primary interest is in sustaining itself with food obtained in the way it is specially equipped to obtain it. If it could obtain the same food and leave the prey alive, it would be just as content. Cruelty is not involved in its act. Neither is there a design to terminate life. There is only a design to prolong life, its own life, that, by nature's laws, is—for each living thing—its primary concern.

A U G U S T 7 .   In a hundred places, lifting the clear yellow of its flower spikes above the gray of old stone walls, adding slender vertical lines to the open pastures, the mullein has begun the long period of its blooming.

In our area it is in bloom from July to September. The individual blooms, packed together along the floral spike, may occupy as much as the final foot of the upright stem. Above their dense rosettes of gray-green woolly basal leaves, many of these stalks rise to a height of six feet or more. From Nova Scotia to Florida and as far west as California, this alien plant, arriving in the New World from Europe, has entrenched itself throughout much of North America.

By means of its deep-penetrating root system, it thrives in dry places—a sun-loving inhabitant of old pastures and wasteland. When it arrived from the other side of the Atlantic, it brought with it a list of more than forty colloquial names, including: velvet plant, candlewick, torches, feltwort, hedge taper, hare's beard, blanket leaf, Jacob's staff, shepherd's club, and Adam's flannel. It was in 1753 that Carl Linnaeus bestowed upon it its present scientific name: *Verbascum thapsus*.

A U G U S T 8 .   Among the many mulleins of our farm, one in particular attracts my attention. It ascends close beside a stone wall. Whenever I come this way, I swing my field glasses in its direction. For I am not the only one with an interest in the massed blooms of its floral spike.

On this day I discover three young Baltimore orioles fluttering around the mullein and alighting among the flowers. Not long ago they were swinging in the woven-basket nest that dangles from the tip of a high limb in our largest hickory tree. I see them pecking steadily among the petals. When I examine the flowers after they are gone, I can find nothing to explain their attraction for the orioles. I suspect that sharper eyes than mine have found small larvae or tiny insects concealed among the blooms.

Chickadees are swarming over the mullein when I return home from my walk. A little later a downy woodpecker holds the center of the stage. As I watch it I see a chickadee flit down, alight, and ride on the tip of the swaying stalk while the woodpecker continues its feeding. I remember another time when a monarch butterfly clung to the tip where the chickadee now rides, basking in the rays of the late-afternoon sunshine. A downy swooped in for a landing and I saw the insect flap away in alarm.

These small woodpeckers are particularly valued in the Iowa cornfields. There, in the wintertime, they hammer their way into the dry cornstocks to extract and consume the pupae of the corn borers. There are never enough downy woodpeckers among the Iowa fields. Even with abundant food, their numbers do not increase greatly. For there are other factors in bird populations than ample food. One is nesting places. And the wide fields of the corn country are relatively treeless.

There are times when the visits of the woodpeckers to our Trail Wood mullein stalk last as long as a quarter of an hour. Now I watch one hammering away as though among the crevices in the bark of a tree. When at last it darts away, I go over the area where it has been working. Already, where the flowers have bloomed, small, compact, rounded pods have formed. Now green, they will, in the course of time, become the dry capsules that are crammed with the dark and minute seeds. Looking through my hand lens, I notice numerous holes peppering the little seedpods. Each punctured capsule is empty. When I break free an unpunctured one and split it open with my thumbnail I discover a small grub feeding within.

This answers the question of what the woodpeckers are getting. But, as is so often the way with unfolding questions in the out-of-doors, it leaves me pondering another. How does this little expert in extracting hidden larvae know the grub is there? Is it unlearned wisdom handed down from generation to generation or is it knowledge obtained from observation when following its parent about during its earliest days after leaving the nesting hole? It may be a combination of both. But I suspect the latter is the more important of the two.

AUGUST 9. Last evening dull overcast shuttered the sky. The threat of rain hung heavy in the air. Yet all across the yard in the unstirring calm winged ants set sail from the grass in a widespread dispersal flight. Young queens and smaller males fluttered weakly into the air on their first and only flights, on the one aerial journey of their lives. The rain held off. When we went to bed the weather was still calm. And by this morning the skies had cleared. How right were the flying ants!

Thinking back, I cannot recall a time when heavy rain or bad weather followed on the heels of such a dispersal flight. I once asked the famous ant authority at the American Museum of Natural History, Dr. T. C. Schnierla, if he had ever observed a case when the ants were mistaken. He thought for a time, then said he could not remember an instance of this ever happening. Few things occur infallibly in nature and there may be exceptions to the rule. But at Trail Wood the summer weather forecasters we pay special attention to are the ants when they take to the air on their dispersal flights.

This morning, of all the gauzy-winged hosts that set forth on the calm air of last evening, the only ones I see are the unfortunates that wavered down onto the surface of the pond. Their floating bodies sparkle across the water with little glints of reflected sunshine. Scattered afar, the others have already mated. In small crannies in the ground—their aerial adventure over, earthbound now for the rest of their lives—the fertilized queens are founding their new colonies.

AUGUST 10. It takes a long time to perfect the art of sauntering. Nellie has perfected it far better than I. She lags behind while I roam aside over more territory, seeing, perhaps, a greater variety of things. But always she has something of special interest to relate when she catches up. On this day we saunter with a broad margin of time. We loiter along the way. We have brought our lunch and can spend hours on our trails.

Everywhere we go, along the Colonial Road, beside Hyla Pond, up the Old Woods Road, around Witch Hazel Hill, we are in a time of little toads. We see them hopping over the decaying carpet of last year's fallen leaves. Each is only about half an inch long and so dark in color it appears black. Our impression is that on this day the whole woods is swarming with toadlets. This is also a day of the hunting wasps. We see them running over the leaves in search of insect prey or digging burrows in the hard-packed soil of the paths.

Walking along the same trails through the same woods at frequent intervals, we are continuing our progressive education. We are instructed in the succession of little events that mark the beginning, the middle, and the end of each of the seasons. We see the small advances and retreats of nature. Isolated encounters take their place in the general plan.

Today, as always, we are leaf turners in the summer woods. Whenever we pause, we look on the underside of the foliage. And thus we are introduced to many a small and unfamiliar creature. Another of our pleasures, at this time of year, is to look up at leaves of varied forms where they hang luminous in the backlighting of shafts of sunshine probing down through openings in the canopy above. Particularly beautiful in such lighting today are the mitten-shaped leaves of the sassafras and the broad, veined ovals of the witch hazel.

It is noon before we reach the Brook Crossing. We sit on the chestnut-plank bench amid the ferns and partridge-berry vines and eat our lunch. The hot woods around us echoes with the shrill singing of the August cicadas. As we eat, we watch the slow, drifting flight over the water of a large crane fly with legs strikingly banded with black and white and with part of each leg swollen almost to globular form. With these legs outspread like the spokes of a wheel, and with its body tilted almost to a vertical position, it gives the impression of riding through the air supported by half a dozen small balloons.

On this afternoon, instead of turning back at the Brook Crossing, we follow the flow of the water of Hampton Brook for a time and then climb to a dry open expanse covered with wiry yellow grass and scattered bayberry bushes. Here, long ago, a farmhouse stood. During the years we have known it the area has changed hardly at all. Deer trails cross it as they have for two generations and more. During our earliest days at Trail Wood, we named it the Deerfield.

From the Deerfield we wander into the more remote portions of our Far North Woods. We encounter tangles and fallen trees, damp little hollows, and a swamp we skirt because of its poison sumac. A sense of brooding solitude surrounds us. It recalls an echo from *Lohengrin:* "In distant land, by ways remote and hidden . . ." We turn aside to visit the largest of all our glacially deposited boulders. A mass of stone as large as a truck, it lies half-buried among the trees, a landmark along our northern border.

Here we discover we are in the home of a red squirrel. The area is *his* home and he resents our intrusion. Peering down at us from a limb, it chirps, churrs, spits, coughs, hiccoughs, sputters. Its tail flips. Its body jumps. Its hind feet seem beating a tattoo. It gives the impression of a small motor running away and likely to explode. A red squirrel uses up more energy staying in one place than a gray fox uses in running. Per ounce and inch, nothing develops more indignation than a chickaree.

Its tirade gradually diminishes behind us as we turn back toward home. The diversity of our woodland trails leads us through three miles of varied nature. Toward the end of this wandering day, late in the afternoon, I find myself counting my steps: one, two, three, and so on to ten, then beginning again for another ten. This is something I have noticed myself doing for years, without conscious thought, when I begin to tire. It breaks things into smaller pieces. I remember once meeting a woman who had climbed to the top of Mt. Kilimanjaro in Africa. Her companions had all given up before they reached the summit. She had gone on, she said, methodically counting ten steps, then resting, then taking the same number of steps again and resting. "I climbed the mountain," she told me, "ten steps at a time."

A U G U S T   11.   I lock the door of my log writing cabin after a morning's work and then pause to look up at the top of the frame of one of the inset windows. Like seven parallel pipes of a pipe organ or seven cartridges placed side by side or seven slender cream puffs, the tube nests of the mud-dauber wasp, *Trypoxylon albitarsis,* are cemented to the wood.

In the way memory often links remote events together, as I stand here looking up I experience a sensation of closeness and dampness, or being surrounded by the smooth, water-worn rock of a limestone cavern. For it was in the spring of the year, in Tennessee, on that special day when Nellie and I rode on an underground river deep into the serpentine of Nickajack Cave, that I first encountered this nest of the pipe-organ wasp. Union soldiers, stationed there during the Civil War, had carved their names on the limestone walls near the entrance of the cave. And among these names I caught sight of a cluster of the parallel tubes of hardened mud. They were long deserted but still intact. We examined them for some time. We had seen such nests before only in pictures.

Now, as I stand below my cabin window, the maker of the tubes comes flying in. The sun glints on the burnished black body of the hunting wasp, on its legs marked with white. Clutched in those legs, it carries its paralyzed prey. Always the prey is the same—a spider. With quick, nervous movements, it drags its victim into one of the tubes. Then it reappears and is airborne in an instant. Within each tube the space is divided by masonry partitions into smaller cells. Beginning at one end, the wasp crams in its prey, lays an egg on the close-packed mass of spiders, then walls them in. This process is repeated until all the space in the tube is occupied.

Within each cell a larva hatches from the deposited egg, gorges itself on spiders, and pupates. The adult that emerges bites its way to freedom through the hard mud shell of the tube. In this process some inherited wisdom comes to its aid, for it always appears from its opening on the exposed side of the tube. It never tries to bite its way through the part that is cemented to the support or attached to an adjoining nest.

One habit among these masonry wasps is described as "exceptional among hymenoptera." This is the way the males stand guard over the partially stocked tubes while the females are away hunting spiders. If the nests were left unguarded at such times, small, parasitic cuckoo wasps of the genus *Chrysis* would dart into the cells and lay their own eggs on the food collected for the larvae.

I remember once observing one of the pipe-organ wasps at work on the stone wall of our garage. It was repairing the mud tubes of a previous year, patching holes with new material. I saw it fly away for a fresh supply of mud. Immediately a smaller wasp, brilliant metallic green, alighted and hurriedly investigated the open tubes. It had come too soon. There were no spiders on which to deposit its eggs. With the buzzing return of the mud dauber, the green interloper darted away.

When cornered, one of these cuckoo wasps will roll itself into a tight ball with only its wings exposed. During such a time, an entomologist once watched a dramatic sequence of events. Unable to sting the smaller wasp to death, the nest builder chewed off its wings. Then it flew away. Mutilated, mortally crippled, but undaunted, the small invader—as soon as the attacker left—began crawling upward. Unable to fly, it still reached the nest and deposited its eggs before it died.

A U G U S T 1 2 .   For two days now, we have been watching the same grackle. Of all the tens of thousands of these birds that have flown over or landed at Trail Wood, it is the one we will remember longest.

I first came upon it near the rustic bridge at the end of the dam. Other birds of the kind stalk over the ground with a slow, rather dignified pace, heads held high, occasionally running or making long hops to pick some insect from the air. This individual plodded doggedly ahead. It never ran. It never flew. But it kept in constant motion. It will remain in our minds as The Walking Grackle.

Later we saw it along the lane, on the slope of Juniper Hill, ascending the path up Firefly Meadow, in the Starfield, around the house, and following the winding paths of my Insect Garden. It appeared thin. Either disease or injury to its wings had robbed it of the power of flight. We encountered it over and over again. The distance it covered on foot was amazing. Once I saw it near the house, then on the dam, then on Juniper Hill, and then—all in the space of less than half an hour—back near the kitchen door. It was as though it could not stop walking, as though it felt safe only when in motion.

We can but guess at how many miles it covered in its ceaseless wandering. We never saw it make any effort to hide. Occasionally it picked among the grass clumps in search of food and then plodded on again. All our attempts to capture it failed. Although thin, it appeared strong. There seemed a chance it might recover and regain the use of its wings.

But last night, a little before midnight, a thunderstorm struck with chill rain and violent gusts. This morning when I walk among the dripping plants of my Insect Garden after breakfast, I come upon the grackle dead not far from the sundial.

As I have noted before, grackles are birds I find it difficult to like. Yet, during its long travail, I developed for this stricken individual a fellow feeling and admiration new to my experience.

AUGUST 13.   In May they rose in white clouds densely clothed with pendant clusters of flowers. Now, in August, the wild cherry trees lift branches laden with the shining black of ripened fruit and shot through with the color and motion of darting songbirds. We follow through our field glasses their ceaseless comings and goings amid the hanging masses of the small cherries that provide an abundant harvest. We see the gray-black of starlings, the slate blue of catbirds, the iridescent hues of grackles intermingling with the eye-catching colors of Baltimore orioles, scarlet tanagers and rose-breasted grosbeaks. Three reds of different shades appear and disappear—the rusty red of the robin, the scarlet red of the tanager, and the rosy red of the grosbeak. We watch the birds hop from limb, alight, and take off in their feeding. Once a flock of fifty grackles comes in like a dark scudding cloud and settles in one of the treetops.

The list of birds that feed on the fruit of the wild black cherry includes more than forty species. Among them are flickers, crows, cardinals, bluebirds, crested flycatchers, cedar waxwings, red-eyed verios, brown thrashers, mockingbirds, evening grosbeaks, towhees, and wood thrushes. After gorging themselves on the plump, shiny-skinned fruit, birds such as robins and starlings have been observed perching in nearby trees regurgitating the pits.

In addition to the birds, such mammals as opossums, raccoons, chipmunks, red and gray squirrels, red and gray foxes feed on the wild cherries. About this time of year, I always find fox droppings in the woods filled with the pits of this abundant fruit. In their *American Wildlife and Plants*, Martin, Zim, and Nelson rank this wild cherry as one of the most important sources of food for wildlife.

The two largest mammals feeding on the fruit are the black bear and man. In former times, clusters of the wild cherries— each of the round fruits about a quarter of an inch in diameter— were gathered extensively for making preserves and jellies. The colloquial name of rum cherry dates from Colonial times when the wild fruit of the black cherry trees was employed in flavoring a drink popular in taverns. On my walks during this season of the year, I compete with the birds and the squirrels, picking handfuls of the small cherries. Combining sweetness and tartness, their juice, on sultry summer days, provides tangy refreshment along the way.

AUGUST 14. Several of us at the village post office this morning fall to reminiscing about dogs. Something reminds me of a cocker spaniel that belonged to a friend of mine. It was a dog that loved flowers. Whenever a bouquet was placed in a vase on a sideboard or table, it would stand on its hind legs, stretch out its neck, and sniff the air, straining to get as close as possible to the source of the floral perfume.

This recalls to someone else's mind two dogs that wanted to do what their owners were doing. When nuts were being cracked, one wanted to crack nuts with its teeth. When trading stamps were being licked and pasted into books, the second dog wanted to lick stamps, too.

A third reminiscence concerns an Airedale that had the run of the house. When no one was watching, it began surreptitiously snatching titbits from the table. Then it added an innovation. When it saw the meal was laid out, it would dash to the front door barking loudly. Thinking a visitor was there, its mistress would go to the door. While she was occupied, the dog selected a choice morsel and gulped it down. This stratagem worked a second time and then no more.

The postmaster, Charlie Fox, recalls a puppy—half poodle and half terrier—that used to tag along at his heels when he was working in the yard. One evening, just as the two came out of the house, a barred owl began its hooting where dusk had come to the edge of a nearby woods. The puppy dashed for the kitchen door and whined to be let in. No sooner was it inside than it turned around and, looking out from the protection of the screen door, set up a loud and defiant barking.

This story, in turn, revives a memory of a dachshund that belonged to Evelyn Estabrooks, Charlie Fox's predecessor at the post office. The men who collected garbage once a week in front of the Estabrooks home on Cedar Swamp Road noticed a vast difference in the attitude of the low-slung dog when someone was home and when everyone was gone. In the former case, the dachshund came rushing out, barking aggressively. In the latter, when it was all alone, it would approach slowly, without a bark, wagging its tail vigorously. It was a watchdog that wanted moral support for its barking.

A U G U S T 15 . While the evening advances, Nellie and I, after a walk over the meadows, sit in the rustic summerhouse, once more viewing everything through the bronze mesh that keeps the mosquitoes at bay.

Out across the still water dimples form and fade away in ripples where the white-bellied tadpoles of the bullfrogs break the surface. Among them, sweeping across the mirror of the water, I see the reflection of a bird with long and slender wings. I think: "nighthawk!" and look up. Seven of these graceful birds are passing overhead. After they have gone, Nellie and I run over in our minds the list of birds—flicker, blue heron, green heron, robin, nighthawk, mourning dove—that we remember recognizing from their reflections in the pond.

After a time a blue and white domestic pigeon flutters down on flat Summerhouse Rock. It walks to the water's edge, lowers its head, and takes a long draught, drinking like a horse at a watering trough. By chance, only a few days ago I received a letter from a man in Revere, Massachusetts, commenting on how the pigeons that came to his back yard drank in this way instead of lifting their heads to let the water run down their throats in the manner of so many other birds.

The next event in this evening of small events comes just before we shut the summerhouse door and turn the wooden button to hold it in place. Nellie spies three tiny seeds lodged in a crack in the floor.

"You," I observe, "would make a good sparrow."

Her reply: "Well—if I *were* a sparrow, I'd want to be a *field* sparrow and not a *house* sparrow!"

A U G U S T 16 . The scales have tipped and the days have more of early fall than early summer in them. Mornings dawn with heavy mists that burn away in time. The annual overturn in sound has already taken place. The chorus of the singing birds has diminished; the music of the singing insects has swelled. Tree swallows, moving south, swirl above the meadows. The tide of the goldenrod is rising. In the woods the beaked hazelnuts are large and plump within their snouted envelopes of green. All across the fields seeds seem more numerous than flowers. That, perhaps, marks the time of the seasonal shift in balance, the time when the number of plants that have produced their seeds outnumbers those that are still in bloom.

Wherever we go, from the far end of the Far North Woods to the hillside with its seven springs close to our southern boundary, from Mulberry Meadow on the east to Broad Beech Crossing on the west, we see changes antedating the arrival of fall. This sense of autumn within summer, of the endless intermingling of the seasons, has been felt by many. But none has expressed it better than the Colorado author Ann Zwinger, one of the finest of modern writers about nature. In *Beyond the Aspen Grove*, speaking of how, in her high country, "part of each season is contained in every other," she writes: "This land is a place of all seasons, for even in winter there is the promise of spring, and in spring, the foretaste of summer. The white of snow becomes the white of summer clouds. . . ."

A U G U S T 1 7 .   It is about three inches long, plump, leaf green, its segmented body decorated with rows of small reddish tubercles. A narrow line of yellow runs down the length of either side. It lies motionless in a wheel track of the lane, fallen from a hickory tree beside Mulberry Meadow—the fully grown larva of a luna moth. This is a caterpillar I see but rarely now. The adult moths, so graceful in form, so delicate in coloring, one of the most beautiful of all animal creations—once so plentiful here—are now encountered infrequently. The only one we have seen this year is the one Nellie found clinging to the jack-in-the-pulpit beside the wall at Pussy Willow Corner in June.

Among the lunas of other years, one individual stands out as an accident-prone insect. I first noticed it on the grass outside the kitchen door. It tried to rise at my approach, staggered into the air, and fell. One of its wings was broken. The outer third hung useless. In a painless operation, I clipped away the dangling part and trimmed the other wing to match. When I released it, it fluttered—in balance—into the air. Its flight was more labored but it remained aloft. I watched it disappear toward the brook and thought my contact with it had ended. But half an hour later Nellie came in to report it was ensnared in a spider web. We freed it, cleaned it off, and launched it into the air again. After that this easily recognized insect was seen no more.

While building up the tissues that are transformed into the adult luna moth, the caterpillar spends its time of growth clinging firmly to twigs and leaves. Neither wind nor storm dislodges it. But when its full larval growth is attained—and a pinkish flush spreads across its green—instincts reverse themselves. The urge to cling is replaced by the urge to let go. It releases its hold and plunges, often for a considerable distance, to the ground. There, wrapping itself in a dead leaf, pulling it close with threads of silk, it spins its cocoon. In this manner, during the winter of its pupal life it is protected by an almost perfect camouflage.

As I look down at the larva in the wheel track that runs on before me, I see that it will produce no silk, will weave no leaf-covered cocoon, will never transform into the pale-green beauty of the adult. It is one of the doomed. It lies in the lane unmoving, close to death. One end of its body is blackened, deliquescing, oozing slowly the liquids of decay. It is the victim of one of those devastating bacterial diseases that sweep, at times, through the population of the caterpillars.

I walk on, following the lane on this warm and pleasant morning. How difficult it is for the human mind to accept the fact that bacteria and viruses, the seeds of disease and death, are also an integral part of nature, as much a part of nature's whole as are the luna moth and the pasture rose. All have come into being, all have reached the present time as part of the same overall relationship. Not the beautiful moth alone, not the terrifying bacteria of contagion and death alone, but both are inherent in this ancient, preman scheme of things.

AUGUST 18.   Misty dawn. Bluebird dawn. Little rabbit dawn. The clinging vapor wraps me around, cool and damp, when I go down the lane. I hear the soft, questioning voices of bluebirds descending from treetops hidden in the mist. Once two pass overhead, above the ground fog, calling as they fly. Ahead of me, toward the sunrise, the vapor glows with silvery luminiscence; behind me, it lies milky and opaque.

Just beyond the stone bridge, I come upon a young cottontail nibbling on a dew-covered leaf at the edge of the lane. Because I loom up, a dim shape shrouded in vapor, it lets me approach until I am only a few yards away. I see its fur is wet. When it hops away, it gives two little jumps into the air, shaking itself and sending the moisture flying. It was this same young cottontail, I believe, that I observed on a recent sunny day rolling like a cat in the dust of the lane. It was engaged in a rabbit dustbath. On that occasion, too, it made a small leap into the air and shook itself. But then it was a fine scattering of dust instead of dewdrops that it sent flying.

When I come back carrying the morning paper and near the same spot again, out of the thinning vapor bursts a ruby-throated hummingbird. All the movements of this winged mite are quick, abrupt. It seems unable to conserve energy by slowing down. Even when resting on a twig, it changes position in sudden twists and jerks. I remember one evening when we sat on the terrace watching a rubythroat that had darted down to perch on a telephone wire. It gave the impression of a twitchy little bird—highstrung, rarely still for long.

Now I watch the hummingbird shift from bloom to bloom among the cardinal flowers near the bridge, hanging momentarily on vibrating wings before each, running its slender bill to the hidden source of nectar. A sip and it reverses itself, backpeddling in the air, and darts to another flower. A spider or an ant lion, immobile, consuming a minimum of energy, can wait for a long time for food to come to it. But a hummingbird, because of its rapid metabolism and its prodigal expenditure of energy, must range widely and swiftly in its feeding. It must be—of necessity—a bird in a hurry.

A U G U S T 1 9 .   Down winding paths, among the weed tangles of my Insect Garden, now run wild, I advance with frequent pauses, descending the gradual slope toward the dense stand of spicebushes, the hidden nook and the bench beside Hampton Brook. Along the way I spend more time standing still than moving ahead. Every few steps I halt to observe some fly basking in the sun, some beetle toiling up a weed stem, some spider waiting for visitors to its web.

Once, where weeds and wide-leaved grasses rise tangled together, I stand motionless for several minutes following with my eyes a little pearl crescent butterfly as it drifts in and out, up and down, winging its way under sweeping arches of the grass blades, exploring the herbaceous canyons and caverns of the tangle. When I move on I am thinking how much fun it would be, at a time like this, to possess the size and the wings of a little pearl crescent.

Then sitting on the bench beside the stream, a century-old book lying unopened beside me, I watch events taking place around me. Where glowing streamers of sunshine angle downward from two openings in the foliage, one holds in its spotlight the brilliant red of a cluster of cardinal flowers, the other the white or pale pink terminal clusters of the blooms of a clethra bush. Sometimes called the white bush or white alder and, more commonly, the sweet-pepper bush, its flowers fill the air with a rich and spicy fragrance. I catch it now drifting across the brook.

Minutes go by. Then a movement on the other side of the water catches my attention. On its bright orange legs, a wood turtle is scrambling in slow motion up the farther bank. Ponderously it pushes itself among the weeds and disappears. I bridge the stream on stepping stones and, bending, lift the wanderer for a closer inspection. It hisses gently. I turn it over and see a small nick on the edge of its shell, a nick I made years ago. It is an old friend of mine, the same wood turtle, with the same great luminous eyes, that I have encountered on successive summers along this stretch of Hampton Brook. When I put it down it takes refuge in the water. I watch it sink to the bottom, fight its way against the current, and find sanctuary beneath the overhang of a flat rock.

For a time all is still except for the shrill drone of the cicadas in the mid-afternoon heat. A damselfly flutters past in silence in the shadows. I pick up the book that lies beside me. It is a gift from Walter Harding, a duplicate copy of the insect guide in Thoreau's library in Concord: *The Life of North American Insects*, by B. Jaeger, published in 1859.

I thumb through it once more without finding much of interest. The pages seem desiccated. Drained away, lost entirely are the living aspects of the outdoor world. Gone are the interest, the beauty, the mystery of the animate insects. As I close the book, my eye is arrested by one line of the title page: "With numerous illustrations from specimens in the cabinet of the author."

Those eleven words underscore the shift in emphasis achieved by natural history guidebooks during the past half century. The dramatic change began in 1935. There appeared in that year the first of the classic Peterson Field Guide Series—Roger Tory Peterson's *A Field Guide to the Birds*. It was based on field marks, on observing the bird in the bush and on the wing instead of in the hand. Increasingly since then, attention has shifted from the dead specimen to the living creature, from the laboratory to the out-of-doors, to nature alive and in action.

A U G U S T 2 0 .   With the going down of the sun at the end of this incandescent day, we walk along the water's edge, circling the pond. The rocks shed turtles as we come down the slope to this water field of ours. Caddis flies dance above the shallows where snails, with gelatinous clumps of extruded eggs adhering to their shells, float on the still surface. We hear a splash and look up to see a kingfisher rise with a sunfish in its bill. Close to the spot a green heron is stalking with deliberate movements along the edge of Azalea Shore. As the kingfisher rushes away unreeling its rattling call, the heron flaps into the air and pursues it as far as the boundary trees. One fish eater resenting competition from another.

Toward the end of our walk, where the pondside path curves around Whippoorwill Cove, we find the tall grass has gone to seed. When we brush against the stems, dry seeds fly out and droop on the water. With them, occasionally, go one or two leafhoppers. They, too, are pale yellow. At first glance, floating leafhoppers and floating seeds are indistinguishable. Only when a seed begins to swim am I sure it is not a seed but a leafhopper. The sunfish, as well as I, are often deceived. A score of times I see bluegills strike at alighting seeds, start to gulp them down, discover their mistake, and spit them out again.

Instead of a leafhopper, on one occasion, it is a small black field cricket that takes fright at our approach. It leaps blindly into the water. But it lands without a splash. By pure chance it descends directly on the head of an enemy, a green frog resting partly submerged in the shallows. There it clings for a moment. Then it gives another leap back toward the shore. The green frog, mouth agape, jumps after it but misses. We see the insect splash down a couple of inches short of the safety of the weeds. Rapid kicks of its jumping legs drive it ahead. A cricket favored by fortune, it climbs out unharmed among the vegetation.

AUGUST 21.   Small green apples lie in the green grass. I watch a chipmunk, foraging for food, arrive under the tree where the sour fruit has fallen. For a moment I look away. When I glance back I see the little animal making short jumps, carrying in its mouth an apple fully as large as its head. I am surprised it can hold it with its teeth. After each five or six short, top-heavy leaps, it stops and rests. Progress is slow, but it keeps going, and I finally watch it disappear among the stones of a wall.

The chipmunk is sampling a new taste. Hopefully it is one that will make it forget the flavor of petunias. For chipmunks seem born with a craving for petunias. Nellie—choosing chipmunks instead of petunias—has substituted other garden flowers. I remember a friend of ours who was baffled by a mystery in her garden. In an elongated bed, she noticed her row of petunias vanished except for a dozen feet at one end. There, near a kennel where a dog was chained, they flourished. Measured by the length of the chain, that area was patrolled and protected and there the chipmunks feared to venture.

Several years ago, friends of ours arrived from Cambridge, England. One of the things they wanted to hear especially was the chipping of a chipmunk. Before they came—and after they left—the striped rodents were all around the house. But during their visit we saw only a quickly appearing and disappearing form, and the loud, familiar chipping we heard not at all.

This is something we have noticed repeatedly. When visitors come—when new forms, new clothes, unrecognized strangers arrive—the wild creatures around the house pull back. They become less evident, more wary, more secretive. Almost at once these nonhuman neighbors of ours thin out and draw farther away. In our own comings and goiings about this old farm, our chipmunks and rabbits and birds know us. We are part of their environment. Many of the creatures appear to consider us as harmless as cottontails. And that, so far as we are concerned, is a compliment.

We could easily tame or make pets of some of these natural inhabitants of Trail Wood. But this we do not do. It would be a source of pleasure for us but it would be a handicap for them. They would lose their natural wariness, would be less fit to survive. They become used to us, are unfrightened when they see us. But they never come too close. They live in an unstable world that can suddenly turn hostile. They maintain a margin of safety. They are as we would have them; they are on their own, as nature intended them to be.

A U G U S T 2 2 . The feather first catches our eye. No more than three-quarters of an inch long, it is tipped with scarlet. Like a tiny, perfect jewel resting on the rich green velvet of a case, it lies on a cushion of moss beside the trail—the body feather dropped by a molting scarlet tanager.

Then we notice, a foot or two away, something else clad in red—this time orange-red. Each about an inch long and shaped like a small, juicy persimmon, three plump objects lift above the moist soil and woodland litter. We recognize the trio. It is comprised of the fruiting bodies of one of the rarest and most interesting of all our fungi, *Colostoma cinnabarinum.*

We first encountered it at the base of an oak tree just south of Old Cabin Hill during our earliest summer at Trail Wood. Farida A. Wiley of the American Museum of Natural History was visiting us at the time. When she returned home, she showed a color photograph I had taken to the curator of the cryptogamic herbarium at the New York Botanical Garden, Dr. Clark T. Rogerson. He identified our mystery growth. On mycological lists, it occupies a place between puffballs and earthstars. The genus, *Colostoma,* is represented in America by only three species. Our species has been recorded from New England to Florida and as far west as Iowa and Texas. But in New England it is rather rare. In the collection of the New York Botanical Garden at that time, there were only two other records for the state of Connecticut, both made near New Haven in the 1890s.

Nellie and I bend over the three fruiting bodies beside this path leading to the Old Colonial Road. We run a thumb and forefinger over their smooth, almost slippery exteriors. The outer wall is thick, viscid, transparent. Visible beneath is a thinner bright vermilion inner layer. In a short time the outer layer will burst open, slough away, and lie in a jellylike mass around the stem. This shedding of the outer layer is comparable to the spreading out of the points of the earthstar. In both instances, what is left is the case containing the maturing spores.

By the time fall is well advanced, the red will have faded from the *cinnabarinum* spore case. It will have contracted, become more globular, assumed a grayish hue. Its outer skin will have hardened and it will perch at the top of a thick and slightly curving stem. From late fall on into spring, a jar or slight pressure will send a puff of pale ochre-yellow spores spurting upward from a small opening at the top.

Almost every year since our original discovery, we have sighted this comparatively rare botanical inhabitant of Trail Wood. So far we have found it in four separate stations—in its original site near Old Cabin Hill and in three other stands, all amid moss along the trails in the vicinity of Fern Brook and Lost Spring Swamp. We continue our walk on this August day debating the possibility that we ourselves may be responsible for the appearance of *Colostoma cinnabarinum* at its three latter stands. Perhaps it has become established here along these trails we follow through spores that, unawares, we have transported on our shoes.

AUGUST 23. Coming home in the late afternoon I cross the meadow to the house. For unhurried hours I have wandered along the Ground Pine Crossing Trail, past the Lost Spring, to the top of Old Cabin Hill, down the ridgetop leading north, through the North Boulderfield, across a narrow stretch of lowland, up the slope where moccasin flowers bloom in the spring, and home by the Old Woods Road.

I return with tautness gone, with delight in simple things heightened, with a sense of health and sanity and well-being. I feel more calm, more capable. I have been in contact with the enduring and the real. Cares have shrunk to proper proportions. For many of us a return to the out-of-doors is more than a pleasure; it is a basic need essential to physical, psychological, and emotional welfare.

Thirty centuries ago, the Greeks repeated the fable of Antaeus, the giant who in mortal combat renewed his strength by contact with the ground, who was invincible so long as he was touching the earth.

A century and a quarter ago, in Concord, Henry Thoreau expressed it this way: "I think I cannot preserve my health and spirits unless I spend four hours a day—and it is commonly more than that—sauntering through the woods and over the hills and fields, absolutely free from worldly engagements."

A quarter of a century ago, Aldo Leopold was putting down his expression of this feeling in words that have been repeated so often they have assumed the character of a kind of modern manifesto: "There are some who can live without live things, and some who cannot. For us of the minority, the opportunity to see geese is more important than television, and the chance to find a pasque-flower is a right as inalienable as free speech."

This feeling is a thread that has run through the ages. It has echoed through the thoughts of men remote from each other. This emotion of drawing calm and strength from contact with the earth is one known in widely different times and expressed in varied ways.

A U G U S T 2 4 . Scattered over all the former pasture fields that lie outspread around us in the August sunshine today, we see the last little splashes of bright golden color that previously have been so numerous. Each floral cluster marks the place where a common Saint-John's-wort has taken root.

Naturalized from Europe, this perennial wild flower has extended its range as far north as Newfoundland and British Columbia and as far south as the limits of the United States and beyond. It blooms as freshly in our New England meadows now as it did in English wasteland when Monday was "moon's day" to the early Saxons. An ancient herb, it was employed for centuries for curing wounds. One of its common English names is "touch-and-heal." During medieval times it was gathered and hung in doors and windows to ward off evil spirits. Its name comes from an old belief in England that it begins to bloom on June 24, the day of St. John the Divine.

A closer look reveals a many-branched plant, stiff of stem and small of leaf. Nellie and I run our fingers down the stem and find that two raised lines extend along its length. This feature, together with small translucent dots—oil-producing cells—that are found on the dusky-green leaves, form the identifying field marks that set the common Saint-John's-wort apart from other species of its kind. These translucent "perforations" in the leaves are the source of the second part of the scientific name Carl Linnaeus bestowed on the plant—*Hypericum perforatum*. We look at the shining yellow flowers through our pocket magnifying glass and note another characteristic of the plant. Along its margin each golden petal is decorated with tiny dots of black.

To us this plant is one of the beautiful wild flowers of summer. To the thrifty farmer it is an irksome weed. Its hold on life, its ability to survive under varied conditions make it one of the tenacious plant successes. Nowhere in America has this success been greater than in its rapid expansion, since about 1900, across the rangeland of northern California near the Klamath River. There this same plant, our common Saint-John's-wort, is the notorious Klamath weed or goat weed that has spread over an estimated 400,000 acres in that state and since has invaded cattle country in Oregon, Washington, Idaho, Nevada, and Montana.

Not only does it impoverish the range by replacing more desirable species of plants, but it is poisonous to livestock. Although the Klamath weed rarely causes death, the cattle that eat it become sore-mouthed and scabby. White patches on their skin become more sensitive to sunlight and blisters form on these unpigmented areas. On ranges where grasses dry rapidly in the spring, the Saint-John's-wort often becomes the most abundant of all the varied plants.

For almost half a century this hardy immigrant to the western states evaded control. Then one of the leaf-feeding *Chrysolina* beetles was imported from abroad. In test plots the insects wiped out the Klamath weed over areas of as much as fifty acres. The beetles now are becoming naturalized. How far they will go in eliminating the tenacious plant and what other, more desirable plants they will begin consuming if the Klamath weed is greatly reduced are chapters still to be written in the New World history of the Saint-John's-wort.

A U G U S T 2 5 .   Four stepping stones, glistening with spray and green with moss, carry me over Hampton Brook a dozen paces downstream from the waterfall. Almost every living thing, except the singing insects, is dozing in the midafternoon heat. Feeling only partially awake, dull witted, drugged by the oppressive warmth, I climb the steep path that angles up the farther bank and start on through the meadow grass toward the gray pile of the hired man's monument.

I have advanced only a few yards when, with the suddenness of an exploding land mine, the air around me is whirring with hurtling forms. My ears are filled with an airy confusion of sound produced by many wings. Nine bobwhite quail shoot in all directions from almost beneath my feet. I hear their little cries of alarm. I see them laboring on their stubby wings up and over the rise and out of sight.

Camouflaged by their stripings of grays and browns and whites and blacks, the nine birds have been clustered in a ring, tails together and heads pointing out in all directions like the rays of a starfish. In winter coveys of as many as thirty birds will take up such a position. By this stratagem, when all are resting, the quail are able to obtain advance warning of danger from all points of the compass. Then, when danger comes—when a dog or fox surprises them—with each bird launching itself in a different direction, most are likely to survive. Had I been a predator, the surprise and distraction of their explosive burst into the air would have prevented instantaneous reaction or my concentrating on one particular bird. I watch the birds go, marveling at this age-old artifice that they employ.

For how many unnumbered generations have the quail been relying upon this simple and effective stratagem? When—and how—did it originate? I know of no other bird that makes use of it.

AUGUST 26. Something Samuel Johnson said in the eighteenth century is responsible for the fact I am sitting on this moss-plated boulder in the shade of this maple tree overlooking this expanse of open fields on this afternoon in August.

Once more I have crossed the four stepping stones and ascended the path to Monument Pasture. This time there is no explosion of quail—only the shimmering heat and the sound of the stridulating insects shrill across the field. I have come on serious business, bearing long-handled clippers, more or less intent on trimming an opening for a new path into the woods. But in the heavy warmth my resolution wavers. It topples completely when a sentence from one of Johnson's conversations with Boswell comes to mind.

"No man," Johnson had declared, "is obliged to do as much as he can do. A man is to have part of his life to himself."

I turn aside. I lay down my clippers. I sit on this shaded, cushioned rock. I decide to enjoy the part of my life that belongs to me.

A catbird mews. A Carolina locust dances in the sunshine. Cumulus clouds drift overhead trailing islands of shade across the woods and fields. The day dreams.

I watch a soaring hawk turning in tight circles, riding a thermal updraft in a spiraling climb up and up into the sky. I try to imagine how, on this day, this familiar scene around me—the valley of Little River, the winding country roads, the fields, the woods, the dappling cloud shadows—must appear from the vantage point of its lofty climb. Then my gaze drops from the soaring hawk. I follow the hurried, fluttering flight from bush to bush, only two or three feet above the ground, of a diminutive field sparrow.

I consider the two—one climbing so high, soaring so free, the other almost scuttling from cover to cover low above the ground. For a moment my mind toys with the idea of the sparrow looking up with envy at the high freedom of the hawk. But I discard it instantly. The answer is obvious. The sparrow is intent only on what concerns its own life, only with its own niche in the world, only in carrying on its own inherited pattern of behavior. Each natural form of wildlife is streamlined to the things vital to it, streamlined to necessity. The hawk soars and the sparrow flutters from bush to weed stem and back again. Each plays but a single role. This is a feature that runs through all nature—with a single exception. I am part of that exception.

Man is discontented with a single role. He wants to play all roles. Questing, unsatisfied, he is perennially the discontented animal. The wants of every other living thing, Henry George pointed out, are uniform and fixed. Man is the only animal whose desires increase as they are fed. His wants forever outstrip his gains.

This restless quest has produced the field glasses through which I watch the hawk and the sparrow. It has produced the materials with which I set down these words. It has also produced the poisons that drench the land and flow down the rivers, the chemical pollution in the air, the abused forests, the extinct wildlife, the threat to man's own survival on earth. What direction this questing, this eternal discontent, will take in the future is of concern to the hawk and to the sparrow—and to me.

AUGUST 27. The moon, now one day from full, clears the black cutouts of the silhouetted trees and spreads the glow of its light across the meadows. Looking up, we note high above us a faint sheet washing over the dark foliage of the hickories. There the long contention of the summer night—the "Katy did!" "Katy didn't!" "She did!" "She didn't!"—has already begun. One caller gives an added flourish: "Katy did! She did! She did!" Where two insects seem in agreement, both repeating "Katy did," the rhythm of their calls suggests to Nellie the words: "Go to bed, sleepy head!"

Each year, about the end of the first week in August, we hear the initial call of a katydid in our hickory trees. With that sound the year has reached one of the small milestones of summer. From then on the wrangling of the contentious insects mounts in volume, goes on and on, continues until the first black frost of fall.

Listening in the shadows we count six different variations of the call: "Katy did!" "Katy didn't!" "Did!" "She did!" "She didn't!" and "Katy she did!" A few times we hear one of the insects continue on in a kind of excited stutter: "Katy did-did-did-did!" We wonder if a female has come close and aroused the calling male. Because these insects are slow-moving creatures, the process of finding a mate no doubt is a deliberate one. The repetitious calling of the male is needed to guide a female to the spot.

Surrounded by the warm moonlit night, with the debate of the katydids coming down from the dark bulk of the trees overhead, we speculate on how territorial these insects are. So far as I can discover, no one has made a thorough study of this aspect of their lives. Because they are found most often high in trees, they are difficult to mark and difficult to observe. But from the nature of their mating, it is logical to assume that the males remain in the same tree and, it may be, on the same branch rather than move about. Night after night I hear a katydid calling here, there in the tree exactly where I heard a katydid calling the night before. All I can say with certainty is that it is *a* katydid. In such instances I have no proof that it is the same individual. But the simpler of the two explanations is that I am hearing the same individual rather than that, by coincidence, I am hearing different insects calling from the same identical location.

More conclusive evidence of a territorial sense in katydids, it seems to me, is provided by something else we have heard on recent nights. One of the hickory trees is inhabited by a katydid with an abnormal call. Perhaps one of its wings hardened out of shape. The sound it produces is less clear, less incisive, more hoarse and muffled. At times it seems echoing itself. We can recognize it among all the others. The striking thing is that whenever we hear this insect the sound comes from the same part of this one tree and never from any other point.

A U G U S T 2 8 . If I were asked to pick the strangest form of animal life we have at Trail Wood, I think I would choose the small primitive creature I have found at the western edge of the pond. It is part of a smooth, brownish, gelatinous mass about the shape and size of half of a large watermelon. The mass adheres to the side of a submerged boulder only a few feet out from the path on which I stand. Its exterior is engraved with markings. Some five-sided, some irregular, these rosette shapes combine to form a pattern that vaguely suggests the outside of a pineapple. Each rosette represents the space occupied by one cluster of the tiny animals that live on the surface of the mass. As the colony grows, the bulk of the jelly—produced by the animals—increases.

I remember one year when such masses, large and small, appeared in a dozen places around our pond. Most were attached to submerged branches of fallen limbs. They were globular, some no larger than small oranges, others the size of a grapefruit, still others as big as a basketball. In addition, at other places in the shallows, they were elongated and attached to rocks. It is always in the month of August that these colonies become large enough to attract attention.

When I first encountered them at Trail Wood, I thought I was discovering some kind of freshwater sponge. In the past such masses have been described as corals, seaweed, and hydroids. Their true nature and identity was finally established less than a century and a half ago. The mass on the submerged rock before me is formed of a colony of thousands of bryozoans or moss animals. These creatures live in fresh water and also in salt water, where they originated. The species in our pond is *Pectinatella magnifica.*

With a stick, I break off a small fragment from one end of the mass and maneuver it within reach. I carry it home, place it in a bowl of water, and Nellie and I examine it with care. Embedded in the solid mass of jelly, invisible except through a magnifying lens, the individual bryozoans appear slender, cylindrical, and partly surrounded by a limy wall which persists after death. As we watch them without moving, we see heads appear surrounded by horseshoe-shaped or circular wreaths of waving tentacles. Each tentacle is covered with cilia that sweep through the water swirling diatoms and other microscopic organisms into the creature's primitive mouth.

Faster than I can follow the movement, when I jar the bowl slightly all the heads, all the nodding tentacles snap out of sight. A minute goes by. Then they reappear. From end to end across the fragment of the colony there are again the exposed heads and the swirl of the moving tentacles.

After a time I carry the bowl back to the pond, emptying its contents into the food-filled water. I watch the little mass of jelly with its freight of living creatures sink back into its submerged habitat among the rocks of the shallows. There, in the course of a few weeks, the individual bryozoans will enter one of the strangest periods of their annual cycle. Nothing is more surprising about these creatures than their method of reproduction.

As fall comes on, thousands of buds, or statoblasts, will appear on the surface of the jelly. The colonies die and the statoblasts—some resembling small doughnuts—wash away. Tough cushions buoy them up. Around their perimeters run circles of strong hooks that anchor them to submerged twigs or trash or to the feet or feathers of the water birds that sometimes transport them for long distances to other ponds. These minute statoblasts function like burs. They are resistant to cold and resistant to drought. When winter is past and spring warms the water of the ponds again, each statoblast is capable of founding a whole new colony of the primitive creatures, a colony that in some instances may grow until, like a city, it contains tens of thousands of individuals.

A U G U S T   29.   Once, twice, a dozen times—first toward the north, then overhead, then toward the south—the sound comes down from the sky. It is a single note, like the twang of a plucked musical string, a sharp, metallic "plink!" Nellie and I stand in the after-sunset light looking up, seeking to make out the little band of southward-flying birds high above us. They pass on—bobolinks in the vanguard of migration. They are leaving New England fields behind them, beginning one of the longest migratory flights made by any songbird.

Now they look down on the fields of Trail Wood. At the end of their journey they may look down on the pampas of Argentina, where W. H. Hudson spent the boyhood he describes in *Far Away and Long Ago*. Their short call note is a sound we catch only in passing. But it is a sound that these journeying birds will hear across the sky all the way to South America.

In my mind's eye I follow them down the map—to southern states, where they are "reed birds" or "rice birds" and where the slaughter by meat hunters was once so great that nearly three-quarters of a million bobolinks were shipped to market from one small town on the South Carolina coast. Then on to Jamaica. There the black and white migrants, with the fuel of surplus fat stored up for the journey ahead, were called "butter birds." And so on across the Caribbean toward their wintering grounds. Most of this migration occurs at night. In New England, as the nocturnal flocks pass overhead, the plucked-string note that holds the traveling birds together is a frequent sound heard in the darkness of the August nights.

Home for the bobolink, no doubt, is where it hatches, where it nests—in the fields of North America. Yet it spends only about three months, one-quarter of the year in its breeding area. For nine months out of twelve it is a bird that is "away from home." Like its near relative, the red-winged blackbird, it alters its feeding habits when autumn comes. Mainly insectivorous in spring and summer, the bobolink turns to the consumption of seeds in the fall.

Nellie and I stand in the advancing dusk looking after the departing birds. Will they—the better part of a year hence—look down on Trail Wood fields again? Will we hear their song in another spring—joyous, rollicking, the musical notes tumbling over each other like the jingling of coins? It is a song we hear less frequently now. Only occasionally, and not every year, bobolinks come to our pastures. For this bird, the "meadow wink" of an earlier New England day, is now a sadly reduced species.

Many factors besides the merciless slaughter of so many years along its migratory route have contributed to its decline. A nester in open, extensive fields of tall grass or grain, it was originally benefited by the cutting of the forests and clearing of land in eastern North America. But during the past century and more, as farms have been deserted and fields have become overgrown, the favored habitat of the birds has steadily decreased.

Moreover, changes in haying methods have reduced the chances of survival of the young. At the time when the bobolink population was at its peak, mowing was done largely by hand and rarely commenced before the first or middle of July. By then the nestlings were safely on the wing. With the introduction of mechanical mowing, however, the harvest began earlier, before the birds were fledged. Mowing machines cut closer to the ground. And even when nestlings escaped the chattering cutting blades, they were often killed by the horse-drawn rakes that scraped the dried hay into windrows.

Another factor in the bird's decline has been the replacement of the horse by the tractor on the farm. The demand for hay decreased. Fewer fields have been devoted to the tall grass the bobolink needs for nesting. Add to this the deadly scourge of DDT and other pesticides in recent years and the picture becomes clear. The little flock that has just passed above us, heading into the night, represents—like the bluebird—a species sorely pressed. This far traveler with its striking black and white plumage and its jubilant song—one of New England's favorite birds— year after year while a century has passed has been fighting a tide that still runs against it.

A U G U S T   3 0 .   On a blur of shining wings beside the flower of a pasture thistle, a red-bodied sphinx moth hangs in the air. It appears newly emerged, immaculate, untouched by the world, a moth as perfect as any flower. Before we turn away we pause to listen to a kingfisher as it makes a last round of the pond, trailing its ratchetlike call behind it. Then, as we come up the path to the house, the white taillight of a rabbit goes bobbing away to disappear through an opening in the wall.

While the light fades, Nellie and I sit on the back steps, where the stones retain some of the warmth of the day. We speak of all the small similar events that, unseen, unrecognized by us, have occurred since the sun rose over our acres. Even at Trail Wood, I am well aware that what I record in this outdoor journal of the year is but a thimbleful dipped from the wide lake of occurrences transpiring unobserved around us. There is more going on than the owner is aware of in every pasture, every swamp, every pond and ditch. What farmer knows a tenth of what takes place in his woodlot?

A U G U S T   3 1 .   At the close of this day that ends the month of August, we sit on a bench beneath an apple tree gazing across the meadow, looking once more toward a summer sunset. As far as we can see, cloud beyond cloud, multitudinous midges climb and plunge in the shimmer and vibration of their mating dance. They rise and fall in a luminous host, each minute insect in its own halo of light. We see perhaps a million tiny moving bodies, all shining like moats drifting in a sunbeam.

Scything through these clouds, each surrounded by a larger glitter of light, dart and hover a hundred or more dragonflies.

They are mainly *Anax junius,* the green darner, and the darker *Epiaeschna heros.* We watch them swinging back and forth through air alive with food. During the late-summer days, these largest, swiftest, most voracious of our North American dragonflies migrate south like birds, journeying for hundreds of miles down the coast. At the end of this day, their migratory movement has brought them to Trail Wood at a propitious hour—an hour of calm when the dance of the midges has produced an airborne feat, edible clouds, on which the dragonflies can dine.

We follow with our eyes these slender insect falcons streaking low above the meadow grass, whirling as high as the treetops, halting in midair, darting on long straightaways as they hawk among their smaller prey. All along the way on their southward movement at this time of year, food is plentiful.

I have seen such migrating dragonflies skimming over the dunes on the south shore of Long Island, weaving between skyscrapers in Manhattan, feeding among the Spanish oaks of Cape May Point on the New Jersey coast. But it is only during years of unusual abundance that their migratory movement is noticed. Usually when large concentrations arrive in an area it is assumed it is comprised of a temporary assemblage of local dragonflies. This conclusion is strengthened by the fact they soon disappear, often on the following day.

This southward journey of these larger dragonflies is unhurried. Like the swallows, they feed in the air as they advance. One year, on the eighteenth of August, dragonflies were everywhere at Trail Wood. A week later, the concentration had reached New York City. Officials were reassuring alarmed citizens in all the boroughs that the invasion consisted of beneficial, not harmful, insects.

During the slow downward drift of the sun, until it reaches the trees on the ridge to the west, we remain engrossed in watching the endless repetition of the same activity—the luminous dragonflies, riding between the twin shimmers of their wings, hurling their open jaws and sweeping the net of their bunched, spined legs through the clouds of haloed midges. Then the sun sinks and the backlighting leaves the meadow. All the insects, large and small, merge with their background. We can glimpse only the highest of the dancers, the uppermost of the sweeping dragonflies. Shorn of their glitter, the multitudinous insects move in silhouette against the still light-flooded sky.

SEPTEMBER 1. The transient beauty of the flowers—that blooming of the plant world that J. H. Fabre, the French naturalist, once referred to as a "frail magnificience"—has now largely left the fields. Today I walk among seeds where once I walked among flowers. Remembering the long succession of the summer blooms, I recall many of their common names, repeating them just for the pleasure of the sound, just for the enjoyment of the picturesque and imaginative designations they have received.

Nowhere else in the realm of nature—except perhaps among the butterflies with their whirlabouts and firery skippers, their spring azures and brown elfins, their anglewings and hairstreaks and metal marks, their satyrs and hop merchants, their mourning cloaks and dusky dreamy wings—has the poetry of naming reached such heights. Walking slowly along, I run over the list of plants whose flowers we see coming in season to Trail Wood. I repeat them aloud: goldthread, windflower, wake robin, gay-wings, thimbleweed, Queen Anne's lace, enchanter's nightshade, pussytoes, turtlehead, blue flag, jack-in-the-pulpit, blue-eyed grass, fleabane, meadow rue, buttercup, boneset, baneberry, ladies tresses, black-eyed Susan.

Was there ever a real Susan whose name was bestowed on this flower of the fields? No one knows. Who first referred to plants as the gaywings, the jack-in-the-pulpit, the windflower, the wake robin? No one can guess. When and where and under what conditions did the inspiration come for such imaginative common names that have survived generation after generation, names as old or older than the scientific designations of the botanists? No one can tell. Lost in the remote past is the source of most of these names we commonly use. Creating such apt and striking designations is a rare art of the imagination. It is part of the poetry of the common people of the past. I continue my walk on this day thinking of the debt we all owe those nameless namers of long ago, the ones who, unrecorded, gave to wild flowers the colloquial names—gay or sinister, poetic or down-to-earth—that have enriched the language of botany.

S E P T E M B E R 2 .   Here is something curious. Nellie and I, following a leisurely path coming home from the Brook Crossing, reach a white oak tree where, yesterday afternoon, Nellie saw one of those striking black and yellow wasplike sawflies, the *Tremex columba*. She is just telling me how it withdrew its stout wood-boring ovipositor, or egg placer—the "horn" at the end of its body that gives it its common name of horntail—and flew away to land on the limb of a nearby sapling when I spot its black and ochre-yellow colors. It is clinging exactly where Nellie saw it alight twenty-four hours ago.

We watch it for several minutes. It remains motionless. I shout. I wave my arms. It does not stir. I shake the sapling. Instead of taking wing, the insect tumbles to the ground. I pick it up and find it is dead. The females of these wood borers not infrequently have difficulty withdrawing their ovipositors when their egg laying is completed. They are then near the end of their lives and may die anchored in place.

Within the wood of the dead or dying tree in which the *Tremex* implants its egg, the cylindrical, large-headed larva hatches into a life of darkness. Slowly eating its way, burrowing through heartwood or sapwood, it leaves behind it a tunnel that grows larger as it increases in size until, after it has transformed into a pupa within a silken cocoon and emerges as an adult early the following summer, it leaves an exit hole about the diameter of a lead pencil. The winged and strikingly marked adult insect that appears is almost two inches long.

During their months of darkness, larvae consume the wood of various trees, including oak, maple, beech, apple, elm, and sycamore. There they are secure from almost all enemies except one, the large ichneumon fly, *Megarhyssa lunator*. This parasite flies slowly through the air, trailing behind it a threadlike ovipositor that is often more than four inches long. The entire length of the insect, including the ovipositor, is nearly half a foot.

A specialist in finding *Tremex* larvae within a tree, it prospects down the length of a branch or trunk, playing the supersensitive smelling organs of its antennae over the bark until it locates a hidden grub. Then, lifting high its abdomen and standing to the full length of its slender legs, it hitches its ovipositor into a vertical position and begins drilling downward. It has the appearance of an oil-field derrick in operation. When it reaches the larva or its tunnel, it deposits its egg. The parasitic grub that hatches from it is uninterested in wood. It is carnivorous. It devours the *Tremex* larva.

In earlier times when wood was seasoned naturally and not dried more quickly with the aid of kilns, the burrowing grubs of the horntail were the subject of stories of miracles in the form of living insects appearing from table legs and chairs that had been in use in a family for many years. The explanation is that in the abnormal dryness of the seasoned wood the growth of the larva was slowed down. It required, often, a considerable number of years, instead of one, to reach its adult stage.

SEPTEMBER 3. For days now, monarch butterflies have been drifting through. While birds are migrating south by night, in starlight and moonlight, these butterflies are moving south by day, in brilliant sunshine and under the shadows of the clouds. The night of the migrating songbird finds its parallel in the day of the migrating monarch.

Everywhere this year people are talking about the abnormal numbers of these orange, brown and black milkweed butterflies. On this September afternoon I count more than a hundred during my meadow walks. A dozen flutter around a single stand of goldenrod at the foot of Firefly Meadow. And where long yellow shoals of the massed flower heads run along the edges of the old pastures, I see a continual procession of butterflies alighting, hanging in the sunshine, drinking nectar, taking off, and alighting again. A favorite resting place for these migrants, I find, is a sheltered corner of the Starfield at the edge of the North Woods. When I push myself through the forests of the goldenrod, a wave of butterflies takes wing before me. The richness and darkness of the orange hue in their markings vary widely with the individuals. It is this hue that originally gave the butterfly the designation of monarch. It was named for William of Orange.

For more than a hundred years it has been known that insect-eating birds tend to avoid monarch butterflies. For most of that time this aversion was attributed to the "bitterness" of their blood. Only in recent years has the true explanation revealed a more important basis for the reaction of the birds. In 1968, chemists in Switzerland reported the isolation of a digitalislike heart poison from such butterflies and also from their black-and-white-and-yellow-banded larvae.

More recently—in the April 4, 1975, issue of *Science*—two research workers at Amherst College in Massachusetts, Lincoln P. Brower and Susan G. Glazier, announced the results of further experiments. The insects, they found, store the cardiac glucosides in various parts of their bodies—even in their wings. One of the first effects of this poison is to produce violent vomiting. Thus birds or animals swallowing a monarch butterfly may get rid of most of the poison and so survive. But they have learned a lesson. Henceforth they avoid the strikingly marked and easily recognized insects.

All this day, from soon after dawn until dusk grows denser, this butterfly parade continues moving through. But there is no sense of haste. The journeying insects pause. They sip nectar. They turn aside. But the general trend of their movement is toward the south. It is long after sunset when I look up in the twilight and see a monarch flying overhead, a small, black, fluttering form against the fading glow in the sky. It is headed south.

S E P T E M B E R   4 .   In *The Narrow Road to the Deep North*, that classic story of the seventeenth-century wanderings of a Japanese poet, Bashō includes descriptions of such excursions as "a moon-viewing trip" and "a snow-viewing party."

On this day, three centuries after Basho, our own wanderings in our own woods might well be called "a fungus-viewing trip." Accompanied by our knowledgeable friend, Farida A. Wiley of the American Museum of Natural History, who has come up for several days, we start out after breakfast.

There are times when Nellie and I have the feeling we are seeing "birds at their best." On this hot and humid day we have the feeling we are seeing fungi at their best. Rising from the damp woodland mold, clustered on decaying stumps, growing on the bark of dying trees, dense on moldering logs—some like parasols, some like clubs, some like shelves, some like dense hair, some like masses of coral—fungi, in their varied forms and sizes and colors, arrest our attention all along the way.

One of the most beautiful is one of the first we see in the woodland. Beside the decaying ruin of an ancient chestnut stump—plated with the vivid green of lichens—it lifts its bright violet parasol. This is the violet cortinarius, *Cortinarius violaceus*. Hardly a dozen steps beyond, where it is sheltered under the scraggy branches of a witch hazel, we bend over a white mass of delicate branching growths bunched closely together. Its name is apt. It is a coral fungus. Later we encounter similar masses, one bright yellow, another tinted a pale shade of lavender.

In diverse forms, other fungi appear around us as we advance: the knobby-surfaced conelike *Boletus* or pinecone mushroom, *Strobilomyces strobilaceus;* the thick-stemmed peppery *Lactarius*, *Lactarius piperatus*, with the milky, pungent fluid it exudes when broken; the parasol mushroom, *Lepiota procera*, at the top of a stem that in some instances reaches a height of half a foot or more; the solid-appearing yellow-capped *Russula foetens;* and a whole charming colony of the tiny wheel mushrooms, *Marasmius rotula*, springing from damp, decaying leaves, each at the top of its threadlike stem and with its broad gills radiating out like the spokes of a wagon wheel. In most cases, Farida recognizes a fungus at a glance. Each time she visits us, we add a few more species to the ones we know.

At the summit of Old Cabin Hill, beside the rough rectangle of stones that once formed the foundation of a small house, we find a large and striking shaggy mass attached to the trunk of a dead wild apple tree. It is white with a tinging of yellow. It resembles a dense beard. What we are seeing is the bear's-head *Hydnum, Hydnum caput-ursi*. Our next species is as beautiful as it is deadly. Graceful in form and immaculate in whiteness, it is the death angel or destroying angel, *Amanita phalloides*. A little latter, in an opening in the Far North Woods, Farida picks a bright crimson mushroom supported by a thick, rough-surfaced stem. This is *Boletus alveolatus*. She turns it in the sun and we notice a sheen running across its satiny undersurface. She breaks the cap open, revealing the firm white flesh within. Slowly, as we watch, we see one of the identifying reactions of this fungus. The white of the flesh turns blue.

Our roundabout circuit of the woods has carried us into the damp, deeply shaded ravine below Whippoorwill Spring when, for Farida, we come upon the find of the day. We are all moving slowly among the delicate fronds of an extensive stand of maidenhair ferns when she spots not only a new fungus for the day but one that—although she recognizes it—she has never seen before. It is the large-club *Clavaria*, the fungus Linnaeus named *Clavaria pistillaris*. Like a thick, short cudgel in miniature, enlarged at its upper end, yellowish-tan in color, spongy within, it rises three or four inches above the ground. When we break it open we discover small insect larvae feeding on the interior.

Ours has been a good day; this has been a good year for fungi; we are fortunate to have made our "fungus-viewing trip" at such a time. For the forms and colors of these growths usually represent a quickly passing show. Nor are we likely to find them growing, like wild flowers, in the same place year after year. Often half a dozen years may pass before we come upon the same type of fungus growing in the same place where we saw it before.

S E P T E M B E R   5 .   When darkness falls we all set out again, this time not on a fungus-viewing trip but on a singing-insect expedition. Carrying flashlights, the three of us head down the lane, our trio of beams running over the weed tangles to our right and left.

Before the rising of the moon, the soft darkness around us rings with the music of the night insects. From the shadowed trees comes the calling of the katydids, from the dark of the meadows the stridulating of the black field crickets, from the dense cover of the massed weeds along the lane the insect music, soft or shrill, of the snowy tree cricket, the green meadow grasshopper, and the cone-headed grasshopper. The game is to trace the sound to the small musician and watch its activity spotlighted in our beams.

Farida is the first to succeed. Probing with her wandering shaft of illumination, she narrows down her search and we all gather around to watch a delicate pale-green creature, snowy and luminous in the concentrated light, moving above its back like two blurring fans the translucent wings that rub together and give off the sweet, mellow, rhythmical music that is the song of the snowy tree cricket. We watch it until one of us, seeking a closer look, accidentally jars a weed stem. Instantly the music ceases.

Moving on, slowly and silently, being careful not to brush against the outer plants of the weed tangles, we advance at a snail's pace down the lane. By the bridge, Nellie traces another insect song—a short, low buzz lasting only two or three seconds—to a different singer. Long and slender and darker in hue, it is a green meadow grasshopper. My first discovery spotlights the source of the shrillest of all these songs of the night. It comes from a leaf-green cone-headed grasshopper with its head rising in a peak like a dunce's cap. Its vibrating wings appear to shake uncontrollably as it rubs them together in a stridulation that fills the surrounding air with a high, steely, ear-piercing buzz like that of a band saw.

These three performers we encounter at other places along the lane. But one of the night musicians, which Nellie picks out with her flashlight beam as we are returning, we hear and see but once. It is the narrow-winged tree cricket, a more slender relative of the snowy. Its song comes to us as a sad, frail, sweet lament. It is often lost in the uproar of the more robust singers of the summer nights. But once recognized, its minor melody—like the song of the white-throated sparrow—has the power to move us deeply.

By the time we reach the house again, moonlight is flooding across the meadows. We stop beside a large clump of catnip bathed in its illumination. Little spots, shining like jewels, are scattered over the leaves and seed heads of the plant. At first we assume we are seeing dew. But when I touch a drop with a fingertip, the thicker fluid smells strongly of catnip. The "dewdrops" are formed of oil exuded by the plant, drops that glitter with reflected light. This is something none of us has ever noticed before—a new form of beauty on a moonlit night.

SEPTEMBER 6.   The red ball of a heat-wave sun rises over the brookside trees. All morning we watch the mercury climb in the thermometer beside the kitchen door. Suddenly we are back in the sweltering heat of mid-July. Farida and Nellie spend part of the day under the apple tree checking in guidebooks the puzzling plants they have seen in the woods and fields. When I glance down at the pond from the shade of the hickory trees I notice four painted turtles lined up on a half-submerged log. They appear to be soaking up warmth for the long winter ahead.

After sunset, in the cooler air of evening, we pack a basket with sandwiches and a thermos bottle of iced tea and descend the slope for a leisurely picnic in the summerhouse by the pond. A small shorebird precedes us along the path at the water's edge. It stops, runs on, stops again. At each pause it bobs its tail in the characteristic teetering motion of a spotted sandpiper. We talk about how, if we could know other birds as well as we know this one, if we could become familiar with all their individual movements and reactions, we could recognize them all at a glance, hardly aware of how we identify them, just as we recognize an old friend coming down the street in the distance by the way he walks or carries his head or swings his arm.

It is almost dark and the sandwiches are gone and the iced tea drunk when three wood ducks come speeding in over the treetops. They circle the pond and then, with wild squealing cries, pitch steeply down to alight on Whippoorwill Cove. As long as we remain unmoving, the waterfowl seem unaware of our presence. Later, when the ducks have taken off again, the three of us follow the darkened path back to the house deep in a discussion of how accurately wild creatures see.

I have noticed that a woodchuck or a grouse apparently observes no difference between an empty hammock and a hammock in which I am lying. The birds we feed in winter, as I have noted before, become accustomed to our coats and hats rather than to us. They react exactly the same if someone else wears our clothing. Weasels, in certain experiments, have exhibited no fear of rattlesnakes when their rattles were removed. In other tests, birds showed no recognition of monarch butterflies when their wings were clipped off. Wild creatures appear to observe things in general rather than in detail.

This was dramatically emphasized by one of the classic experiments conducted by the European student of animal behavior Niko Tinbergen. Above a chicken yard, he towed the cutout of a bird silhouette. It appeared to have a long neck and short tail when towed in one direction; a short neck and long tail when towed in the other. In the first passage it suggested a goose in flight, in the second a hawk passing overhead. The poultry paid no attention when the cutout passed above them in the goose position. But they became greatly excited when it was reversed and moved over them in the hawk position. They were reacting to a quick impression rather than to a detailed observation. In the wild, creatures that wait long enough to see sharp details may wait too long.

SEPTEMBER 7. Red, white, and blue—the red of the cardinal flower, the white of the turtlehead, the blue of the bottle gentian—we see them rising almost side by side along the brook when we look down from the bridge. The air is fresh, the heat less oppressive on this morning when Farida, Nellie, and I start out after breakfast on an unhurried walk over the open fields.

At the edge of Monument Pasture we reach a dense stand of high goldenrod. In these September days, the rough-stemmed goldenrod, *Solidago rugosa,* which sometimes attains a height of seven feet, is coming into its prime. Here it rises above my shoulders, and when five-foot Nellie pushes her way through a stand she disappears entirely. Once in a little bay or glade among the tall goldenrod, we pause to watch a small skipper butterfly whirl around the opening, apparently pursuing a honeybee.

As we come back up the lane, toward noon, I pick up a small piece of windblown paper and stick it in my pocket. This reminds Farida of a woman who went on one of the famous bird trips that—even after she has reached the age of ninety—she still leads around New York City. The woman began tossing away pieces of paper all along the way. When called to account, she explained: "I *always* throw out litter. I think it is my *duty.* It makes *work* for people."

We have almost reached the kitchen door when we discover one of the most interesting things of the day, where the tires of cars and trucks have packed the ground of the lane until it is almost as hard as cement. A red-legged locust is walking slowly about with its antennae lowered. It appears to be hunting for something. Then we see it stop, lower the tip of its abdomen, and work it deeper and deeper into the ground. There follows a pumping motion of the abdomen as a cluster of eggs is deposited. Why does the insect pick a spot where the earth is packed so hard? Is it to protect the eggs from enemies or is it to lessen the chances of their being washed away in the heavy rains of spring? While the process of egg laying continues, the locust supports itself entirely by its abdomen, which is planted in the ground like a post. All the time, its red jumping legs wave in the air.

As we watch it, Nellie tells Farida of the experience of a friend of ours. When she emerged from a visit to the Louisa May Alcott home in Concord, Massachusetts, she noticed a woman on her hands and knees in the grass of the front yard. She thought she had lost something. But it developed that the woman was really engaged in one of the most specialized avocations in the world. Her hobby, she said, was collecting a grasshopper from every famous home or historic spot she visited.

SEPTEMBER 8. On this last day of Farida Wiley's visit, she and Nellie spend hours in the woods trying to add a new fern to the nearly thirty kinds we have recorded for Trail Wood. As author of the simplest and easiest-to-use fern guide I know, *Ferns of the Northeastern United States,* Farida is the perfect scout for such an expedition. Wandering into secluded nooks in the woods, along the edges of swamps, beside half-buried stone walls, among glacial boulders and at the headwaters of tiny streams that wind away between mossy banks, they check on thousands of fronds, hoping to find a Clinton's or a Boott's, a climbing fern or a Massachusetts—possible for our area but ones we have never been able to locate here.

Everywhere they go they see ferns—royal ferns, cinnamon ferns, maidenhair ferns, broad beech ferns, Christmas ferns, long beech ferns, grape ferns, lady ferns, fragile ferns, sensitive ferns—but all ferns they have seen here before. Some species are represented by only a few plants, some grow in isolated colonies, some are scattered all through the woods.

It is late in the afternoon, both are tired, and Nellie's foot is hurting, when they turn toward home after their unsuccessful search. They have enjoyed the beauty of many ferns; but they have missed the thrill of adding a new name to the Trail Wood list.

At the place where the Fern Brook Trail makes its last turn before joining the path across the meadow to the house, they swing aside for a final exploration of a restricted area along the serpentine of Hampton Brook. Here they are among Christmas and New York and spinulose shield ferns. One stretch of the stream bank presents a succession of hollows and hummocks. They have just passed the edge of one depression when Nellie first hears Farida say: "Whoa up, Wiley!"

And then: "Come back, Nellie. Here is something to see."

She is bending over a fern that Nellie has almost stepped on in passing. It grows among Christmas and New York ferns. At first glance it looks like one of the familiar shield ferns of the woods. But Farida points out subtle differences. It is neither a crested shield nor an American shield but a cross between the two. It is a natural hybrid, the rare Boott's fern, *Dryopteris Boottii*.

The two come home, where I am working at my desk, with their big news. We all troop back to examine the new fern. Nellie and Farida, in the success that has come at the very end of their hours of searching, find they are no longer tired. And Nellie, reacting to the excitement of the discovery, makes another discovery. Her foot has stopped hurting.

SEPTEMBER 9.   In the dawn I come upon three little rabbits feeding among the dewy grass. They are still, in these early days of September, hardly more than half grown. One races away as soon as I appear. Another waits until I head directly toward it. The third remains quietly watching me until I am less than a dozen feet away. Three little rabbits, each with a different temperament, each with a different "flight distance." As in the trio of young foxes we watched in the long twilight of the July evening, their varying degrees of wariness appear part of their natures from the beginning. They distinguish them as surely as though they were marked red and blue and yellow.

I remember a wildlife photographer I once talked to in the West. He had just come down from the high Rockies after making a motion-picture film of mountain sheep. He explained that on every expedition of the kind he spends the first few days not in photographing but in carefully studying the temperament of the different animals. Then, when he begins shooting, he wastes no time stalking the more nervous and restless ones. He concentrates all his efforts on the one or two or three that are more relaxed, less easily alarmed, more likely to let him approach within camera range.

S E P T E M B E R   1 0 .   Stars lit our way when we walked down the lane last night. But about two o'clock this morning, I awoke to hear the steady drumming of rain angling against the windowpane beside my bed. The stars glittered unseen beyond blanketing clouds. Now, as I get dressed in the milky light of this sluggish dawn, rain still falls and heavy mist fills the hollows and clings among the trees along the brook.

All down the lane, the calling of birds in the mist accompanies me. A little later, when I come back, the vapor has thinned slightly. I glimpse birds everywhere along the brook, birds chasing each other, birds plunging in and out of bushes, birds flying from tree to tree. In the dark before the dawn, flocks of varied migrants, winging their way south under the stars, had encountered the rain and fog. Forced down, they had landed pell-mell among our Trail Wood trees.

I stand by the bridge listening to the voices of the travelers, the calls and snatches of song uttered by the smaller birds. Their little forms dart and flutter, shuttling in quick dashes among the misty trees and bushes. A wave of warblers has ended its night flight along our lane. I see the nearer birds pause, cling to a twig, lift their heads, and with quivering throats send forth their fragments of song. Among them I recognize the voices of prairie warblers, black-throated green warblers, chestnut-sided warblers, yellowthroats, and black-and-white warblers. I hear the sharp "chick!" that is the call note of a yellow-rumped myrtle warbler. Each year, in our area, it is the myrtles that form the rear guard of the southward movement of the autumn warblers. Stragglers go through after all other migrants among these small birds have left the region.

Catbirds mew and flap their tails as I pass by. Once I catch the call note of an olive-backed thrush; another time that of a wood thrush. But loudest of all the bird sounds I hear, cutting through the excited voices of the warblers, are the ringing notes of the robins. They call back and forth continually as they dart from tree to tree—now in an ash, now in a maple, now in a butternut.

All around Trail Wood, probably over an area covering many square miles, the migrants have come to ground. As the morning advances and the mist burns away and the sunshine breaks through the clouds, the light will no doubt reveal other birds of passage among other trees, beside other brooks, along the edges of other woods. Abruptly the southward movement of the migrants has been halted. An emergency has brought them down. By chance they reached these Hampton trees.

SEPTEMBER 11.   In the apple tree beside the terrace there lives a special katydid. Each evening when dusk falls, we recognize it from all the others by its more deliberate, more distinct enunciation. Tonight as we stand in the warm darkness with the wrangling of the insects coming from the woods, from the lane, from trees that rise above the ground mist, from the hickories beside the house, our ears pick out the sound its wings produce. Instead of the usual "katydid"—a call that led Alexander Wilson to spell the name of the insect "ka-te-did"—it spaces out, enunciates more slowly, as though speaking to a child, four syllables: "ka-te-she-did." We have never heard it utter its call in any other way.

On the hottest nights of late summer, the increasing tempo of the calling of the katydids blends into a kind of uproarious chorus, a shouting melee of sound. The stridulation of the insects becomes so rapid we "can hardly make out what they say." Individual calls are lost in the general tumult as the katydids seem bent on shouting each other down. But on most nights we recognize several individuals from the out-of-the-ordinary calls they produce. One that both Nellie and I have noticed stresses the last syllable. It seems to enunciate it with special clarity, to end its "katy- *did*" with added emphasis. Another apparently has one abnormal wing. Its call ends in a kind of "yip! yip!" A third, mentioned before, has a curious muffled call which at times seems echoing itself among the foliage of one of the hickory trees.

Almost always the oval-winged katydids—the true katydids, the ones that "speak our language"—are only voices in the darkness. They live mainly in the upper stories of the trees. The angular-winged katydid that produces only a "tzeet!" we encounter in weed tangles and bushes; but the oval-winged we almost never see. Even after the killing frosts of autumn, when I have hunted in the grass beneath the trees where the insects have debated so long on summer nights, I have never found a fallen katydid. Veined and colored like leaves, their wings provide their almost perfect camouflage. In the tropics, relatives of our katydids carry their camouflage to even more amazing lengths. William Beebe, after returning from an expedition to South America, once showed me a katydid he had brought back from the jungles of Venezuela. Not only had its wings the color and veins of leaves but evolution had decorated them with imitation patches of fungus and imitation drops of dew.

SEPTEMBER 12. A turkey vulture, tossing and rocking in the unstable air, hugs the ridge to the west. A cicada commences the shrill, soaring "bur-r-r" of its call at the edge of the North Woods. A woodchuck halts its hurried feeding to sit up and sweep its eyes in a swift, vigilant circle around the meadow. Beside the pond, a bullfrog broadcasts its hoarse "trump-p-p." And all around me from the meadow grass, as I lean against the plated bark of this tilting wild cherry tree, rises the sustained chorus of the black field crickets.

The life of the vulture, the life of the cicada, the life of the woodchuck, the life of the bullfrog, the life of the field cricket— how dissimilar are they all from the life of the human being, this life that we experience and know so well. If my fate has been more fortunate than these others, it is nothing, I know, for which I can claim credit. So far as my control of events is concerned, I might have come into the world a woodchuck or a cicada.

As I stand in this shade in the midst of this pasture, the evidence is that among all the forms of life around me I alone am considering such things in my mind. For the others life is a pencil sharpened to a point. They are concerning themselves almost entirely with their own affairs, with what affects them directly, with the moment in which they exist. But the human mind covers a wider span, ranges over more territory, scouts far into the future, works back far into the past.

Each of us in our own fashion—turkey vulture, cicada, wood-chuck, bullfrog, cricket, man—represents a kind of mountaintop, the tip of a peak produced by aeons of trial and error, by ages of slow evolving. In this long ascent, man has eclipsed the rest of living things that, in the main, follow patterns of behavior little changed for ages. Many wild creatures possess inherited, special-ized abilities we cannot equal. But it is mankind alone that has developed an inventiveness, a breadth of investigation, a capacity for thought and reflection, an interrelation of coordinated effect, that sets him apart from all else that has evolved on earth. He is the animal that has altered and advanced the fastest and the far-thest. But my good fortune in being born a human being is no cause for contempt for lesser creatures; we are each part of the vast chain of life that includes all from the highest to the lowest.

"I have never understood," Dr. Karl von Frisch, the Nobel Prize-winning student of animal behavior, wrote when thinking along a similar line, "why many people find the idea of an animal origin so disagreeable. Does not the thought of a development to greater perfection, of the gradual development of the spirit far beyond all other living creatures on earth provide more satisfac-tion than the idea of a creation which from the very start gave man a privileged position in the world?"

It was Thomas Hardy's conviction that the ethical implication of evolution is just as momentous as its scientific importance. It links us more closely to all the humbler creatures. It emphasizes our kinship with all the life of the world. It provides "a magic thread of fellow feeling" that unites our lives with theirs.

SEPTEMBER 13.   Burnished skies and goldenrod at its height. A shimmering pond and a slight haze above the meadows. This is the setting in which, on this day, Nellie becomes one year older.

During much of the morning we wander in the sunlit woods. Already the foliage is noticeably thinning. Additional sunshine comes flooding in. Woolly bears are out, humping along the paths. Flickers are moving through. We encounter half a dozen hunting for ants along the Old Colonial Road. At one place we catch the rich, musky perfume of ripe fox grapes. But most of the

time we find our enjoyment in little things. We notice how the
shadow of a fern frond forms zebra stripes of light and dark on the
smooth bark of a young red maple. We observe how low water in
the brook has concentrated the water striders into sheltered pools
among the mossy rocks. We catch isolated spots of crimson that
mark the earliest leaves to change color on the swamp maples.
We see small dragonflies dip and dip again as they deposit their
eggs in the saturated moss of submerged stones. And we note
how the lateness of the season is reflected in the singing of the
tree crickets, no longer waiting for the coming of dusk.

Toward the end of the afternoon, as we are descending the
slope toward the summerhouse, Nellie receives the special
present of the day. We catch sight of a bird, smaller than the
mallard ducks, disappearing underwater near Azalea Shore. It
pops up and dives and pops up again. But these momentary
glimpses are enough. It is a newcomer, a bird we have never seen
on the pond before. At the beginning of its long flight south, a
pied-billed grebe has stopped to rest and feed at the small sanctuary of our acre of water. In doing so, the little waterfowl has contributed, on this special day, a new addition to our Trail Wood
list.

As we sit in the summerhouse, watching the diving bird appear and disappear, a letter of recent days comes to mind. An elderly lady in the Midwest wrote that every year she reads again
the four books about our trips through the four American seasons,
traveling with us anew. And the wonderful part about it, she concluded, is that neither Nellie nor I get any older. At least not in
the pages of the books. For while our lives have moved on, that
part of our lives has remained unchanged.

SEPTEMBER 14. I lie on my back in the meadow
grass and look up into the clear sky. I notice something that I
have not been aware of before. High above me, small winged
forms are swooping, circling, veering in the afternoon sunshine. A
little band of tree swallows, moving south, is feeding on flying insects far above the earth. A sign of fair weather.

I leave the meadow and sit on a log in the woods and close my
eyes. I concentrate on my ears and the little sounds they pick up,

mentally trying to trace each to its source. A tiny ticking or tapping comes from a drying leaf that swings in the slight breeze and strikes a nearby twig. A soft plumping sound is made by a falling acorn landing on damp leaves. A long sigh, faint and thin, is the voice of the breeze among the slender needle leaves of a nearby pine.

I walk down a trail and notice how the silverrod has come into bloom; how the Canada mayflower has gone to seed; how the ground all around a clump of light tan mushrooms appears covered with scattered flour, the white dust of the spores of the fungi. I catch the rank smell of a fox in the woods and see a red-bodied daddy longlegs carrying a dead fly over the brown carpet of the decaying leaves. Where the silvered stub of the great chestnut tree is moldering away, I find a ruffed grouse has dusted in the decayed wood at its base. There it has left a rounded depression a foot and a half across.

I stand on the moist ground of a little hollow beside Hampton Brook and gaze around me. Leaves of the wild sarsaparilla have all turned to golden yellow. Each year they are among the earliest of our woodland plants to lose their summer green. Like brilliant flowers of autumn, the lacquered scarlet of the massed jack-in-the-pulpit berries and the red stems of the white baneberry—where the "doll's eyes" of the fruit have fallen away or been eaten by birds—shine out amid the greens and browns of their surroundings.

When I turn toward home, following the trail from the woods across the meadow, I notice it contains a path within a path. Running down the center of the wider trail our feet have trodden is a kind of groove that follows all its turnings. It is the narrower path made by the woodchucks and skunks and foxes and raccoons that have followed our footsteps across the Starfield.

SEPTEMBER 15.   In the dark that comes so noticeably sooner now, Nellie and I, a little after nine o'clock this evening, follow the path up onto the open, higher ground of the Starfield. To our south, our white house rises, a vaguely glimmering shape under the black bulk of the hickory trees. To our north, the edge of the woods lifts in an ebony wall against the sky. The air is limpid and from black horizon to black horizon points of light—big stars, little stars, bright stars, faint stars—all are burning with a special intensity.

As the air grows chill, we note how the calling of the katydids becomes slower and slower. They drawl, drag out each syllable. At times we wonder if they will finish what they begin to say. They all speak deliberately, as though enunciating for listeners who are hard of hearing.

While we are standing here beneath the multitude of the stars, our ears catch other sounds—the little cheeps and calls of migrants passing overhead. By now many of our birds have slipped away almost unnoticed. Where are the kingbirds of summer days and where the orioles that hatched in the nest in the hickory tree? Almost in surprise, we realize they are gone, that we will not see them again until spring. And now, night after night, birds from farther north—like these unknown wayfarers traveling under the stars above Trail Wood—are streaming southward.

We gaze after the sounds when they have moved away beyond reach of our ears. How beautiful the night! How wonderful a time for flying! What a great adventure lies before these small travelers whose lives are touching ours only once, only through faint sounds in this starlit night. Standing here motionless, wingless, earthbound, Nellie and I almost feel the tug of migration ourselves.

SEPTEMBER 16. Close to the threshold of autumn, this is one of those days when the air, pure and transparent, appears to shimmer over the open fields. Across Firefly Meadow and the Starfield, through Monument Pasture, Mulberry Meadow, and the pasture where the woodcock sings, I wander from field to field as the afternoon advances.

Amid a jungle of weeds close to the stone wall that bounds one side of Monument Pasture, I find two brownish walkingsticks in the midst of mating. The male is hardly half the size of the female. Like the Galapagos turtles, these creatures, moving with such deliberation, live a life that to us seems dim and almost inert. How infinitely removed it is from the speed of the falcon on a windy sky or the intense alertness of a fox in the snow of a winter night.

SEPTEMBER 17. In the morning light, white globes shine out like snowballs lying on the still rich green of the pasture grass. Puffball time has come to the meadows. Each of these roughly spherical fungi has enlarged in a silent explosion of growth.

This swift, now visible expansion has been preceded by a long, invisible preparation. Most of the life of the puffball is spent unseen underground, getting food from decaying vegetable matter through an extensive network of white threads. It is in its latter days, when it is ready to mature its spores, that the globular form we see rises above the ground.

I walk among the dew-covered globes. Some are lighted by their first rising sun. I break off several, as large or larger than my fist, and bring them home when I return from my before-breakfast walk. Nellie washes them, scrapes off the skin, slices them into thin sections, and fries them lightly in vegetable oil. They provide us with the delicacy of "mushroom fritters." There are other ways of cooking the slices—dipped in egg batter, fried with bacon or with onion juice added. We prefer them prepared in their simplest form, seasoned only with a slight addition of salt and pepper. Thus we enjoy the wild mushroom flavor to the full.

The life of the late-summer puffball, once it has appeared, is a story of continual change. Its color alters. The white fades. The tinging of brown grows deeper. At first, as the spores mature, the interior is wet and soggy. Then it loses its moisture and becomes arid and powdery. In its final stage, the puffball resembles a leathery bag filled with dust. This dust is composed of those light and infinitesimally small particles, the spores, which are carried on the wind. Strike such a puffball with a stick or kick it with the toe of your shoe and a smokelike puff of brown spores shoots into the air. Hence the name puffball. "Smokeball" and "Devil's snuffbox" are two other folk names for the globular fungi. Because the spores mature within the cavity of the sphere, these mushrooms are sometimes classed as "pouch" or "stomach" fungi.

In days before matches came into general use, American pioneers gathered the dry threads of the underground network of the puffballs to use as tinder to catch sparks from flintstones in igniting fireplace fires. Physicians at one time employed the spore dust in staunching the flow of blood from wounds. In their varying forms, different groups of puffballs have been designated as skull-shaped, hard-skinned, engraved, gem-studded, pear-shaped, long-stemmed, and giant puffballs.

Only once have we encountered one of these giant puffballs at Trail Wood. It rose among weeds and went unnoticed until it had grown into a cocoa-brown ball nearly a foot in diameter. Its skin, in time, split and peeled back in patches, revealing the solid mass of the spores within. One morning I tested something I had read about how the first raindrops in a storm aid such puffballs in the distribution of their spores. Dipping one hand in water, I let drops dribble off my fingertips onto the open patches. Each drop that struck produced a little puff—like smoke billowing up from a miniature explosion.

In extreme cases, the giant puffball, *Calvatia gigantea,* may attain a circumference of as much as six feet and may weigh three and a half pounds. One British mycologist, Dr. A. H. R. Buller, made a study of the spore content of one of these giants of the fungi world. He calculated that the dry "dust" of its interior consisted of at least seven trillion spores.

S E P T E M B E R 18. Facing the sun, after a night of ground mist, I see thin, taut, shining lines stretching from side to side across the lane. Some are high, some only a foot or so above my head. Each, like a burnished-silver thread, extends from a bush or tree on one side of the wheeltracks to a bush or tree on the other side. All are formed of spider silk.

Devoid of wings, how do these small creatures stretch their strands through fifteen feet of thin air and anchor them to a twig on the opposite side of the lane? It is obviously impossible for the spider, without getting its silk entangled on the way, to anchor one end, then descend among the weeds and bushes and, trailing its thread, cross the lane to a tree on the other side. Somehow the silk must be transported straight across the intervening space. How does the spider do it? At first glance, the accomplishment seems as baffling as the rope trick of the Indian fakir.

But its explanation is relatively simple. It lies in the lightness of the thread of silk. Clinging to a twig on the upwind side of the lane, the spider commences spinning and paying out its slender filament. So weightless is it that it floats on even the faintest breeze. Steadily, minute after minute, the spider extends this air-borne line. Eventually the thread, buoyed up and drawn out by the flowing currents of air, is long enough to reach the other side. There the waving end becomes firmly entangled on a twig. The spider pulls the thread taut and secures it and the feat is accomplished.

Once, when I encounter a particularly low strand of such silk this morning, I put a forefinger against the middle and pull the taut thread more than two feet toward me. The filament does not break. And when I release it, it returns to its former position. Thin as it is, the silken line has surprising strength. Once it is secured, it provides a dependable bridge across which its maker can tightrope above the lane and safely reach the other side.

S E P T E M B E R 1 9 . Yesterday it was spider silk. Today it is burs. At the end of my walk this afternoon, when I come to the barway by the shed at the edge of the meadow, I lift one foot and then the other to the lowest pole. Laboriously, I pick off the small burs attached to the legs of my corduroy trousers. These days I can tell where I have been by the burs I collect.

From the walks I come home bearing clinging seeds of the agrimony that borders the lane or the enchanter's nightshade that grows in the moist soil under the old apple tree on the path to the summerhouse or the tick trefoil of the Fern Brook Trail or the sticktights or devil's-pitchforks—the *Bidens frondosa*—of the edges of the fields.

With their minute hooks, they cling tenaciously in place. One by one I pick them off and cast them aside. In that act I am fulfilling the purpose for which the hooks of the burs were created. I am aiding in the dissemination of the seeds, assisting them to reach the ground in some new place removed from the plant that produced them.

The one great employment of the plants in these waning summer days is the spreading and perpetuation of the species. Varied are the means, in varied seasons of the year, that are employed in the world of flowering herbs and bushes. The dandelion, the milkweed, the thistle, the cattail trust to the wind for their seeds' dispersal; the cardinal flower to the moving water; the witch hazel to the capsules that, through tensions, shoot out and scatter the seeds like hard black bullets. Over wide areas, birds deposit the seeds of berries they have eaten. The wild grape the fox consumes and the wild grape that dries up on the vine both, by different routes, play their part in nature's plan. The seeds expelled by the digestive system of the fox and the seeds in the dried grape shaken from the vine by autumn winds arrive at the same journey's end, the soil of the earth for which they were designed.

In this dispersal of the seeds, the berry and the bur employ opposite means to achieve the same end. The berry entices, offers something in exchange, pays its way. The bur forces us willy-nilly to carry on the work of its dissemination.

S E P T E M B E R   2 0 .   This evening culminates in a glow-worm walk. About nine thirty, Nellie and I start down the lane. I look at the thermometer before we go. It stands at seventy degrees Fahrenheit. The night is soft and velvety; the air moist from a shower in the afternoon. From the trees and from the open fields the serenade of the night insects reaches our ears in swelling, overlapping waves of sound. Intermittently a sluggish breeze awakens, drifts up the slope of Firefly Meadow from the south, then sinks away. Off to the north, heat lightning soundlessly flickers on and off in stabs of quickly ebbing light. Ours is a Florida night.

Hardly have we started when we begin to come upon greenish sparks of life—glowworms in the still wet grass. From then on we are rarely out of sight of these glowing spots of heatless light. They shine steadily in the darkness. Sometimes they are isolated. Sometimes several glow in unison. Most often they are in the edging of grass but on occasion we find them among the compressed gravel of the wheel tracks. So we continue our walk with glowworms shining beneath us and katydids calling in the trees above us.

On our way down the lane, Nellie and I make a tally of the glowworms. The number reaches fifty-one. Beginning over again when we turn back at Kenyon Road, we count, on our return, fifty-two.

Where the soil is moister, near the bridge, the green-tinted lights are most numerous of all. We count a dozen glowing within the space of a few square feet. Nellie clicks on her flashlight. As its beam centers on one of the greenish spots at the lane edge, the glow is eclipsed. It disappears, overpowered by the more intense illumination of the electric torch. What is left is a small, light-colored creature with the form of a beetle larva. Nellie extinguishes her light. The silent glimmer of green returns again.

Among the *Lympyridae,* not only the wingless females of the fireflies glow, but also, in numerous cases, the earthbound larvae and even the eggs. In the air over the darkened fields, we glimpse only two or three winged fireflies switching on and off their moving lights. For these insects, the mating season is virtually over. Because of this absence of flying males to be attracted by the lights of the females, we suspect that the majority of the glowing lights we find along the lane tonight are those of the luminous larvae wandering in search of the small crustaceans, snails, earthworms, and insects that comprise their nocturnal food.

SEPTEMBER 21. Sitting at the top of Juniper Hill, with the wide, variegated view across the valley of the river opening up before us, Nellie and I find ourselves discussing, this afternoon, a particular aspect of the beauty around us. These are the days of glowing mist and the beginning of the tinted trees.

We are always fascinated by the way nature achieves her ends with color and music and fragrance, with graceful motion, with forms and patterns, with beauty that is ever changing. In his Presidential Address before The Royal Photographic Society of Great Britain in London in 1976, Vernon Harrison spoke of beauty. "Beauty," he noted, "does not enter into the equations of the physicist and does not seem necessary for the running of the universe. It is, as it were, a bonus thrown in to make life worthwhile."

Yet, as we sit here, we wonder how far we are influenced in our ideas of beauty and what is beautiful by what we have been surrounded by, what we have become accustomed to, what we have been taught, and what our emotional associations lead us to accept as beautiful.

I remember a man who once wrote a book in which he pointed out how the human skin can stand no more than a certain amount of specific rays coming from the sun. He went on to note how wonderful it is that nature has adjusted things so exactly that this precise amount of the rays reaches the surface of the earth. But he had the cart before the horse. The adjustment was the other way around. Man adapts himself to nature; not nature to man.

So when we look around us and marvel how nature has placed us in surroundings of beauty, making life more rich and pleasant, how can we be sure we are not using, as our own yardstick of beauty, merely the things with which we have become familiar? In some future time, in a world divorced from nature, will a brick instead of a flower, the whir of an electric fan instead of the song of a thrush, the straight line of a skyscraper instead of the curve of a winding brook be considered examples of the beautiful? We find our beauty in the things we know. New lives, lived in new surroundings—will they bring new yardsticks of what is attractive, what is beautiful? The beauty we appreciate today and the beauty of some remote future—who can tell how they will differ?

S E P T E M B E R 2 2 . Dancing grasshoppers are now part of every walk on a sunny day. In the heat of this early afternoon, I stop to watch a dust-gray Carolina locust, *Dissosteira carolina,* again and again hover five or six feet above the ground amid the crackling flutter of its wings. It rises and falls irregularly as though jiggled at the end of a rubber band. All the time I see its black, fanlike, yellow-bordered hind wings flickering like thin parchment in the air. Then suddenly it drops into the grass or onto some sunburned patch of open ground. Throughout our pastureland, wherever I go, I come upon these gray grasshoppers rising, hanging, crackling, descending.

A lighter sound, higher pitched, surrounds the smaller red-legged locust, *Melanoplus femur-rubrum,* the most common short-horned grasshopper throughout the United States. As it hovers, the roughened thighs of its jumping legs and sharp-edged veins in its wings scrape together producing the dry music of its serenade. Such sounds do not accompany the ordinary flights of grasshoppers. They are a feature of the late-summer mating displays of these insects.

So loud is the noise produced by one locust inhabiting the high Sierras of California that it is known locally as the firecracker grasshopper. In our travels through the mountain country of the West, Nellie and I often noticed how hovering grasshoppers hang in the air beside a cliff. The rock acts as a sounding board, projecting outward the dry clatter of the mating performances.

Providing my amusement today, this dancing of the grasshoppers is one of many signs I note that summer has almost reached its end. The orb webs of the golden garden spiders are larger and more common among the weed tangles. And when I lie back in the sunshine on a pasture hillside, looking up at the drifting clouds, my ears catch a faint stridulated trilling among the grass blades. I am hearing that smallest of all our crickets of the fields, the little *Nemobius*. In making its music, its tiny wings vibrate above a brownish body hardly half an inch long.

Each year, it is in the latter days of summer that the *Orthoptera* come into their own. Theirs is the dominant music of the woods and fields. It ranges from the calling of the katydids through the clatter of the locusts and the endless chirping of the black field crickets to the mellow pulsating rhythm of the snowy tree crickets and the long trilling of the narrow-winged tree cricket—that frail sound filled with gentle melancholy that each year seems to me, more than any other, the voice of summer's ending.

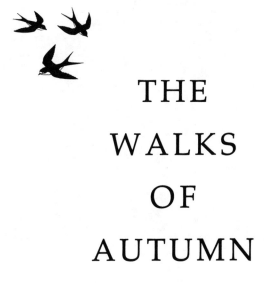

THE
WALKS
OF
AUTUMN

SEPTEMBER 23. On this day of the autumnal equinox the seasons change before dawn arrives.

When I start out after breakfast, I see robins running over the lawn and out across the meadows. I see the white rump patches of flickers where the woodpeckers take off and land along the path across the Starfield. Flocks from the north have dropped down at daybreak. The earthworm hunters, the robins, and the ant hunters, the flickers, are searching the ground for food together on this first morning of fall.

More than once during my walk, I watch a flicker abandon its pursuit of ants and alight in a pokeberry plant to consume the juicy fruit—black, flattened on top, with purple juices and stems that are reddish or raspberry colored. Ripening in immense numbers, this wild fruit is also one of the favorite foods of the autumn thrushes. I stop beside one pokeberry with fourteen stalks rising higher than my head. It is decorated with more than 500 elongated clusters of shining berries. Counting the number in several clusters and taking the average, I calculate that this one plant is supplying more than 25,000 large, plump, juicy berries for the migrating birds to consume. One common name for them is inkberries. Another is pigeon berries. That dates back to the earlier years when hordes of passenger pigeons stripped the plants of their fruit. It was their favorite food during the migration south.

During the day I notice the number of bluejays increases, then decreases again. They, too, in their more secretive fashion, are making an autumnal movement toward the south.

A song of spring brings me up short when I am out toward sunset. Loud and ringing, it is the "tea-leaf" song of two robins clinging to treetop twigs along the lane. I rest an elbow on the stone wall, listen to the singers, and watch other robins on the ground as they run and halt and run again, as they take off in sudden flights and pursuits, in twisting chases, calling continually.

Judging by their excitement, the birds around me might be reversing their journey, going north in spring instead of south in autumn. Their elation bridges the winter, advances the calendar. But I am well aware that now, even while robins are everywhere, we are nearer than ever to the days without robins.

SEPTEMBER 24. Part way across the open meadow this morning, I pause and look back toward the house. In so doing, I notice something I have missed before. Only a foot or two from the path, a couple of yards behind me, a woodchuck is flattened in the grass. I have passed it without realizing it is there. And it has remained fixed, without the slightest twitch that might have caught my eye. But now, the instant it sees me looking directly at it, the moment it realizes I have caught sight of it and recognize it, it bolts away through the grass.

I remember once walking about in open ground no more than fifty yards from a crow perched at the tip of a tall tree. It remained there until I looked directly at it. Then it flew. Another time, among the glacial rocks of the boulderfield north of Old Cabin Hill, I passed within four feet of the base of a gray birch tree. When I was coming back, the glint of something bright amid the tan carpet of fallen leaves at the base of the tree caught my eye. A hen grouse, camouflaged among the leaves, was brooding her eggs on a nest. While my eyes slid over her, she remained unmoving. But when I saw the glint of her eye at last, peered intently, looked with recognition, she shot into the air and disappeared among the trees.

Wild creatures watch our eyes. They note our reactions. The sudden halt, the tenseness of surprise, the change in position or expression—these wild things are quick to observe. For upon them their lives may depend. For uncounted centuries their species have been studying the ways of their enemies.

In one of Roger Tory Peterson's nature films, I remember seeing how the wild hunting dogs of Africa hold their heads high when they are well fed and uninterested in prey but how they invariably lower their heads when they run in packs like wolves, intent on killing. The herds of the plains, the grazing animals that are their prey are aware of this. They pay little attention to the wild dogs, even those passing close by, when their heads are held high. But they become wary or take flight at once when they see these hunting dogs running with their heads lowered.

S E P T E M B E R 2 5 .   Coming back from the north boulder-field toward the end of this afternoon, Nellie and I swing aside to visit Wild Apple Glade. There, in a diversion of other years, I used to sling my canvas hammock under a low, outstretched limb and, lying in it, spend summer hours absorbed in observing wildlife unobserved.

As we step out from the edge of the opening, a ruffed grouse takes off with a windy rush and threads away at high speed through the needle eyes of small openings in the underbrush. Undisturbed, a white-breasted nuthatch continues, with starts and stops, down the trunk of the decaying apple tree. Its "yank! yank!" changes in tempo, speeding up and slowing down as, with its slender up-turned bill, it investigates grooves and recesses and loose fragments of the bark.

This part of our West Woods, almost within sight of the house across the meadow, seems as wild as when the Indians lived here. All our woodland—north, west, and south—continues year in and year out untouched, unchanged, except as windstorm and ice-storm, flood and lightning, insect and woodpecker and beaver alter it. In the main, our woods are changed only as natural events affected wooded areas here before the Europeans reached the New World. Here we are in contact with a nature that, except for the trails we have cut and the fallen branches we gather for fireplace kindling in the fall, remains almost undisturbed by human hands.

When we are coming home over the wooded higher ground above the western end of the pond, we pass beneath an oak tree. I glance up. In a crotch twenty or twenty-five feet above the ground, a raccoon is curled up fast asleep. Its ringed tail hangs down like the tail on a Daniel Boone coonskin cap. Only the tail and one of its two ears are visible from directly below. The raccoon sleeps on. It never knows we have passed beneath it. How many times in how many years, I wonder, will we come this way and, arriving at this same oak tree, look up, expecting to see a raccoon?

SEPTEMBER 26. In the blaze of the midday sun, where the trail descends the slope of Firefly Meadow to the pond, I come upon a large black field cricket. To all appearances it lies dead in the middle of my path. Undisturbed by my approach, unaroused by the vibrations of my footfalls, it remains motionless even when I stand above it. Half-convinced that it really is lifeless, I touch the insect with the toe of my shoe. It explodes into the air, arcing away in a leap that carries it out of sight among the weeds and grasses. I have come upon a cricket in the profound sleep of a midday siesta.

Across the warm meadows, other crickets, crickets in uncounted numbers, are wide awake. The air pulsates with the shrill sound of their stridulating wings. The ringing, singing fields of late September!

Rising and falling, the sound of this insect music is dominant in my ears as I stand, a little later, watching a butterfly migrant—a lone monarch, perhaps the last of the year. The journeying insect drifts by in the sunshine, coming from somewhere farther to the north, probably from Canada, perhaps from as far as the land below Hudson Bay. So late on its long journey, it moves without haste, with no indication of urgency. I follow its flight with my eyes. It crosses the pond and rises over the trees. It disappears and leaves me behind. I feel suddenly closer to the months of cold.

SEPTEMBER 27. A bug in the sun. A flat bug. A brownish bug. A long-legged bug. A squash bug soaking up warmth from the morning rays after the chill of the night. I stand looking down at it. We share the warmth together. The sunshine, the fresh air, the green of growing plants are its as well as mine. The world was made for it—and for the chipmunks and the bluebirds and the water striders—as well as for me. We share it all in common.

At the Old Manse beside the Concord River in the summer of 1843, Nathaniel Hawthorne came in from inspecting his growing vegetables and wrote in his journal: "The garden looks well now; the potatoes flourish; the early corn waves in the wind. I am forced, however, to carry on continual warfare with the squash bugs, who, were I to let them alone for a day, would perhaps quite destroy the prospects of the whole summer. It is impossible not to feel angry with these unconscionable insects, who scruple not to do such excessive mischief to me. Why is it, I wonder, that nature has provided such a host of enemies for every useful esculent? It is truly a mystery."

Long before it became a cornerstone in the thinking of modern ecologists such as Aldo Leopold, John Muir expressed the idea that "the universe would be incomplete without man; but it would also be incomplete without the smallest transmicroscopic creature that dwells beyond our conceitful eyes and knowledge."

For those who start with the assumption that from the beginning our welfare has been nature's primary concern, there come a thousand variations of Hawthorne's question. Why are *they* here? Why the squash bugs, why the mosquitoes, why the wasps, the tigers, and the botflies? Why are they part of a world made for man?

Unless we accept the evidence that we are but part of the complex whole, that each creature on earth is here to live its own life, without any more regard for man's wishes or his welfare than for the wishes and welfare of any other dweller on the planet, we are destined to live and die without being able to reconcile what seems to us *should be* with what *is*.

SEPTEMBER 28. The sequence of the falling of the leaves is like the sequence of the blooming of the wild flowers. It repeats itself each year. The time of its leaf-fall is a characteristic of the tree. At Trail Wood we see first of all the downward drift of those elongated leaves with delicate tintings of yellow and salmon and purple that begin descending from the white ash trees in the last of the September days. They come down in slow swirls or sudden leaf showers. They are followed, as the weeks move into October, by the descent of the crimson of the red maples, the gold of the hickories, the wine red of the tupelos, the pale yellow of the aspens.

Already, as I go down the lane this morning, my feet scuff through the initial layer of the fallen ash leaves. Here and there I am surrounded by the leisurely drift of spent foliage descending in the only journey of its life from the twigs of the ash trees to the lane below. The slender leaves are like the first deliberate flakes of a snowstorm.

And when I go to the woods in the afternoon, I notice already a perceptible increase in the widening of my view. New vistas are opening up. Secret places are revealed. In this bush I glimpse the nest where a brood was raised unobserved. On that tree limb I see another nest that has been screened by leaves before. The mystery of where birds we met a hundred times along these trails cared for their young becomes, with even a slight thinning of the clothing of the trees, a mystery no longer. Like the opening pages of a book, the woodland scene is spreading out. From now on, as we go along these paths, we will learn progressively something new about something old as we see revealed what the dense foliage of summer has hidden before.

SEPTEMBER 29.   Why do we smile when Nellie and I pause for a moment beside a heap of wood ashes in the meadow? We have seen it there since last winter. The reason goes back half a century to a schoolteacher in Hampton who told her pupils as a fact of natural history that if they placed wood ashes in a meadow they would attract whippoorwills.

That idea, no doubt, originated with someone who heard a whippoorwill calling where ashes had been dumped and leaped to an unjustified cause-and-effect relationship between the two. I know how ridiculous the idea is. I am aware it can have no basis in fact. Yet, somehow, ever since I heard the story, some of our wood ashes seem to get dumped in the meadow. My defense to Nellie is that I have to dump them *somewhere*.

But each time I carried a pailful from the fireplace, I remember, I recalled a visit I once made to the old Ward's Natural History Establishment in Rochester, New York. One of the men in charge told me of an early order the museum supply house received. It came from a businessman in Havana, Cuba. He requested a stuffed white weasel to put in the window of his store because, he explained, it is a well-known fact that as long as a white weasel remains on the premises a business will never fail. The staff at Ward's laughed heartily at that idea. However, it was noticed in the succeeding years that, somehow, at this scientific supply house, a white weasel was always kept in stock.

Legion are the superstitions and misconceptions related to the field of natural history. Probably most of them originated in the same combination of circumstances—an observation made and a wrong conclusion derived from it. Accuracy in nature observation is a twofold problem. No one has expressed this better than the famous Cornell botanist, Liberty Hyde Bailey. He used to admonish his students: "See what you look at and draw proper conclusions from what you see."

S E P T E M B E R 3 0 . I hear a continual movement, a scurrying this way and that over fallen leaves. The lane this morning seems alive with running mice. A flight of small sparrows, after journeying through the night, has landed at daybreak. Some are field sparrows, others chipping sparrows, others song sparrows, a few are juncos, and one or two are white throats. Active, well, filled with energy at the beginning of migration, they dart about, hunting for food. They represent an early installment of the uncountable numbers of these small seed feeders that will stream south in the coming weeks. In the course of every fall migration at Trail Wood, October is our month of sparrows.

During the hours of this day I wander widely. I note with intense interest all the small events around me—a single cardinal flower still blooming in a wet declivity by the pond; dark bumblebees sleeping afield in Mulberry Meadow in this time of the breaking up of the colonies; slender ash leaves revolving and sweeping in serpentines on the currents of Hampton Brook. Near the beaver pond in the Far North Woods, I pause to listen to the hubbub of crows mobbing another owl beyond Witch Hazel Hill. And on the Big Grapevine Trail, in the South Woods, I become aware of low-pitched pulsations in the air. A grouse is drumming in an autumn recrudescence of its springtime activity.

Along the top of Lichen Ridge and on the slope of Juniper Hill, my ears catch tiny scraping and rustling sounds, while in all the open fields I am aware of the thin complaint of parched leaves rubbing together in the breeze. I am surrounded by the little voices of autumn dryness. They are sounds unassociated with the lush and sap-filled days of only a month ago. I stop to enjoy those flowers of the latter days—the graceful wreath goldenrod and the creamy-white heath aster. They are now at the peak of their blooming. But around them how tarnished are the pastures! How far the tide of the wild flowers has ebbed! Largely, the colors of summer have left the fields. Now, the colors of autumn, mounting to their annual climax, are investing the trees of the woodland.

OCTOBER 1.   As lightly as a dry fly cast by an expert fisherman, the falling leaf drifts down through the still air and touches the surface of the brook. I watch it turn in the current, bob on little ripples. I follow it with my eyes as it curves deliberately around a mossy rock, gains momentum, and is swept away downstream.

Where I sit on this wooden bench set amid ferns beside the brook and overarched by the sprawling, outspread branches of the spicebushes, I am less than 200 yards from the house. Yet here I am as in some remote and secret place. An early junco discovers me. It gives its small repeated ticking call as it moves among the alders. It tires of this in time. Then everything is still except for the liquid solo of the stream coursing over its stony bed almost at my feet.

Leaning back, I look up among the spicebush branches. I notice how the leaves are beginning to alter from the green of summer to the bright yellow that fills our lowland woods each fall with waves of golden light. The leaves, the bark, the fruit of this member of the laurel family all are strongly aromatic. This spicy benzoin fragrance has given to the shrub one of its common names: wild allspice. Along so many of our woodland trails it is the most companionable of our bushes. We often stop beside it, pluck and crush a leaf, then stand enjoying the simple pleasure its wild incense imparts.

I reach up and pick one of the yellowing leaves above my head. I rub it between a thumb and forefinger. When I lift it to my nose I catch only a faint remnant of the scent that is so strong in the sap-filled tissues when the leaves are green. Looking at the maze of crisscrossing branches overhead, I recall two things about them. They are among the few branches I know whose wood burns well when it is still green. And they are so brittle, so easily broken, that another name for the spicebush, *Lindera benzoin,* is snapwood.

Only here and there, widely scattered over the canopy of the branches, do I catch sight of the brilliant lacquered red of the elongated berries that form the autumn fruit of the spicebush. In former days on New England farms, this fruit was sometimes collected and, with the large single seeds squeezed out, stirred into batter to add its individual flavor to homemade cookies.

Each fall I pick a few of the berries and eat the pulp with its pungent juices. The first effect is an immediate and copious flow of saliva. It is like putting a slice of lemon in my mouth. But now the berries are largely gone. Songbirds of various kinds, storing up fat for migration, have feasted on them. Especially fond of this fruit are wood thrushes, robins, red-eyed vireos, great-crested flycatchers, kingbirds, and veeries.

O C T O B E R 2 . Aesop recorded it in his fables nearly twenty-five centuries ago. It is repeating itself on this misty morning in autumn at Trail Wood. The old rivalry goes on. The crow and the fox—the cunning bird and the cunning animal—are once more in opposition. The uproar it produces draws me to the far end of the pond.

I see everything dimly. Trees and bushes are shrouded shapes, pale gray mixed strongly with silver. Vaguely I distinguish the dark forms of half a dozen crows perching, peering down, fluttering from limb to limb among the pond-side trees. They all are cawing in a raucous din that drowns out the other sounds of the morning. Moving stealthily from bush to bush, drawing closer cautiously, a little at a time, I advance until I can see the cause of the birds' excitement.

A red fox is sitting in the open on the slope above Summerhouse Rock. As long as the animal remains motionless, the crows assail it with voice alone. But as I watch the animal stands up, turns to one side, begins following the edge of the pond, nosing, as it advances, among grass clumps and fallen leaves. No sooner is it in motion than the cawing is redoubled. It increases into a gale of sound. The crows take wing, swirl around the fox, dive on it, strafe it, harry it as it trots along. The black birds are continuing the age-old feud of Aesop's time.

To the fox's nose, this moist air no doubt is filled with delicious and beguiling scents. Ignoring the hubbub of the crows— and even the diving birds except when they sweep too close—it winds in and out among the trees and bushes bordering Veery Lane. The pursuing birds follow its every twist and turn. They are like foxhounds of the air. The keenness of their eyes matches the sensitiveness of the noses of the earthbound hounds. The fox is never completely out of their sight. Sometimes when a fox has made a kill and is carrying its prey back to its den, crows will mob it with redoubled intensity, apparently in the vain hope that their old enemy will become distracted and drop its booty.

The better to see what is taking place, I make a slight movement behind my bush. One that sees better than I sees me instead. The red fox catches sight of that small movement. It wheels and races for the woods. In the space of no more than three ticks of a watch, it has vanished. The animal is lost to my sight but not to the sight of the crows. I listen for some time to their pursuing clamor receding and growing fainter among the trees.

O C T O B E R 3 .   Several times in our walk today we stop to listen to the lonely piping calls of autumn hylas. Only during the warmer hours of the day now do we hear their voices. Twice we see one of these smallest of our frogs, tiny, light-colored, bearing a cross of darker hue on its back. Each time it is making its way over the fallen leaves in diminutive jumps. It seems so frail, so unprotected, so ill-fitted to survive in winter's cold, so unlikely ever to join the great peeper chorus of another springtime. It appears more naked to the blasts than King Lear wandering on the moors in Shakespeare's play.

To us, on this October day, the future of this small creature seems bleak, an unfair contest with the elements. To us its months of dormancy appear a time of hazard, the outcome uncertain, the chances of survival scant. We fancy it is facing, as autumn draws on, a great ordeal. Rather, in truth, it faces what it is

fitted by nature to endure. When it burrows deep into the woodland leaf cover and settles down to hibernation, what lies ahead
for it is no more an ordeal than a familiar time of sleep—this time
a deeper sleep and winter long.

OCTOBER 4. The autumn bumblebees drone from lavender-blue flower to lavender-blue flower where the heart-leaved
aster grows. Because of its late nectar, one of the common names
for this wild flower, *Aster cordifolius,* is bee weed. We watch the
furry black-and-gold insects rising and descending, winding
among stems and leaves accompanied by the glittering blur of
their transparent wings.

As always on these bright autumn days, from the woods we
hear the continual chipping of the chipmunks. Each now has
stores underground—the chipmunk equivalent of money in the
bank. Each now has a proprietary interest, a territory to protect,
property to guard, and its vocal "no trespassing" signs ring out
from morning till night.

At the top of the slope overlooking the western end of the
pond we meet the most interesting creature our walk provides
today. It is ascending among the leaves of a ground cherry. Its
body is round, flattened on the bottom, convex on top. Its head is
almost entirely hidden. It has the appearance of a tiny, brilliantly
colored turtle. When we move our viewpoint, iridescent hues
play along its body. We recognize this metallically beautiful creature as one of the small tortoise beetles of the family *Chrysomeli-
dae.* Some species of these leaf feeders are brilliant red, others
are satiny green. One, found on bindweed, appears plated with
gold. It was this latter insect that gave Edgar Allan Poe the idea
for his famous short story, *The Gold Bug.*

The striking beauty of this small living jewel we see on the
ground cherry represents the denouement of a kind of entomological version of the ugly duckling of Hans Christian Andersen's fairy tale. During its larval life, the tortoise beetle's body
ends at the rear in a forklike tail. Onto this appendage it heaps
excrement and cast skins, binding them in place with fine threads
of silk. When this mass is hoisted over its back, the larva is so
well camouflaged enemies have difficulty distinguishing it from a
birddropping or a bit of mud. Because the mass resembles a pack

carried on its back, this immature beetle long ago was given its common name: "the peddler."

OCTOBER 5.   How easily our minds and bodies separate! How far away, oftentimes, are our minds; how anchored fast our bodies! These words are put down as, cushioned by moss and fallen leaves, I sit in the autumn woods, my back against the trunk of an oak tree. To all appearances I have been—mind and body—at rest. Yet in fact my mind has roamed a thousand miles away. I am reliving, once more, as I sit here motionless in this New England woods, a long summer day when I was in my twenties. It is a time I will always remember as "The Day the Trains Didn't Stop."

On that morning, I had walked for a mile and a half along a sandy dune-country road in northern Indiana to reach a small, weathered shelter set beside the right-of-way of the Chicago and South Bend interurban railroad. To the north a wide green belt of swampland extended between me and the shining chain of the dunes, hiding the Lake Michigan shore beyond. A change I was unaware of had been made in the schedule of the trains. They no longer made their customary stop at this isolated station on their way to Michigan City. Train after train—in spite of my signals—hurtled by.

The elements of the scene that day—the feel of the sun, the shriek of the train whistles, the vibration of the ground as the cars rushed by, the scent of the breeze combining the fresh smell of the lake and the dank smell of the swamp—they all return with vividness and clarity. Expecting the next train to stop, I waited on and on. Hours passed. But those hours were least of all wasted. In retrospect that day seems one of special importance in my life.

When I had started out, I had slipped a small book in my pocket. It was a compact edition of Edward FitzGerald's *The Rubáiyát of Omar Khayyám*. Now, with nothing else to do, I read over and over the classic quatrains that FitzGerald—living his secluded life amid his gardens on the east coast of Suffolk, in England—had fashioned with such care. As the hot hours dragged slowly by, I sat on the plank seat in the shade of the open-faced shelter reading first the quatrains from beginning to end and then, half a hundred times, reading aloud the ones that meant most to me. Nobody came. I was all alone. It was a strange and wonderful time, that day when no trains stopped. Echoes of it have reverberated through all the succeeding years.

What has set in motion this sequence of remembrance? What has sent my mind bridging the gap between this New England farm and the dune country of those earlier years? Nothing of which I am conscious. Was it some scent in the air, some sound in the woods? Was it the sight, as I crossed the pasture field, of the cloudless heavens—"that inverted bowl they call the sky"? Was it some departing swallow—"Ah whence, and whither flown again, who knows!"? Was it some flower gone to seed that brought an echo of the past—"The Flower that once has blown forever dies."? I will never know. What I do know is that as I sit here suddenly both the surroundings and the mood of that long-gone day live again. The mood of those hours becomes as vivid in my mind as the details of the remembered scene that lay around me then. Unspoken, difficult to put into words, the moods of our remembered past may lie deeper than ordinary memory and nearer to the heart.

OCTOBER 6. Its feet in the air, the short-tailed shrew lies on its back in the meadow path. Sometime during the night, a hunting fox has caught and killed it and left it uneaten. Perhaps because of its strong and musky smell, each time I come upon a dead shrew, I discover it has been discarded by its captor. At first glance, this smaller relative of the mole could easily be mistaken for a mouse. But a closer inspection reveals the long, pointed, flexible snout which in life keeps twitching constantly.

Beginning life in a form so tiny they have been described as looking like wrinkled pink honeybees, shrews live lives of intense, nervous activity. When in an excited condition, one species of these *Insectivora* was found to breathe as fast as 850 times a minute and its heart to beat up to 800 times a minute—at least 150 beats a minute in excess of the rate of a hummingbird. The voice of the short-tailed shrew consists of a high-pitched rapid twitter. Because it spends its days underground, often in the tunnels of moles and meadow mice, appearing aboveground only after dark, we see such creatures more often dead than alive.

In hunting its prey, which ranges from worms and slugs and

insects to animals as large as mice, the short-tailed shrew is aided by a nerve poison secreted from glands into its saliva. This immobilizes such smaller prey as the slugs and snails it sometimes stores underground for later use. But this providing for the future is the exception. For the little shrew is driven by such a voracious appetite it must consume its weight in food every twenty-four hours just to sustain life. Individual shrews have been recorded as eating more than three times their weight in the course of one day. Like everything else about these strenuous creatures, metabolism is extremely high. Deprived of all food, a shrew may die of starvation in hardly more than a single day.

Like numerous other forms of life, its days are spent on a razor's edge. Certain fish in the sea live on the borderline of their ability to endure salt. Ants that inhabit Death Valley have been shown to exist almost at the limits of their tolerance for heat. The fastest serpents in North America are certain diurnal racers inhabiting the dry, hot regions of the Southwest. They streak from shade to shade, spending a minimum of time in the direct rays of the deadly desert sun. Water striders skate over the surface film less than a hairsbreadth from the water in which they would drown. And, in the hot springs of Yellowstone, small insects spend their whole lives in water that is within a few degrees of the maximum temperatures they can survive. All these varied forms of life—like the voracious, vulnerable little shrews—endure and thrive through a lifetime of days spent on the verge of disaster.

O C T O B E R   7 .   Four miles away, on the other side of the village, Hilltop Farm was once the scene of a curious friendship between two oddly assorted animals. Among a score or more of young cattle kept in one of the pastures, a reddish steer stood out because of its appearance. Its hair grew in swirls on its body. Because of these roughened places in its coat, the reddish animal was named "Royal Fudge."

For some unknown reason this particular steer and a cocker spaniel living on the farm became close friends. The little dog followed its larger companion about the field. The two lay down close together in the shade. Day after day they engaged in a running game that was repeated over and over. The dog would chase the steer across the field, barking at its heels. Then the pair would turn around and the steer would chase the dog back again.

One morning when Bill Stocking, manager of the farm, was watching the two animals, he saw the chase come within seconds of tragedy. Just as they made an abrupt turn, with the dog racing ahead and the galloping steer close on its heels, the cocker spaniel's feet shot out from under it. It fell heavily on its side. The steer, its head lowered, was on it in an instant. The age-old impulse to crush a smaller fallen rival seemed about to assert itself. But just in time it was superseded by another impulse that spared the spaniel's life. The steer stopped with its heavy head only a few inches from the dog. It nudged its defenseless companion with its muzzle. The spaniel leaped to its feet and the chase began again.

O C T O B E R 8 .   I have just returned, late in the afternoon, from a necessary journey to a big city. I have changed my clothes, and Nellie and I are out on the green path that follows the brook to the waterfall. The sunshine is mellow. Frail gnats dance in sheltered nooks along the way. After the noise and hurry and pressures of the metropolis, I sink into this Trail Wood life as easily as does the green frog I see—a lone sojourner—on a pad of dripping moss close beside the waterfall.

In the midst of this tranquil scene, my mind is filled with a mood of calm content, of deep and quiet pleasure. Someone else—who was it?—once recorded in the pages of a book just such a moment, just such a return to the quiet of the out-of-doors, just such an emotion as I am feeling. I vaguely remember the paragraph, but the words refuse to come into focus in my mind. When we come back, half an hour later, I take down book after book in my library. Finally I discover the half-remembered sentences. I find them in Sarah Orne Jewett's *The Country of the Pointed Firs*. Near the middle of the book, recalling a day when she left a big city behind and returned by water to the coast of Maine, she recorded her sensations:

"But the first salt wind from the east, the first sight of a lighthouse set boldly on its outer rock, the flash of a gull, the waiting procession of seaward-bound firs on an island made me feel solid and definite again, instead of a poor, incoherent being. Life was resumed and anxious living blew away as if it had not been. I could not breathe deep enough or long enough. It was a return to happiness."

OCTOBER 9. The Geese-Going Days. So this time of bright October weather was known to the Indians. We remember this, late this morning, as we come home from Broad Beech Crossing along the rough trail that follows the windings of a stone wall through the woods past Wild Apple Glade.

Our ears catch a stirring, faraway sound. It is the clamor of approaching Canada geese. Their voices grow stronger, draw nearer. We look up and through a gap in the treetops glimpse a long wedge of flying birds, an arrowhead pointing south—an arrowhead without a shaft. The two legs of the V are of unequal length. Before the journeying birds, traveling perhaps forty-five miles an hour, disappear from sight, I count between seventy-five and eighty geese.

Ever since the last week in September, when one or two small flocks went through, the migratory movement of these northern waterfowl has been building up. Last evening in the dusk, we heard the calling of geese along the ridge that runs west of Trail Wood. In the misty dawn this morning, we made out the vague V of another flock forging southward over the Little River valley. Reversing itself twice a year, fall and spring, the inherited compass these birds carry within their bodies leads them to and from their distant southern wintering grounds.

Today the weather is bright and warm. But the wild geese sense an impending change. The nip of cold has come to their northern lakes and marshes. With the steady, rhythmical beating of their wings running back along each leg of the wedge of flying birds, the flocks cleave through the air hour after hour. When the migratory urge is at its height, they may cover hundreds of miles in a day.

In one of the early books on American ornithology, these wild geese are referred to as "the forerunners of winter and the harbingers of spring." They still are. All along the way, their passing has special import for country people. These days at the post office in the village friends trade reports on the size of the flocks and the number of geese they have seen. So far the record is held by a man who spent the day shingling his house. During the time he was working, he counted 400 Canada geese passing within sight of his rooftop.

O C T O B E R 1 0 . One day when Henry Thoreau was digging for fish worms at Walden Pond, he uncovered a string of tubers he called "the potato of the aborigines, a sort of fabulous fruit"—the groundnut, *Apios tuberosa*. This same groundnut or Indian potato or traveler's delight—but now with its scientific name changed to *Apios Americana*—I find rooted in moist ground in several places at Trail Wood.

During late-summer days, I see its slender vines entwined in the higher vegetation. At times they extend to a length of seven or eight feet. I sometimes catch in passing the strong perfume of the brownish-purple flowers. And toward the end of summer I notice small pealike pods maturing where the flowers had been.

All this aboveground activity is obvious. But unseen, below-ground, there is the development of a succession of tubers strung along the roots like round pearls on a necklace. These are the so-called nuts of the groundnut. The plant is able to reproduce either through the tubers or through the seeds within the pealike pods.

Late on this October morning I start out with a pail and a small spade to harvest a crop I have not planted. Near the rustic bridge over Stepping Stone Brook, groundnut vines are wound in and out among the sweet fern. There I begin my digging. Only a few inches below the surface I unearth the first of several strings of tubers. Each tuber, looking like a small rounded potato, is an inch or an inch and a half in diameter. A few are twice that size. With a quart or more in my pail, I climb the path to the house.

Nellie washes and boils them and we dine at noon on this favorite food of the New England Indians, the same wild tubers that helped the Pilgrims survive their first bitter winter in America. Eaten hot, salted and buttered and mashed like potatoes, the groundnuts have a pleasing flavor of their own, perhaps a trifle turniplike. Sometimes they are eaten roasted, sometimes raw, sometimes sliced and fried in butter, sometimes cooked with other vegetables. The Indians and early pioneers dried them and stored them for winter. Also they collected the pods, the seeds being boiled and eaten like peas. The groundnut is, in fact, a member of the pea family. Some of its other common names are trailing pea, ground pea, and potato pea.

In Virginia, colonists who preceded the Pilgrims found in the groundnut a staple and welcome food. Captain John Smith refers to: "Grounds nuts as large as Egges, and as good as Potatoes, and forty to a string not two inches under ground." As early as 1635, attempts were made to introduce this New World vine into Europe. But these efforts to establish it as an agricultural crop were soon abandoned. The roots lie so close to the ground cultivation is difficult, and the tubers are so slow growing, two or three years are required to attain a size suitable for eating.

Whenever Nellie slices one of the raw tubers in two, we notice a milky, viscid fluid within. When I nibble at one of the pieces, I find this juice leaves a thin coating on my teeth and the roof of my mouth. And after the meal, Nellie discovers when she washes the dishes that the interior of the pot in which the groundnuts were boiled is covered with a rubbery film of residue, a gluelike deposit apparently left by the viscid fluids. To eliminate it requires a strong cleaning powder and a metal scouring pad. When the tubers are roasted instead of being boiled, this difficulty is avoided.

Throughout the world, there are only five species in the genus to which our groundnut belongs. Two are found in the United States. The other three are widely scattered. One is native to the region of the Himalaya Mountains, the other two to China. The giant of them all is the highly restricted American species, *Apios priceana*. Its range is confined to parts of Kentucky and Tennessee. Instead of a necklace of small tubers along the roots, this groundnut produces a single irregularly shaped turniplike tuber. In extreme instances, it has attained a diameter of as much as half a foot.

OCTOBER 11. Fields that yesterday were tan and green this morning are fields of silver. Grasses, weeds, ferns, boulders, bushes, the bark of the lower tree trunks, the rustic fence, the stone walls, the path around the pond, the roof of the log cabin are all plated with the cold and shining beauty of the frost. Once or twice before this month we have seen the grass whitened in colder spots. But this is the first heavy frost of our autumn. Under the rising sun, I walk through the glitter of a scene clothed in fragile crystalline beauty.

In its steady ascent beyond the brookside trees, the sun throws long streamers of its rays across the frosted fields. Looking toward the sun, I see tussocks of grass and islands of goldenrod plants transformed into low-lying luminous clouds. Now the glory of the goldenrod is largely gone from the fields. But beauty goes and other beauty succeeds it.

The long slope of Firefly Meadow is shining silver from end to end except where it meets the woods. There a band of green extends along the pasture edge. It testifies to the equalizing effect of the trees, to the slightly warmer air of the woodland. When I come upon a place where a chipmunk has burrowed its tunnel, I find the entrance of the hole encircled by an unfrosted ring of green grass. Warmer air flowing from the underground passages has kept the frost away.

I see, when I bend close, how each leaflet of a climbing rose is bordered with frost, the autumn counterpart of the dewdrops of summer dawns. The feathery leaves of yarrow are thick with silver rime and dry thistleheads rise like goblets plated with silver catching the sun. Where the rosettes of the mullein hug the ground, their furry leaves that even in midsummer have a slightly frosty appearance now are densely clothed with crystals. And everywhere the seeds of fall, seeds in all their varied forms, appear as though suddenly preserved within a sculptured shell of precious metal.

As I return home I surprise a chipmunk half in and half out of its hole. It is sitting in the sunshine at the entrance of its burrow. Bright of stripe and bright of eye, it looks about it. On every side it is surrounded by the silvering of the frost.

The morning advances. The sun gradually mounts above the treetops. I see the long shadows of the dawn pull back across the meadows, making a slow retreat toward the trees of the brookside. As they shorten, the increasing warmth of the sunshine melts the frost in the area newly exposed to the rays. So the thin white coverlet of the fields is drawn back, giving the impression of being dragged steadily toward the east. By eleven o'clock, I see this first of our heavy frosts making its last stand along the brook. By high noon the silver show is over.

OCTOBER 12. The sweet smell of woodsmoke drifts down the slope. Looking from across the pond I see a skein of bluish vapor trailing from the fireplace chimney and losing itself in the blue of the sky. On the hearth below, the first small fire of the year is blazing.

Farther down my side of the pond I notice a painted turtle on a log. It is basking in the warmth of the sun on this day in October, that jewel among the latter months that round out every year. Only a comparatively few times will I see this turtle again before, leaving light and warmth and consciousness behind, it will bury itself in the mud and, for the better part of half a year, so far as we can tell, will be as lost to sensations as a stone. I, with much labor, store up wood for my fireplace and pay for fuel oil for my furnace while this small turtle reduces its wants almost to zero. So we both survive until spring.

OCTOBER 13. Before we bought Trail Wood, Margaret Marcus, from whom we purchased the farm, told us of the golden light that fills the rooms with windows facing south at this time in fall. Today this light comes flooding in from the high hickories across the lane. They are clouded in the brightest yellow of their autumn foliage. Now, more than ever, the pignut hickory deserves its common name of "the shining hickory." The high point of its fall foliage, the time when the sun shines through the gold of its brightest leaves, varies from year to year. But it is an event, amid all the flooding tide of color sweeping across the woodlands, to which we look forward.

Whenever it comes, whenever we pause to enjoy it once more, we remember how much Margaret Marcus also enjoyed such things, how much these same acres meant to her. She walked in the woods. She looked forward to the spring wild flowers. The bird and animal neighbors were as interesting to her as they are interesting to us. At the time when she was thinking about the possibility of having to sell the farm, she made up her mind to sell it to no one who would not appreciate it as much as she did.

Once, years after we had settled in at Trail Wood, she recalled how it had seemed to her on the day when we appeared at her back door and she found our love for the land was as great as hers.

"It was," she said, "just as though I had been looking for you and you had been looking for me."

OCTOBER 14. Once seen, never forgotten. As easily recognized as a zebra or an ostrich or a panda. Imagine a creature with a snout, slender and downcurving, projecting out to a length almost twice that of all the rest of its body. Such is the dull-colored, plump little dweller in our South Woods we come upon today under our largest white oak tree beside the Big Grapevine Trail.

It is clinging to a fallen acorn. One of more than 20,000 species belonging to that immensely large family, the *Curculionidae*, this particular snout beetle is an insect specialist. It is interested only in acorns, and its common name is acorn curculio.

Carefully I pick up the acorn. The insect rides along, anchored in place. At the extremity of its slender proboscis lie its biting mouthparts. When we look closely we see they have been at work, for a tiny rounded hole penetrates the hard shell. And the tip of the parabola of the thin beak is thrust through the hole into the interior. But the beetle is intent not on feeding but on excavating a tunnel into the meat within. When it is completed, it will drop an egg through the hole and painstakingly push it with its proboscis to the bottom of the tunnel and then seal up the hole with a plug of excrement. The larva that hatches from the egg will spend its immature life feeding on the meat within the acorn.

Squirrels, it has been found, are particularly fond of acorns containing such larvae. In tests, the animals always chose infested acorns, as long as they were available, and discarded those in which no curculio larvae were present.

The beetle we observe is absorbed in its work. It pays no attention to us even when we breathe on it with our faces only a few inches away. Unlike the mouthparts of many insects, such as ants and wasps and praying mantises, its jaws bite not horizontally, inward from the sides, but vertically like the jaws of a dog or cat or human being. Particularly strong, they easily gouge their way through the hardest shell protecting an acorn. One of its many relatives, a snout beetle in the tropics, is even able to eat its way into tagua nuts, so hard they are called "vegetable ivory." It is believed by some entomologists, although it has not yet been proved, that these beetles secrete a fluid that softens the shell of the nut at the point where they are working.

I put down the acorn and the curculio. The insect—with its single-track mind—has not been disturbed in the least by its ride through space. It is absorbed in a pattern of behavior as old as the Oligocene. For at the famous fossil beds of that period at Florissant, Colorado, numerous species of curculio beetles have been found perfectly preserved within the thin sheets of volcanic ash that had been transformed into stone.

OCTOBER 15. It is curious how close we feel to someone—even someone we have never met, even someone who lived in a remote period in the past and in a far-distant country—when we find that he, too, experienced the same outlook, the same feelings we have known.

For me, each year, at this time of the beauty of the tinted leaves, this most beautiful period of our northern autumn, I come to my own personal Independence Day, my own Fourth of July in fall. For it was on this fifteenth day of October that I escaped to a freedom I have never left. On that day, after thirteen years in a New York magazine office, I left behind my regular salary to launch myself into the hazards of a freelance life as a photographer of and writer about the world of nature. Everything good that has followed—the books I have written, the journeys across the land and through the seasons we have made, these Trail Wood acres where now we are walking through the year—began on this autumn day which I commemorate each year.

And when this Freedom Day comes round, I always remember T'ao Ch'ien, the fourteenth-century Chinese poet. He, too, escaped after thirteen years among city streets. In one of his poems, T'ao Ch'ien writes: "Even as a young man/ I was out of tune with ordinary pleasures./ It was my nature to love the rooted hills,/ The high hills which look upon the four edges of heaven./ What folly to spend one's life like a dropped leaf/ Snared under the dust of city streets,/ But for thirteen years it was so I lived."

Ta'o Ch'ien—so far away, so long ago—knew the same attitudes, the same emotions that have been mine. And in this poem of his, "Once More Fields and Gardens," he set them down on paper half a thousand years before I was born.

OCTOBER 16.   As long as the light lasts in these mid-October days, we want to spend each sunlit moment out-of-doors. The silent tide of the autumn colors that has been rising during the earlier part of the month has held for several days now at its peak. Whichever way we turn this morning, we are surrounded by the most abundant color, in the most brilliant hues of the year, the million tinted leaves that clothe our woodlands—scarlet and crimson and yellow on the swamp maples, wine red on the tupelos, bright, clear red on the highbush blueberries, minted gold on the spicebushes, scarlet and rose red and lavender on the sumacs, yellows of varying shades on the aspens, the birches, the wild cherries, the tulip trees, and the hickories.

This color runs in walls along our pasture edges. It mushrooms up in the clouds of the higher treetops. When we stand on the hillside above the pond and let our eyes slowly follow the curve of the bordering trees, we see the greatest variety and richness in the glory of these autumn hues. And we see it all in double beauty, see it twice, a second time reflected in the mirror of the water. In the course of the day, while the apparent movement of the sun carries it up and over and down the other side of the zenith, we mark the subtle changes brought by the altered angle of illumination.

We note again today what we observe each year—how trees of the same species differ in the richness of their colors and the time when the tinting of their leaves is at its peak. One red maple near the bridge, when October comes each fall, shines out with later and more brilliant foliage than any of the other nearby maples bordering the brook. And beside the house, rising side by side, rooted in the same soil and growing at the same elevation, two of our shining hickories, year after year, have different times of attaining the brightest yellow of their autumn leaves. One is only partially turned when the other is clad in gold.

In this annual pageant of the autumn foliage that extends away across nearly a hundred acres of our land, which leaves are the most beautiful of all? This is difficult to answer. I consider the scarlet of the maples, the gold of the hickories, the delicate, almost indescribable shadings of the white ash foliage in the fall. My choice in the end seems a curious one—the compound leaves of a small tree of the lowland, a growth I dislike to encounter, leaves I always avoid. It is a tree rarely seen or recognized until autumn comes. Then its foliage stands out in swampy areas—so rich, so rare, so varied, so delicate in its intermingling of reds and yellows, sometimes exhibiting metallic hues such as are found in heaps of slag that when we come upon such a low tree in these October days we stop to enjoy its outstanding beauty while keeping our distance. For it is that arboreal relative of the poison ivy, the poison sumac.

In southern states, spring and summer bring a greater abundance of flowering trees and shrubs than in the north. But it is only farther north or in the mountains, where autumn brings sudden plunges of the mercury when evening comes, that we find the full splendor of the colored leaves of fall. On such a day as this everything is beautiful and pensive at once. There is a hint of sadness in the transient glory of these soon-departing colors. This is the culmination of the beauty of our northern year. Now in a relatively few days, in a comparatively swift retreat, will come the rain of colors as, dropping singly or descending in showers, the leaves drift down.

OCTOBER 17.   With a basket on my arm—a flat market basket of split ash—Nellie and I set off on a circuit of our woodland trails this afternoon. The air is tranquil; the sky cloudless, brilliant blue. We are joining the squirrels and the chipmunks. We are going a-nutting.

For us such an expedition to the woods in bright October weather is made almost as much to see what we can see as to find what we can find. We startle a chipmunk carrying home a shelled pignut in its mouth, a gray squirrel burying nuts in the loam. Bluejays are getting acorns among the upper branches of the oaks. Once one drops an acorn and we hear it plunging down, striking limbs and ricocheting off through the leaves. A little later I step on a dry fallen branch that breaks with a crack as loud as a rifle shot. It silences the chipping of the chipmunks and sets the bluejays screaming.

For a hundred feet or so along one swampside trail, the wild crop we gather consists entirely of beaked hazelnuts, each enclosed in a dense, bristly, leaflike covering that terminates in a lengthy tubular projection like a snout or beak. At other places we stoop to retrieve nuts fallen from the branches of the pignut hickories. And widely scattered, growing fewer year by year, our remaining butternut trees have matured the rich and oily meats within their elongated shells. Pioneers collected these nuts in fall not only for food but for the yellow dye obtained from their shucks. Sometimes in the spring they tapped the trunks of butternuts, like those of sugar maples, to obtain sweet-flavored sap.

It is beyond Broad Beech Crossing, following the path through the woods to Juniper Hill, that we add to the mounting harvest in the basket the sweetest-tasting nuts of all we gather— the product of the trees that give the path its name—Shagbark Hickory Trail. The principal hickory nuts sold on the market come from shagbark or shellbark trees. Among these trees we pause to listen to a low-pitched sound reverberating in the air. At some indeterminate distance to the rear, a ruffed grouse is again engaged in its sporadic drumming, perhaps for a last time this fall. Around us, colored leaves lose their hold and descend, swinging back and forth in the still air.

A little later, coming home by way of the pond, we pause to listen to a squirrel making a great noise as it advances in a series of leaps through the fallen leaves along the margin of Azalea Shore. It, and the noise it makes, pass close to the painted turtle sunning itself for another time on the log projecting out into the shallow water. It remains undisturbed. But the moment it catches sight of us, even in the distance, it slides hurriedly into the water and disappears. We are the only large animal it sees walking on two legs. We have a "different" appearance as far as we can be seen. It recognizes us. And its reaction reflects, we are sad to note, the distrust our kind inspires.

Before we climb the slope to the house, I set my basket in the grass. One by one, I remove and discard the richly colored autumn leaves that have collected inside while we walked beneath the trees. Nellie counts them where they lie on the grass. They number more than fifty.

OCTOBER 18. In another month of October, on just such a day as this, the Swiss philosopher, Henri Amiel, wrote in his *Journal Intime:* "One feels the hours gently slipping by, and time, instead of flying, seems to hover."

This sense of time standing still, of hours lengthening, of a slower ticking of the clock—this mood of the lingering days that is peculiar to October—is everywhere around us as we advance, in sunshine and shadow, along our trails today. All is warm and still. Winter seems far away. The insects of the meadow sing on as though they would sing forever. This is the plateau before the mountain climb, the still pool before the rapids, the lull before the storm. It is a time rich in beauty before a time of bleakness. It is a drifting time before the great reversal.

I know that once in May I chose those days of spring as the finest of the year. And I may think so again when I am in the midst of another spring. But now it seems to me it is these few lingering days of October that must be the finest of all. In them, as in the days of spring, there is beauty, sunshine, genial conditions. But here there is an added quality, a sense of maturity, of having experienced more, a greater sense of knowing, a sense of ripening, of fulfillment, of acceptance. October is the culmination of the alterations of the year. This happiness, this deep content, that comes in the serenity of these few latter days, is based on all that has gone before, is heightened by the proximity to change.

OCTOBER 19. In a khaki knapsack, slung over my shoulder, are nails, a hammer, and a stack of bright red-and-white "No Hunting" signs. At intervals I secure a new sign to a tree along our boundary. In only a few days more, in this time of autumn beauty, for many a wild creature the booming of guns will herald a sudden reversal in its days of peace. To give each form of wildlife a chance to live out its span, to protect all that lies within the boundaries of a sanctuary farm, is one of the satisfactions of ownership.

I remember when *The Sea Around Us* brought in the money for Rachel Carson to realize her long dream of owning a house on a headland of the Maine coast, her close friend, Dorothy Freeman Rand, asked her how it felt finally to own a place by the sea. Rachel responded: "Oh, I don't own it. It is only on loan for me to care for."

This is the feeling that is shared by an increasing number of people. Land is more than a commodity to buy and sell; it is something to protect and enjoy and understand. In her book *Beyond the Aspen Grove*, Ann Zwinger tells of her sense of finding a new world, her sense of discovery in learning about her high-country surroundings. And then she adds: "But somewhere in the learning came commitment, the realization that in the understanding of this natural world comes the maintainance of it, that with knowledge comes responsibility."

Just to enjoy the birds and appreciate the wild flowers is no longer enough. No longer can we view either with the single-minded innocence of earlier years. In this time of dying lakes and poisoned rivers, abused land, destroyed wildlife, and polluted air, we need not only to appreciate nature but to appreciate it enough to join privately and in cooperation with groups in—so far as we are able—the protection of what we enjoy. At Trail Wood, we protect our sanctuary farm. But it is part of a greater whole. So the whole is also our concern. Two elements in the credo of the Sierra Club are ". . . to enjoy, to protect. . . ." Enjoyment in the present, in our immediate surroundings, and concern for the future, for the greater whole—both are needed.

O C T O B E R 2 0 .   In the very year—1806—when our Trail Wood house was built, a captain in the British navy—later Rear Admiral Sir Francis Beaufort—devised his famous Beaufort scale for recording the strength of the wind. Warships then sailed under canvas. Beaufort lived to see his scale accepted by the British Admiralty for all vessels at sea, but he died too soon to know that the International Meteorological Committee would adopt his idea for reporting weather throughout the world.

As employed today, Beaufort's wind scale embraces thirteen designations: 0. *Calm.* Smoke rises vertically; the sea is mirror smooth; movement of the air is less than one mile an hour. 1. *Light air.* One to three miles an hour. The drift of smoke indicates the direction of the breeze. 2. *Light breeze.* Four to seven miles an hour. Leaves begin to rustle. 3. *Gentle breeze.* Eight to twelve miles an hour. Leaves and twigs in motion; crests on waves begin to break. 4. *Moderate breeze.* Thirteen to eighteen miles an hour. Small branches move; dust rises; many whitecaps on waves at sea. 5. *Fresh breeze.* Nineteen to twenty-four miles an hour. Small trees in leaf begin to sway. 6. *Strong breeze.* Twenty-five to thirty-one miles an hour. Large branches begin moving. 7. *Moderate Gale.* Thirty-two to thirty-eight miles an hour. Whole trees in motion. 8. *Fresh gale.* Forty-seven to fifty-four miles an hour. Twigs break off. 9. *Strong gale.* Forty-seven to fifty-four miles an hour. Foam blows in dense streaks across the water at sea. 10. *Whole gale.* Fifty-five to sixty-three miles an hour. Trees uprooted; huge waves build up with overhanging crests. 11. *Storm.* Sixty-four to seventy-two miles an hour. 12. *Hurricane.* Wind velocities above seventy-two miles an hour.

On this October morning, when we set forth, the wind is gusting among our trees. Its strength, we decided, rates five on the Beaufort scale—a fresh breeze with a velocity of about twenty miles an hour. But it blows unevenly, coming in blasts, then easing off. At each long gust, we see the leaves whirling away. To Beaufort, more than a century and a half ago, the main importance of the wind was its effect on water and canvas. To us, on this fall day, its chief importance at Trail Wood is its effect on the remaining foliage of autumn around us.

Whenever we are downwind near or among the trees, we are in the midst of leaf showers or leaf blizzards according to the strength of the gusts. Hour after hour the long succession of blasts continues stripping away the colorful foliage. Near the hickories the wind is a yellow wind; near the maples, a red wind; elsewhere among the varied hardwoods, a multicolored wind. The air appears, at times, clad in its own foliage.

Even during the time of our walk we observe how swiftly the work of the gusts is transforming the colorful foliage above our heads into a colorful carpet of leaves beneath our feet. The woodland trails are paved with them. We walk as on a mosaic of many colors.

OCTOBER 21. We drift back and forth across the meadow on this crisp, bright morning. On each return we bring home a load of dry sticks broken from fallen branches in the woods. We add them to the pile accumulating in the entry shed to provide kindling for the fireplace fires of winter. This is always one of the most congenial tasks of the year. It is work that seems half play. If all the other work at Trail Wood were as enjoyable, Nellie observes, we could rename our place and call it "Happy Chores Farm."

So far as I can recall, I have encountered only one reference in literature concerning this unhurried harvest under the trees, amid the ferns and moss and carpets of the autumn leaves, this work and play that occupies us so pleasantly on this late-October day. In *The Alfoxden Journal* that Dorothy Wordsworth kept in 1798 when she and her brother William lived in the south of England near Samuel Taylor Coleridge, there are such recurring mentions of gathering sticks as: "William gathered sticks . . . ," "Gathered sticks with William in the woods . . . ," "Gathered sticks in the further woods. The dell green with moss and brambles. . . ."

In our "gathering sticks" today, we watch the play of sunlight among the tree trunks. We stop to breathe in the heavy fragrance of the autumn woods. We halt our work to follow with our eyes the scurrying progress of a chipmunk or a gray squirrel. We all—chipmunk, gray squirrel, Nellie, and I—are in the mainstream, part of the great movement of fall. We all are preparing for winter. These are the provident days of autumn. Each in its own way—the chipmunk storing nuts in its burrow, the birds, now far away, winging farther and farther toward the south, the *Polistes* wasps and the cluster flies seeking to squeeze themselves through cracks into the protection of buildings, the wood turtle burrowing into the mud for its long hibernation, Nellie and I bringing in sticks—so we all, bird and rodent and insect and reptile and human, make our autumn preparations.

OCTOBER 22. October is the month when summer trees become autumn trees and autumn trees become winter trees. With the exception of the oaks, with their heavy russet leaves and the beech trees with their fluttering tan foliage—both hanging on far into the winter—the month ends, usually, with foliage gone, with our woods almost entirely stripped of leaves. During the nearly six months of our winter trees, the songs of the summer birds and the music of the insects will be replaced by the varied sounds—low or loud, crooning or shrieking—that make up the music of the wind among the leafless twigs and branches.

A chill rain sweeps over us this morning hastening the descent of the remaining leaves. Wherever we go through the dripping woods today, walking on the soggy carpet of fallen foliage, the air is filled with the primal, deeply moving odor of wet autumn leaves, of tissue—alive so short a time ago—beginning its slow process of oxidation and decay, its transformation into woodland mold.

Leaf-fall, in these October days, represents one of the major landmarks of the year. It marks the end of the time of growth, the end—for the deciduous trees—of chlorophyll-making, the end of the green months, the coming of the great change to the predominantly gray and white months. On the wet woodland trails of this day, we have the feeling of walking a ridgetop between the seasons.

O C T O B E R 2 3 . In starlight we follow the twin parallel streaks of the wheel tracks down the length of the lane. The chill increases rapidly with the setting of the sun these later October days and we are clad in jackets. Almost day by day now, the tendency of the thermometer is downward. More and more, we hear the katydids and tree crickets calling during the hours of sunshine.

Now, at nine o'clock, all the insect music we hear is slow and labored. The temperature stands at slightly above fifty degrees Fahrenheit. The movement of the mercury these days, its rise and fall, represents the great conductor's baton governing the tempo of the music produced by the orchestra of the night insects.

On this still, chill evening, we listen for a long time to the katydids in the trees that border the lane. Gone is the storm of contention that filled the summer nights; gone the quick, loud emphatic repartee. Then the insects snapped off their "Katy *did!*" and "Katy *didn't!*" Now the sounds they produce seem less certain, more deliberate. They come to our ears in a lower pitch.

So the long debate of summer and early fall is nearing its conclusion. An agreement of sorts, unanimity after a fashion, appears to be emerging. As the chill of the evenings grows deeper, the katydids scrape their wings together more often in the shorter, abbreviated call, "She did," rather than the longer "Katy did" or "Katy didn't" or "Katy she did." At times such as this, when the mercury hovers close to the fifty-degree mark, "She did" is the dominant sound we hear the insects make. As though tiring of debate, they seem approaching a consensus at the end. This may be the final conclusion of the interminable argument of the earlier nights as to whether she did or whether she didn't: Katy *did*.

For our Trail Wood katydids, fifty degrees forms a dividing line. When the mercury descends lower, most of these insects fall silent. But the mellow rhythm of the snowy tree crickets continues on, slowed in tempo, but filling the darkness until the temperature has dropped into the lower forties. Then they, too, become silent.

How few are the days for all these musicians! How soon the cold will cut them down! The time of the katydids has only a little way to run. Not only is the chill of the autumn nights increasing, but change in the autumn foliage in the recent weeks left them without the green leaves of their previous diet and robbed them of the camouflage their leaflike forms and coloring provided amid the trees of summer. Although they show no awareness of it, the end of their whole generation is close at hand. Painfully slow now among the remaining leaves, the musicians play on. Each year at this time I am profoundly moved by this unconscious lament, a lingering, sad farewell by those who do not know they are going.

O C T O B E R 2 4 . On either side of us, as we follow the Old Woods Road today, tree trunks sway and bend as their tops, with a clamor of sound, heel over as gust follows gust. This is the exercise of the trees. It is an important part of their lives. It adds to the strength of their roots and fibers as exercise adds to the strength of the muscles of an animal. In the hurricane of 1938, it was found at the Pine Acres Tree Farm that in the protected center of large stands of red pines the trees went down like cornstalks, all facing the same way. But where trees of the same species grew singly or in outlying rows—trees that over the years had been strengthened by being subjected to the full force of lesser storms—they all remained erect.

How silent are the woods when the air is still; how filled with sighs and murmurs when a slight breeze springs up; how echoing with the boom and shriek and wail of the treetops when a great wind blows! On this day of tumultuous gusts, the buffeted trees respond to the wind like a vast orchestra of aeolian harps. In a thousand variations, the twigs, the branches, the individual forms of the bare treetops contribute different strains, different tones to the roaring medley that rises and falls around us.

All along the way, as we come home, we find small fallen branches scattered on our path. We toss them aside, remembering another wind among the trees, a far greater wind in a far greater woods.

On the last day of August, in 1954, Nellie and I were staying in a log cabin overlooking Crocker Lake and the Mooseback Mountains of the Canadian line, miles back in the forest of northern Maine. The radio in the lodge of the camp was blaring out frequent warnings of a hurricane sweeping northward. We started down a familiar trail for a quick last walk before the storm. Already gusts were rocking the evergreens. At one point, our mossy path dipped into a small moist stretch of lowland floored with sphagnum moss and the interlacing roots of halsam firs. On all sides of us, with the whitish undersides of their needles appearing and disappearing, the firs tossed in the growing violence of the blasts. As their shallow roots pulled taut and relaxed again, the ground lifted and descended beneath our feet. We walked on land as though on rising and falling waves of water.

Later, back at the cabin, Nellie and I watched the drama of the greater storm unfold as the tail end of the hurricane struck. With a velocity of at least seventy-five miles an hour, gusts exploded out of coves and flailed the water of the lake. Watching from the rustic porch of the cabin, we could see the darkness of the driven water spreading out fanwise before these successive onslaughts of the wind. At times the gusts whirled on an erratic course, flicking the fans first to one side and then to the other. Through the rain that streamed across the lake, we glimpsed aquatic dust devils, miniature waterspouts, once three at a time, spinning over the surface. I can recall no other time when the wind hurled itself in so many directions at once.

With every blast, birch leaves and twigs of the evergreens went streaming past the cabin and far out over the lake. Once we glimpsed what appeared to be green grass thrust above the water. Our field glasses showed the "grass" we were seeing was, in reality, the long needles of a large, partially submerged branch broken from a white pine tree. Where coves were rimmed with jumbled stumps—silvered by years of weathering—fragments broke free in the storm. We saw them drifting down the lake before the wind, some with twisted roots lifted above the water like the antlers of swimming deer.

The gale continued on into the gathering dusk. We were startled occasionally by a sound like the crack of a high-powered rifle when one of the solid green cones of a white pine struck the roof. Another sound—duller, heavier, accompanied by a jar that sent a tremor through the boards beneath our feet—came just before darkness fell. I hurried out into the rain and wind and discovered a thirty-foot black spruce had been uprooted by the gale and had dropped directly across the ridgepole of the cabin.

Although the damage was negligible, when we measured the tree the next day we found its circumference was thirty-two and a half inches. Its growth rings showed it had been rooted beside the cabin for more than twenty years. And when we returned to the hollow in the forest where we had walked on the rising and falling ground as on waves of water, we discovered that one of the largest of the balsam firs had crashed down across the path that we had followed.

OCTOBER 25. Someone, whom I cannot now remember, once wrote that a hundred sheep are no more instructive than one sheep. But is this entirely true? Had the author known sheep better, he would have understood how individual is each lamb and ewe, how varied in experiences and attitudes and capacities they are. The general pattern may be the same but the specific variations are infinite. It is the *individual* life that interests us most.

Friends of ours, coming back from a trip, once told us of meeting a couple who had read my books. Their first question was: "What are Edwin and Nellie Teale *really* like?"

So far as we can tell, we are really like we appear in the books. But this query represents a universal outlook. We all wonder what the people we meet are *really* like, what life is *really* like for them. And so Nellie and I wonder what life is *really* like for the birds we watch, for all the varied forms of existence we encounter along our trails or beside our pond. This is a secret query that never dies. It is specific, not general; it refers to individuals, not masses.

Sitting row on row, an audience in a theater, a congregation in church, a crowd at a political rally seems to form a unified whole. Yet that whole is like sand contained within a cup. As soon as the sand pours from the cup—as soon as the audience or congregation or crowd arises—it falls apart. No two in a multitude are precisely alike. In outlook, ideas, desires, background, ambition, condition of health, they vary subtly or widely. Two, sitting side by side, may be separated by distances almost unbridgeable. What is life *really* like for each? We always keep coming back to that question when thinking of both human beings and the wild forms of life we meet in the out-of-doors.

As for ourselves, for each of us, the most interesting book in the world is the one we are writing by living. It is the book of our *own* experiences, our *own* journey through the world. It is a personal and unpublished volume. We alone read it in its entirety. Memory turns the pages. It tells us what our own lives are *really* like. Or so we think.

OCTOBER 26.   Nothing creeps up silently on anything in the woods these days. The carpet of fallen leaves, daily becoming more crisp and dry, provides an early warning system for wildlife. No fox can stalk its prey noiselessly here. No Indian could walk in silence through these woods. Stealth is impossible. Even a chipmunk broadcasts its every hop as it moves over the woodland floor. A gray squirrel burying a nut among the leaves sounds at least as large as a woodchuck. And so loud is our own progress as Nellie and I follow the trails or wander aside among the tree trunks that we seem giants scuffing through the leaves. The uproar we produce drowns out our voices. We have to stand still when we want to hear each other speak.

A young raccoon claws its way over an obstruction as it
starts on its nocturnal wandering and foraging for food.

Parachutes of milkweed fluff, tightly packed in the pod,
are released by the autumn wind to broadcast the seeds.

The mirror of swamp water reflects, amid floating fallen
leaves, the image of trees growing bare as autumn advances.

The spore-spikes of ground pine are covered with yellow
living dust known commercially as lycopodium powder.

Looking toward Beaver Rock along the pond's north side.
Near here a chipmunk started to swim to the other shore.

The clustered fruit of staghorn sumac dries and remains in place for months, a favorite food for overwintering robins.

Sunflower seeds attract purple finches to Trail Wood feeders.
This handsomely striped female clings to a silvered stub.

Late autumn sunshine lights the lane leading to Kenyon
Road. It is now bordered by trees that are largely bare.

Once when we halt not far from a dense stand of bushes, an unseen grouse that has kept our approach under surveillance takes off, the explosive suddenness of its bursting from the bushes being magnified by the noise of the dry leaves sent flying by the violent downdraft of its beating wings. These game birds have a special stake in the dryness of leaf cover in the autumn woods. During years when the hunting season begins in particularly dry weather, the fallen leaves broadcast the movement of approaching dogs and hunters. It is when the leaves are wet from rain or heavy mist, when they are soggy and silent underfoot, that the grouse are more likely to be taken unawares.

For a time, on the way to the Brook Crossing, we turn aside along a wall and begin stooping and brushing away the dry leaves, searching for plump red berries that shine out amid the dark glossy green of the three-leaved wintergreen plants. Not long ago this fruit was white; since then it has ripened, turned crimson, and become filled with flavor. Each with a handful of berries, we sit on a hummock among the trees.

Looking up through the lacework of the black twigs of the treetops, we see glimpses of the hard blue of the sky. So brilliantly blue is it, this morning, that it gives the impression of having been polished and scoured by the high wind of two days ago. Looking down, we become aware of the surprising number of wandering ants and small winged insects walking over the leaves. They remind us of all the dense, unseen population hidden in the leafcover. During much of the way home, we both are reflecting on this invisible host. In the sudden silences when we stop, we discuss the beetles, the larvae, the millipeds, the predators, and the preyed-upon that find a home in the layers of this carpet laid down by the leaf-fall of October.

O C T O B E R 2 7 . The breeze sweeps through the dry ferns around me. It draws, in its passing, frail, whispering music from the tangled mass of the brittle fronds. Here, under the dead elm at the far side of the pond, as I stand listening to these elfin strains, I notice something that has escaped my attention during all the summer months I have been following the trail that leads close to the elm and this lacy maze of New York ferns massed beneath it.

Set in the midst of the frond tangle, I spy a little cup formed of grasses and vegetable fibers and fine threads of bark. I observe how the material has been woven in and out among the ferns, attaching the diminutive cup firmly in place half a dozen inches above the ground. I finger the delicate nest. What tiny bird was its architect? In size and shape and materials employed it suggests a warbler's nest. But I know of none of these birds that constructs its nest among ferns.

How many times, as she brooded her tiny eggs, had the unknown builder looked out from amid the ferns to watch me as I loitered by, pausing often, so close at hand? Baffled by the identity of the maker, I stand here picturing the life of the nestlings that spent their earliest days cradled in this cup of grass. How frail appears the support provided by the supple, feathery ferns of spring. How wildly must the nest have gyrated in gusts and windstorms. For the diminutive birds beginning life here, the world into which they had hatched appears as unstable as the world of the young orioles swinging in their high, hanging home. I stand looking down on this deserted nest among the dry interlacing ferns on this day near October's end. So small, so delicate, amid such insubstantial surroundings, it seems a kind of elfin home set in the midst of an elfin scene.

OCTOBER 28. I emerge into the steely light this morning as into the silence of a great battlefield where armies lie slain. Toward sunset last evening the wind shifted. Hour after hour, under the glitter of a wide, cold sky, it blew from the north. When sometime after midnight it died down, the mercury continued its descent. It stands at seven degrees below freezing, twenty-five degrees Fahrenheit, when I look at the thermometer outside the kitchen door at dawn. The night of the killer frost, coming a little late this year, has arrived, as it arrives each fall, with its sudden finality.

White are the brookside bushes, white the walls, white the meadows. Heavy frost overlays all. My feet make a faint crunching sound in the frost-laden meadow, and where I step the imprint of my shoe is recorded by grass that remains flattened. Wandering in the chill of this dawn, my mind is filled with an awareness of all that has been lost in a night. Farewell to the katydids! Farewell to the snowy tree crickets! Farewell to so much—to so many plants, to the singing of so many insects of the early autumn days! With a single night of freezing cold, nature has virtually wiped clean the slate for forms of life at once frail and tenacious.

Perhaps under flat pasture rocks, small clusters of black field crickets have survived. Their chirping is the last of the insect sounds we hear in fall. But on this morning, from the frigid hours of the night has come a sudden silence, the silence of winter extending across our fields. For more than eight months out of the twelve we will miss the debating of the katydids, the rhythmic melody pulsing from the sliding pale-green wings of the snowy tree cricket, all the music from bush and tree and grass tangle that filled the warm, ringing nights of the earlier fall.

It is a somber time, this walk in the dawn at this hour of farewell. We will have our warm days. We will know our Indian Summer. But the great change has set in. The tide of winter cold is rolling nearer. During these hours of darkness, we have tasted its power. I walk about now taking stock of all the small forms of life, forms of beauty and interest, that have been wiped out in this single night of sudden death.

OCTOBER 29. The dark robins have come again.

About this time each year forest thrushes from the evergreen woodlands of the far north reach our area. They come in waves—all noticeably darker than the earlier migrants. People here call them "Newfoundland" robins. When they leave, the fall migration of their kind—speckled new robins, light-colored farm robins, dusky forest robins—is at an end. They trail through, the rear guard of the movement.

Half a hundred of these darker-backed, red-breasted birds are running—with characteristic starts and stops—over the frosty meadows when I step out this morning. The birds and a second heavy frost have arrived together. Walking over the silvered grass that has lost its spring, I watch these last of the migrating robins calling excitedly, landing and taking off again. I see them fluttering among the Japanese barberry bushes to pluck the crimson berries. Where were the robins yesterday? Where did their long flight begin?

According to the *Check-List of North American Birds*, published by the American Ornithologists' Union, the breeding range of *Turdus migratorius*, that favorite and common bird, the robin, extends from southern Mexico to the limit of trees in northern Alaska, northern Canada, and Newfoundland. One subspecies, *Turdus migratorius nigrideus*, is the dominant robin of Newfoundland. The *nigrideus* of its scientific name takes note of its darker plumage. This is a characteristic of birds whose habitat lies in the deep shade of dense forests. It is apparently the robins of this subspecies that have traveled so far to reach our fields.

Of all the thousands of these darker birds that I have seen bringing up the rear guard of migration during nearly twenty successive autumns, one bird still stands out distinctly in my mind. I was carrying in fireplace logs from the center shed when I first sighted it. This was during our fifth October at Trail Wood. Looking out across the Starfield, I was watching the coming and going of a wave of the "Newfoundland" robins when I noticed one that was strikingly different. White feathers drew shining lines along the darker plumage of its wings. It appeared to be a red-breasted bird with wing stripes. I forgot all about the fireplace logs and followed the movements of this partial albino as long as it remained in sight. Each time it flew, the upper surface of its wings presented a dark-and-white pattern. It was as though a shorebird were keeping company with the dark robins from the north.

OCTOBER 30. Violent gusts strike in sudden pounces, scour across the open pastures, and go floundering away through the thicker woodland. In the exposed fields we feel their pressure against our backs, hurrying us along. But in all their movement as they rush toward us, swirl around us, stream beyond us, they exert a force invisible. We trace their progress through their effects, through the movements they produce, through such things as bending bushes and windblown leaves.

On this day, the inanimate travelers that ride on the successive gusts most buoyantly are the parachuted seeds of a familiar plant, a plant that at various times has been called wild cotton, Virginia silk, silky swallowwort—the common milkweed, *Asclepias syriaca*. In wasteland along the meadow edges in recent days, Nellie and I have noted the opening of the milkweed pods.

As each dry pod, bulging and tapered and about four inches long, has split lengthwise, we have glimpsed in the interior that masterwork of symmetry and efficiency, the placing of the close-packed, overlapping seeds and the fine silken threads of the parachutes that will support them in the air. Looking back on this windy day, we see the white of the milkweed silk crowding the openings in the sides of the pods like shining masses of overflowing foam. The airborne seeds have been waiting for the wind and the wind has come. We observe them, one by one, or in small white flocks, break free and whirl away on their journey to a landing place unknown, to a destination that will be decided by chance.

Near the top of Nighthawk Hill, we stand looking back again. All across the field a host of other milkweed parachutes—each racing swiftly, each shining in the sun—hurry toward us. Some stream past overhead, others scud by close to the ground. The milkweeds are scattering their seeds on the wind and in so doing they are sending forth delicate tracer objects that reveal each subtle change in the pathway of the gusts.

We follow with our eyes the flight of one airborne seed as it traverses the Starfield. We watch it race toward the dense mass of a barberry bush. Just before it collides, it zooms up and over the obstruction. The race and the zoom are the race and the zoom of the wind. A little later, we concentrate on another of these airborne seeds as it approaches the woods. It rides on a collision course directly toward the trunk of a large red maple tree. At what seems the last possible instant, we see it whirl to one side, circling around the tree in a swift sidestep to continue on its way. To us, the outstanding event of this October day is this parade of the autumn seeds whose silken parachutes make the invisible track of the wind visible to us.

OCTOBER 31. I once held in my hand a worn copy of Emerson's essays, the same copy that John Muir, in the 1880s, read and reread beneath the sequoias and in his cabin beside the Merced River when he was living in Yosemite Valley. In the margins of a number of the pages, he had written in pencil his own thoughts and reactions to what he read.

I remember once where Emerson says: "There are in the

woods a certain enticement and flattery, together with a failure to yield a present satisfaction." Muir had demurred: "No—always we find more than we expect." Again where Emerson says that "nature takes no thought for the morrow," Muir quieried: "Are not buds and seeds thoughts for the morrow?"

On this ending day of October, on the opposite side of the continent from Yosemite, we note along all the trails we follow— across the pastures, through the woods, beside the brooks—signs of nature's thoughts for the morrow. At no other time of the year are they so obvious. Fall is the provident, the forward-looking season.

Where the leaves of another year are compressed within the compact buds of trees, Nellie and I run them between thumb and finger. We feel their textures, observe their varied forms and sizes, note the waxy coating on some that will weatherproof them until spring. We walk on carpets of ground pine beside the beaver pond and see the new mud and sticks added to the lodge. In more than one place in swampy lowland, we glimpse the tips of pale-green cones thrust above the mud and water. They are the spear points of the tightly rolled leaves of next year's skunk cabbage. Nearly half a year will pass before the broad leaves will expand under a returning sun. But already, so far in advance, the plants have prepared for that distant time.

Yesterday we saw the seeds of the milkweed riding away on the wind. Today, stopping all along the way, we note on plants beside the trails other seeds, endlessly different seeds, each in its own way a tiny thought for the morrow. Returning home after hours in the open, our impression, on this day, is that the world has gone to seed. This—for the plans of nature's year—is a manifestation of success.

NOVEMBER 1. Winter-silver of pelt, autumn-fat of body, the gray squirrel comes nosing over the ground. I follow its progress as it jumps this way, then that, then comes to a stop and begins to dig. Its forepaws blur in its haste and the earth comes flying out as the little pit deepens. Its pointed nose disappears underground and reappears. The squirrel has a hickory nut in its mouth. It hops three or four feet away. Again its forepaws speed into action and it swiftly excavates a new hole. Once more its nose disappears as it deposits the nut at the bottom. Then it rakes in the loose earth and carefully pats it down. Three times I see this performance repeated. Then the little animal dashes to a nearby hickory tree. It leaves me in a quandary, uncertain what I have observed. Are the nuts it has dug up and reburied ones it has previously buried? Why move them to a new hole? Is it like a man transferring some of his savings to a new place of deposit? Or is it just having fun burying nuts? Or is it pirating the stores of another squirrel? And if it is, will they be pirated again in turn?

N O V E M B E R 2 .   All this gray, wet November day, with windless air and low-hanging sky, the hours advance slowly. Everything is relaxed. Action is suspended. We are surrounded, wherever we go, by misty, rainy, whitethroat weather.

From the wild plum tangle, from the dripping bushes along the lane the minor lament of these sparrows, the same pensive song we have heard among the black spruce forests of northern Maine and by Hermit Lake, high on the side of Mount Washington, carries through the moist air around us. The half-light of this leaden and unpromising day, perhaps—as I have noted before—reminding them of the mist clinging to the trees in the chill of the forest dawns of their northern home, appears to stimulate them to song.

Some of the voices are obviously those of young males, birds that hatched in this year's nests. They seem practicing fragments of the melodic sequence that forms the ancient pattern of their songs. They repeat over and over, beginning and stopping abruptly, not the continuous songs but portions with which they appear to be having difficulty.

I remember a young whitethroat Nellie and I encountered on a second of September along a mossy trail after we had paddled the length of Crocker Lake in Maine. It seemed just beginning to use its voice. It floundered, sang off-key, tried over and over again. At a certain point it always sang flat. A score of times it repeated this fragment of the whitethroat's song. The start of each attempt was perfect—then it got off the track. It suggested a human vocalist practicing scales and portions of a song. We wondered if such a bird would ever learn. Would it ever become a finished singer?

The answer appears to depend on two things—whether it hears other males that are expert singers and at what time in its life it is subjected to such examples. Studies at the California Institute of Technology reported in 1975 show that while some species of birds sing perfectly even when they are isolated from others of their kind, many songbirds imitate the songs they hear. In addition, they have a critical period of song learning that extends approximately between the second and seventh week after hatching.

In the case of such birds as the related white-crowned sparrows, the songs may vary slightly in different localities as do the dialects of human inhabitants. The young in each area learn from older singers the avian dialect of its local song pattern.

Even individual singers, especially among the sparrows, may produce variations in the normal song. The analysis of sound recordings made in the field show that, so far as has been discovered, the most diverse of all is the common song sparrow. A single individual produced twenty variations. And the total number of variations recorded in the songs of such sparrows is close to 900.

N O V E M B E R 3 .   The whitethroats are singing again this morning. As I stand listening to them repeat their songs, noting how the singing of one male sets off the singing of another, a memory awakens in my mind concerning an occurrence that took place a few years ago on the west coast of Florida. There a friend of our, Charles W. Lawrence of St. Petersburg, observed an interesting instance in connection with the response of birds to specific songs. One afternoon he and his wife, Helen Mary, were listening to a bird-song record of the voices of North American thrushes. For several days, an olive-backed thrush, overwintering in the south, had remained among the shrubbery of the yard. On this afternoon it paid no attention to the recorded songs of the veery, wood thrush, hermit thrush, and gray-cheeked thrush. But when the song of the oliveback carried out across the yard, it immediately became excited, bursting into a competitive song.

N O V E M B E R 4 . Clothed in the yellow grass of fall, sheltered by trees from the north wind, here where the earliest rays of the morning sun bring the earliest warmth of these November days, this curving western border of the pond descends steeply to the water's edge. Standing near the top of the slope, Nellie and I look down on the activity below us as from a seat in an amphitheater.

On this slanting ground, insect life at Trail Wood finds one of its final strongholds of autumn. Here we listen to the black field crickets chirping long after most of the meadowland is silent. Here we see some of the final grasshoppers of the year clinging to withered stems. Here we watch the glint and flash of small dragonfly wings when the life of the *Odonata* has almost drawn to a close.

On this day, it is those autumnal dragonflies of modest size, the bright red males of *Sympetrum semicinctum,* that hold our attention. Almost suddenly each year, just about the time fall begins, they appear in increasing numbers. For weeks now, we have come upon them hawking along the water's edge, weaving among the branch tips of the alders, and hanging on quivering wings above the path that curves up to the door of my log cabin. On recent days we have encountered them sunning themselves on the weathered rails of the rustic bridge and clinging to twigs overhanging Whippoorwill Brook, where it tumbles down its steep descent to reach the pond. Now they have concentrated along the warmer curve of this western shore.

It has been our experience that the little *Sympetrum* dragonflies permit us to come closer than any other species. This afternoon, while we stand on Summerhouse Rock watching a flotilla of bass go sliding by, one of these insects alights on Nellie's shoulder. As long as we remain motionless, it basks there in the sun. At times we find the red dragonflies anchored to twigs, apparently deep in sleep. On such occasions, I have taken close-up pictures with the lens of my camera no farther away than the length of a forefinger. We notice how these insects, in resting, let their wings droop instead of holding them out rigidly. In some areas, immense concentrations of these dragonflies build up as autumn advances. I once read an account in an entomological journal by a scientist who had the experience of coming upon nearly a mile of *Sympetrum* dragonflies all resting side by side on a telephone wire. The thousands of red bodies were all parallel; the thousands of insects all faced in the same direction.

Among these brilliant little dragonflies, one, haunting the warmer amphitheater at this western end of the pond, has come to seem a special acquaintance of ours. We recognize it by its notched and ragged wings. Usually we find it resting on pickerelweed close to Summerhouse Rock or patrolling the edges of Whippoorwill Cove. We see it now swaying in the breeze at the top of a rush stem.

One recent morning, after a night of descending mercury, we discovered it hanging to a grass stem, apparently lifeless. But as the day advanced and the sun climbed higher, animation returned to the durable insect. Only yesterday Nellie found it floating on the surface of the pond. That seemed the end of all its adventures. But the breeze was blowing toward the shore. It drifted nearer and she was able to fish it out on a twig. She found it bedraggled but still alive. For a long time, while it dried out in the sun, it clung to a weed without moving. Then, abruptly, it rose on its ragged wings, ascended to a height of ten or fifteen feet, descended again, and commenced patrolling along the water's edge. A dragonfly with nine lives!

NOVEMBER 5. Our first flock of evening grosbeaks alights in the apple tree below the terrace this morning. The return of these winter birds forms one of those small markings on the sundial of our natural year. Each such little first-of-its-kind event records an advance in the seasons—the first "okalee" of a returning redwing in the last days of February, the first song of a woodcock in March, the first bluet of spring in April, the first rasp of a katydid's wing in August, the first heath aster blooming in September, the first red maple leaf drifting down Hampton Brook, the first skim of ice along the edges of the pond, the first falling snowflake. They all are small indications of the wider, profounder changes taking place in the sky and on the land.

It was the recurrence of such natural events that provided the world's earliest calendars. For primitive men, the sprouting of plants, the ripening of fruit, the falling of leaves marked the divisions of the advance of the year. That time in spring when oak leaves become as large as a mouse's ear formed, for the Iroquois Indians, an important milestone in the year, the time to plant their maize.

Hesiod, the eighth-century Greek poet, records classical examples of how in his time the events of the seasons were used to determine different stages in the year. When the cry of the migrating cranes descends from the sky, he notes, it is time for plowing and sowing. When the snail climbs up the vines, digging in the vineyards should cease. When the thistle blossoms, summer has come. And when the fig tree sprouts, mariners will find conditions propitious for voyages at sea.

Long before the work of astronomers had refined the present method of dividing time into months and weeks and days and hours and minutes and seconds, the cycle of the seasons was broken up in this way into smaller units. They were based on an earlier, closer relationship with nature.

And here at Trail Wood—where we, too, find ourselves closer to nature—we discover something of the kind taking place. On our daily walks we observe the recurrence of an endless succession of little changes that signify the passing of time. Following in annual sequence, they combine to form for us the natural calendar of our outdoor year.

N O V E M B E R 6 .   About two o'clock this afternoon I load a market basket with camera equipment and climb the slope of Juniper Hill. In a small open space near the top of the ascent, a feathery stand of Virginia beard grass, *Andropogon virginicus*, appears to have concentrated all the available light into one shining pool set amid the dark masses of the evergreen shrubs. Grasses have received many picturesque common names. None is more fitting than that bestowed on this plant among the junipers with its sun-catching heads leaning down with their beards of hairlike filaments haloed in light.

Later, with the glowing heads of the grass stems recorded on film, as I am packing up my equipment I glance down the slope. Beside the lower wall a yellow spot stands out against the gray of the stone—the excavated earth of a woodchuck burrow. Lost now in its long hibernation is the excavator. How many days have passed since I saw a woodchuck? I try to remember. But I recall that days after the last of these hibernators had vanished belowground, Nellie and I heard the shrill, explosive whistle the marmots give. It was repeated a second, a third time. But each time it came to us not from the ground but from a treetop. Once again, one of our starlings was imitating a summer sound. How many times in the days when the woodchucks were abroad had we heard the bird and thought we were hearing the animal? That we will never know. For it was only after all the burrowers had disappeared that we became aware that this sound that reached our ears was a counterfeit call, the imitation whistle of the starling.

NOVEMBER 7 . By five o'clock these November afternoons, the sun has set, the glow is fading from the sky and twilight is a deepening dusk in all the hollows. A little after this hour, Nellie and I are descending into the bowl of land that contains our pond. We are homeward bound, laden with dry sticks gathered in the woods to add to the fireplace kindling. Under these conditions we encounter a little adventure close to the edge of the water.

At the foot of the slope that leads to the western end of the pond we set down our bundles to rest. For several minutes we stand there watching the last tinted light reflected from the sky slowly fading from the mirror of the water. Then in the dusk we see what at first looks like a bat flying erratically over the pond toward us. But this is far too late in the season for bats. Seen against the sky, as it zigzags closer, we recognize the chunky body, the long, downpointing bill, the short, broad wings of a flying woodcock.

The bird circles around us as a small brown bat sometimes wheels around us on summer evenings. Then it turns out over the water and flies back and forth past us on a wandering course. It seems uncertain. It appears unable to make up its mind. Apparently, without realizing it, we are occupying the exact muddy portion of the shore where it had intended to land. Before we can move, we see it swing away, cross to the margin of Azalea Shore, and drop down helicopterwise in an almost vertical descent. When motion ceases, it disappears. In brown and somber plumage, the woodcock is well disguised for its life as a bird of the dusk.

We pick up our bundles and begin the ascent to the house. Halfway up the slope, we look back and glimpse, in the swiftly fading light, the dark form of the bird taking off, flying low and fast, and descending where we had stood.

About this same time of day, when the chill and the dusk come hand in hand, Nellie, a few evenings ago, had another woodcock adventure of her own. She had been coming down from Lichen Ridge carrying her gleanings of dry and fallen branches. Beside a small oak tree overlooking the shallow flow of Stepping Stone Brook, she paused and set down her load. As she was clapping and rubbing together her hands to warm them, she saw a thickset bird come speeding in over the pond. Fluttering down in an abrupt descent, it alighted at the edge of the brook. Her hands held out before her, Nellie froze into immobility. Only fifteen feet away, the bird commenced taking a vigorous bath, splashing, shaking its wings, ducking its head.

Even while she was absorbed in this performance, a second woodcock flew in unnoticed. The first she became aware of it was when she heard a new splashing coming from a little farther along the stream. The two birds, ignoring her motionless form, continued their baths for several minutes. When they finished they rose simultaneously and on whistling wings disappeared into the deeper darkness of the lowland woods close by.

NOVEMBER 8.   At the end of a log stranded on Azalea Shore close to the spot where the woodcock alighted, we find a mound of leaves and twigs and muddy aquatic vegetation. It is more than a foot high. It suggests a muskrat house being built by mistake on solid ground. So far, none of our muskrats has constructed a house at the pond. Instead, they all have lived in burrows in the bank, burrows having underwater entrances.

When we examine the mound, we see it is flat-topped, trampled down, with fragments of shoots and roots and other vegetable food scattered about. What the muskrats have constructed here is what these animals occasionally form—a feeding platform to which they tow loads of food collected at other places along the edges of the water.

For these humble water rats, life—as a species—has been a great biological success. I can think of no other mammal that has extended its range so widely in North America. It thrives from the Atlantic to the Pacific and from the Gulf of Mexico to the Arctic Ocean. Mainly it is an inhabitant of marshes and muddy ponds where waterweeds and cattails and similar fare provide it with vegetable food.

How the muskrat came to settle in this specialized habitat is the subject of a legend of the American Indians which Ernest Thompson Seton recalls in his *Lives of the Game Animals.* The god Nanabojou, rewarding the muskrat for its services in the Time of the Great Flood, decreed it might live in any part of the country it chose. It selected the deep lakes. But the next day it returned and said it had made a mistake. It wanted, instead, to live on the grassy banks where there was ample food. Its wish was granted. But the following day it came back again. It had discovered the banks offered no place to swim and asked for the deep water once more. Nanabojou replied that one day it wanted the land, the next the water. It didn't know its own mind; so he would decide for it. Henceforth it should inhabit the in-between country of the marsh—neither land nor water. So here, the legend concludes, the muskrat has lived ever since.

When I have occasion to look up anything in the fourth volume of Seton's multivolumed work, I always turn to the ending of his treatment of the muskrat to read with pleasure the shortest entry in the book. After setting forth the details of the animal's life and habits under such headings as "Races," "Tunnels," "Home Range," and "Haunts and Migration," he ends with a section entitled: "How to Trap the Muskrat." In its entirety, it reads: "I decline to make any statement."

NOVEMBER 9. Our footfalls are soundless along the trails of this misty morning. After rain in the night, we tread on soggy leaves all the way to the Brook Crossing. Wherever we go, we are in the still woods of November. Wherever we breathe in the moist air, it is laden with the scent of gentle decay.

Following the last of the silent paths, on the way home, we skirt the wooded, glacially deposited knoll we call Twig Hill. As we near its farther side, we halt, arrested by a harsh mingling of many voices. A grating medley rises from the level floor of the woodland where it stretches from the base of the hill to Hampton Brook. We advance cautiously, take a few slow steps, then pause. And so we come within sight of the open ground. All across it we see scattered a flock of grackles—birds that here are one of the last to migrate. As nearly as we can calculate in a hasty count, they number between 200 and 300. Spread out, searching for food, they continually turn over leaves or flip them aside, snapping up bits of food as they advance.

There is only a slight lag. Almost at the same time we catch sight of them, they catch sight of us. With a windy sound of laboring wings and a jarring cacophony of strident voices they all take off at once. After they have disappeared among the trees, Nellie and I walk out across the level stretch where the birds have been working. From end to end it is torn up. The ground appears plowed where the wet leaves have been tossed about in the search for insects and spiders and whatever appealed to the omnivorous birds as food. Who can calculate the number of small inhabitants of the leaf cover that are consumed in this gleaning by the sharp-eyed birds of such a flock?

NOVEMBER 10. Late this morning, when I leave Ground Pine Crossing Trail and come out into the open fields, the sun has melted away the dawn's thin shining coverlet of the frost. The plants of summer, now dead and dry, mingle in their varied shades of yellow and brown—russet and chocolate and tan. I wander aimlessly this way and that, recognizing plants that I stopped beside when they were green, when they were in bloom and visited by bees, when they were just commencing the development of their seeds. All are crisp and brittle. Each plant has completed the beginning, the middle, and the end of its annual life.

In walking among the plants, I am sometimes tempted to view with a tinging of envy the simplicity of their lives. Without straining, without aspiring to more than they can achieve, they grow and bloom and form their seed. The calmness of the plants seems the embodiment of a timeless confidence.

A belief held by some in pre-Darwinian days, three centuries ago, maintained that the life of the plants predates the stars and is older than the sun. In London in 1656, an English writer on botany, William Coles, published a treatise on the origin of plants. Commenting on the antiquity of their world, he wrote: "It is a subject as ancient as creation; yea more ancient than the Sunne or Moone or Stars, they having been created on the fourth daie whereas Plantes were third."

At the American Museum of Natural History, I once talked to a scientific explorer who had returned from the jungles of South America. For several weeks he had camped on a remote stretch of the Orinoco River. One morning, as he was collecting specimens among the rocks of the riverbed at a time when the stream was abnormally low, he suddenly became aware that one of the native Indians had approached noiselessly and was standing at his side.

The jungle dweller was following the river upstream to a distant town. Looking around him, he used one of the few English words in his vocabulary and indicated he had been there "yesterday." The explorer had seen no one for weeks. The Indian added that he would be back this way again "tomorrow." The explorer knew it would be impossible for him to walk to the town and back in one day. Then he realized that in using the words "yesterday" and "tomorrow" the Indian was employing them with a different meaning than is understood when we use them. For him, "yesterday" implied all the time that had gone before, "tomorrow" all the time that lay ahead.

As for the Indian of the Orinoco, so for the species of the plants. Both their yesterday and their tomorrow are indefinite periods. The uncertain span of their past represents their yesterday; the uncertain span of their future their tomorrow. They have existed so much longer than we have that what has come before and what lies ahead seem gauged by a different kind of time— slower, more measured—than ours.

NOVEMBER 11. I come back from the village post office this afternoon bringing a package and a story. The package contains books; the story concerns the teamwork of two unusual pets that once lived at the 1740 colonial home of the postmaster. After I have signed for the package, Charlie Fox recalls the two, a tabby cat and a yellow bantam hen. In another of those odd friendships, the hen kept close to the cat. It followed it around the yard. It settled down beside it when it curled up to sleep. But it was during one particularly cold winter that the most curious aspect of this friendship came to light. Then the hen and the cat cooperated in catching mice.

During the bitterest days, the two found a warm nook in a corner of the kitchen. There, at times, they dozed together side by side. But the bantam hen, less given to daytime naps, was always the first to become active. It would wander off on an inspection tour of the other rooms of the house. During that particular winter, the mouse population in the Hampton region reached one of its periodic peaks. The little animals kept finding their way into the house. Whenever the bantam, on its tour of the other rooms, spied a mouse, it would set up a loud, peculiar clucking sound, reminiscent of a hen that has scratched up a worm and clucks to call its chicks. In an instant, the dozing cat would be wide awake. It would come racing and pounce on the mouse. This teamwork between the hen and the cat was repeated a dozen times or more before the winter was over.

It occurs to me that we have a somewhat similar situation at Trail Wood. Nellie and I take the place of the bantam hen; downy woodpeckers the place of the cat. It is our custom on cold winter mornings to hammer pieces of suet into knotholes on the trunk of the apple tree below the terrace. The far-carrying noise of the pounding always brings the same results. Woodpeckers come flying from the woods. The sound of our hammer calls "suet" to the birds just as the clucking of the hen called "mouse" to the dozing cat.

NOVEMBER 12.   The wind comes up with the sun. The wind is cold, a foretaste of winter. The sunshine is warm, an after-taste of summer. I brush away the fallen leaves from the weathered plank bench in its sheltered place among the spicebushes beside the brook. It is pleasant sitting here watching the leaves—now rusting into brown—go drifting by.

For only a little while, for only a few times more, will I be able to sit here among these bare-branched shrubs looking down on this parade sliding away on the running water. We have, almost suddenly, in these early November days, taken a long stride toward winter. Summer is overthrown; autumn is retreating. The genial, easy days are slipping away. Variety will continue. But change now runs in a different direction. Warmer days will be the exception.

"I cannot endure to waste anything as precious as autumn sunshine by staying in the house. So I spend almost all the daylight hours in the open air." Everyone abroad in such sunny November days as this has known this same emotion that Nathaniel Hawthorne set down beside another stream—a larger, more sluggish stream—the Concord River.

NOVEMBER 13. My feet, as I look down, resemble the feet of some immense giant planted in the midst of a dense forest of evergreens. For a quarter of an acre around me the woodland floor is carpeted with the thick pile of the small, deep-green, treelike masses of the ground pine, *Lycopodium complanatum*. This is the ground pine of our Ground Pine Crossing Trail.

Known variously as festoon pine, creeping Jennie, Christmas green, hogbed, and crowfoot, it is one of the earliest plants to appear on earth, a plant so primitive it is placed just after the horsetails in botanical listings. In Europe, Asia, and North America, it favors open woodlands and soils containing little lime. On this continent it grows as far south as the mountains of Georgia and as far north as Newfoundland and Alaska. Resistant to cold, it is, at Trail Wood, one of the green plants of our northern winters.

On this morning, all across the carpet, bright yellow spikes thrust upward. Each is about the length of my little finger. Sometimes three or four rise side by side like candles in a candelabrum. During the latter days of summer and the earlier days of fall, the slender cylinders were green. But, progressively, I have seen them grow more golden as the microscopic dust of the spores matured.

Now, on this November day, at each stride as I advance across the green expanse little yellow clouds of living dust spurt up around my feet and hang in the air. Looking back I see my trail, the record of my passage, registered in this drift of floating particles. These are the days when the life-producing spores are being carried on the wind to replenish and spread the species.

So light they ride on the faintest breeze, they embody special properties and have many uses. So uniform in size are the particles that at one time they were employed as a standard in microscopic measurements. So fine and water-repellent are they that when sprinkled densely on the surface of water in a glass, they coat a finger thrust into the water and permit it to be withdrawn completely dry. So nearly weightless are they that when placed in a glass tube against which a tuning fork is pressed the powder redistributes itself into a wave pattern within the container. And so inflammable are they that in earlier times the dust was known as "vegetable brimstone."

When William Shakespeare's plays were first performed in seventeenth-century England, the brilliant stabs of stage lightning were produced by igniting the spore dust of the ground pine. Once, at this spot in our woods, Nellie and I shook forty or fifty of the spore-bearing spikes over a glass jar. When we came home, I threw successive handfuls of the golden dust into the fireplace. Instantaneous bursts of blue-tinged light stabbed across our living room just as three centuries before similar light had flicked across Shakespeare's stage.

Commercially, the spores of the ground pine are known as lycopodium powder. For a long time it was employed in the manufacture of fireworks. It was the source of illumination in the earliest flashlight photography. It was an important ingredient in some of the most expensive toilet powders of an earlier age. In the field of medicine, the dust was used as a coating for pills to prevent them from sticking together, as an absorbent in surgery, and as a soothing powder for chafes and wounds. Cannel coal is now believed to have come into being through prehistoric deposits of lycopodium powder.

At the farther edge of this living woodland carpet that I have just crossed, I stand looking back at the drift of the particles. The days when I will come home from the woods with my shoes coated with yellow are nearly over. The time of the lycopodium powder is almost at an end. Before long the last of all the uncounted billions of spores will have ended their journey, scattered through the woods, deposited on the leaf-covered loam where—for better or worse, success or failure—their destiny as links with the future will be determined.

NOVEMBER 14. A couple of days ago I watched a gray squirrel playing like a kitten in the windblown leaves. It whirled and twisted and leaped into the air. It tossed the fallen leaves about, batted at them, and caught them in its forepaws. So fast were its movements, so high its spirits that at times it appeared surrounded by a whirlwind of leaves. I had come upon it at a high-point moment of well-being in its autumn life on this New England farm.

At this time of year, when warm and sunny days are growing fewer, I encounter gray squirrels everywhere. Their population is at a peak. I see them in a hundred places hastily digging holes, burying nuts and acorns, patting down the ground, looking around suspiciously for observers, then racing up the bark of tree trunks to continue their autumn harvest. They are spending their days in a kind of provident frenzy.

Along the backcountry roads, today I come upon the crushed bodies of three squirrels killed by cars. Their quick dodging, their erratic veering, their sudden halts and abrupt zigzags—the very course that saves their lives when pursued by their natural enemies—bring death to many of these animals. Instead of racing at top speed to the other side of the road, instead of getting across in the quickest possible time, they seem possessed to start and stop, to make unpredictable changes in direction that too often carry them beneath the wheels. I once counted eighteen gray squirrels killed by automobiles along the fifteen-mile route from Trail Wood to the nearest town.

For wildlife, probably the three great disaster inventions produced by man are the match, gunpowder, and the gasoline engine. For the gray squirrels of these New England autumn days, the most lethal of the three is the gasoline engine that sends speeding cars charging down all the country roads.

NOVEMBER 15. As far as we can see before us, as far as we can see behind us, the dark river of birds rolls across the sky. It pours out of the northeast and flows away into the southwest. Minute after minute we stand here, our feet planted among the grays and tans and russets of the dead meadow grass, our heads tilted back, gazing up at the streaming multitudes above. We hear a constantly repeated sound, the monosyllabic "chuck" of red-winged blackbirds passing in undulating flight. This is the sound that always accompanies flocks of these birds when they are on the move. Mingled with the redwings, laboring along in level flight are the largely silent common grackles. Thousands, tens of thousands of birds are streaming over Trail Wood. And so they have been doing during the early mornings of a number of days. But these late redwings and grackles are not migrating. Their movement is part of the ebb and flow of a diurnal rhythm.

For toward sunset each day, the movement of this avian stream reverses itself. The birds cross above Trail Wood flying in the opposite direction, returning into the northeast. Like the current of a stream that descends over a varying bed, the current of this river of life speeds up and slows down. The birds go by strung out in loose flocks, then in dense concentrations. But they keep coming. Some days half an hour ticks by before the flow begins to dry up and only hurrying stragglers hasten by overhead.

Sometimes the flocks advance high above us, the forms smaller, the "chucks" fainter; at other times, the dark birds are only a hundred feet above our heads. Although on some occasions the line of movement is to the east of the house, on other occasions over the house, on still other occasions to the west of the house, the course of the journeying birds always follows a direct line angling across our fields.

Never before has the late-autumn concentration of these birds been so large as it is this year. Why on this particular fall are their numbers so great? Only farther south, over the wide marshes of western Louisiana, in Florida, and at Cape May, on the New Jersey coast, have Nellie and I witnessed anything comparable to this movement of birds cutting twice daily across our sky. Somewhere to the north and east lies the roost from which these myriad birds appear in the morning and to which they return in the evening after spreading out during the day to feed. Where do they go in this gleaning before the coming of the first snows? How many miles do they travel in their daily feeding? Where do they congregate, roosting together during the hours of darkness? Their passage above us is bounded by a double mystery.

At first I assumed they spent the night in the dense vegetation around the Hampton Reservoir a couple of miles to the north. But a friend of ours followed one evening flight for ten miles and the birds, veering a little more to the north, were continuing on as far as he could see them with the aid of his field glasses. Although the redwings and grackles have lingered behind after most of the other birds have disappeared, they will not remain much longer. When they move south, the flow of their river will extend down the eastern seaboard for a thousand miles or more. And that flow will not reverse itself until another season comes.

The solution of the mystery of where the birds are spending their nights to the north of us, when I learn it later on, emphasizes the distance even the smaller birds will fly to and from their feeding grounds. The nighttime sanctuary, the roost where the vast concentration of redwings and grackles gather together and sleep when darkness comes, is located nearly twenty-five miles away, near Oxford, Massachusetts. Just flying round trip as far as Trail Wood each day means covering a distance of almost fifty miles. And I see the dark birds stream away and out of sight, journeying how much farther I can only guess—just as I can only guess how many individuals are in these hosts advancing together above us in the mornings and evenings of these November days.

No one can ever supply an exact answer to this latter problem. But a survey made by the Massachusetts Audubon Society of the Oxford roost suggests that well over 1,000,000 birds have been passing twice daily above the fields of Trail Wood. For an estimated count of the grackles alone puts their numbers at 800,000. And the redwings are more numerous than the grackles.

NOVEMBER 16.   (1) I hear the scream of a red-shouldered hawk. But there are no red-shouldered hawks. They left for the south weeks ago. I look around for bluejays that mimick the cry. None is in sight. I hear the shrill, echoing call again and this time trace it to a starling. This leaves me with a riddle I cannot answer: Is the starling imitating the call of the hawk or is it imitating the imitation of the call of the hawk so often given by the bluejays?

(2) Close beside the path across the Starfield, I stop to examine fragments of insect paper scattered over the grass. A skunk has clawed a hole in the ground and torn up a yellow-jacket nest in search of food. All the fiery stings the nest once harbored have been subdued by the cold. But the skunk's discovery reveals that

all summer long, unknowingly, Nellie and I have been living dangerously, trudging back and forth within a couple of feet of this underground fortress of the armed and belligerent wasps.

(3) Deer tracks wander past my cabin. I see where two of the animals have come in the night to scrape aside the leaves with raking forefeet and uncover the small, sour fruit that lies scattered about under a wild apple tree. For a little way I follow their trail in its descent of the slope, noting how one hoof has scored a straight groove a foot or more in length where it slipped on a tilted patch of clay.

(4) I find a wasp hibernating beneath a pumpkin. When I move the large pumpkin we have set just outside the kitchen door this afternoon, I uncover the wasp, one of the *Polistes* paper makers that produce the flat, pendent nests beneath the eaves of our middle shed. It is a fertilized, overwintering queen settled down beneath the ineffective protection of the pumpkin for its winter dormancy. I transport it to a jumble of trash within the shed, reflecting that in the matter of picking a snug hibernaculum I am a better selector than the wasp.

(5) Starlight and woodsmoke. This is the story of our walk down the lane before we go to bed tonight. The frosty air tingles in our nostrils. The sky arches above us unclouded and luminous. Our white house with its dark shutters glows in the starlight. Trailing from the fireplace chimney, the smoke from hickory logs scents the cold night air.

Small and random, such events as these have enlivened the hours and have added their interest to this November day.

NOVEMBER 17. For a long time I keep looking up at the sky this morning. A bluejay wings its way in silence overhead. Two crows, one following the other, veer away in a wide detour around me. That is all. For the dark river of the dark birds, that vast stream of redwings and grackles for which I watch, has ceased its flowing.

On this morning they are crossing other skies. The immense flocks that have held back their departure for the south so long have left at last. The wide sky suddenly seems deserted. With the monosyllabic calls of the passing multitudes no longer heard, the silence that arrives with November becomes more manifest.

Where now are all those hundreds of thousands of moving wings catching the morning light? How far south have the dark concentrations of the birds moved down the flyways of their migration? One of the dramatic events of this particular fall, an event unequaled in any other year at Trail Wood, has come to its end.

When—after how many winter winds, how many winter storms—the earliest flocks of the returning male red-winged blackbirds fill the treetops along the brook, Nellie and I will stand watching and listening and wondering if they number among them birds that we have followed across the sky during these November days. Of course, no one can ever say. But their return will represent, once more, all the unwritten history that migrating birds live on their immense journeys. Although the birds sing their age-old songs when they return, all the adventures they meet along their migratory way remain unsung. If they could talk, if they could write, and we could understand, what sagas those travels would provide.

NOVEMBER 18. Goose-feather snow. Low-hanging clouds. The slow light of a later dawning. These greet me when I step out on my way to the middle shed where our store of fireplace wood is stacked row on row. I stop to watch the leisurely descent of the clustered flakes, these constellations of crystals. They drift down so deliberately, appear so buoyed up, at times they seem to hang motionless in the air.

Each time I bring in a log on my shoulder or an armload of smaller firewood, I notice how the little clumps of the massed flakes form feathery clusters on the wood. But within the heated house, before I reach the fireplace hearth, the clinging flakes have melted, returning to the moisture from which they were formed.

By the time I have heaped sufficient fuel for the day and the evening beside the fireplace, I see that this early snow already has transformed the shakes of my cabin roof into a smooth blanket of shining white. And all across the meadows, the soft snow has gathered into clotted masses like cotton batting entangled in the grass.

Later in the morning the mercury drops. The falling snow

grows smaller, harder. Its descent becomes more direct and swift. The wind rises. Scudding and icy, the small, hard flakes strike and sting my face. By lunchtime all the open ground is touched with white. But afternoon has hardly begun before the wind diminishes to a moderate breeze and the mercury starts to lift again.

First softening, then melting, the snow begins to fade away. Hour by hour, across the fields, the white coverlet becomes more threadbare. The weathered shakes of the cabin roof reappear. We watch the clouds change from a solid, leaden-hued lid to the lighter gray of the woodsmoke's hue and then begin breaking apart. For the rest of the day the sun returns with intermittent shining. Even before the early dusk closes in, all this first snow of the season has disappeared. Our Squaw Winter has come and gone—preceding the Indian Summer of our fall.

N O V E M B E R 1 9.   In the increasing chill of this late afternoon, I lock the door of my cabin. It probably will remain locked now through the winter. For it is in the warmer months that I write most often here among the aspens.

The click of the lock as I turn the key sets in motion a train of thought that continues with me as I walk down the curving path and cross the bridge. If I were the last man on earth, how far back into the history of human ingenuity and invention the world would slip! I do not know what happens in a lock when I turn the key. I have only a vague idea of the sequence of events that transpires in my car when I start it. I do not know what takes place within a radio or a TV or an electronic calculator. I could not make a clock or a pump or an electric light bulb. I probably would go back to stick fires and dipping water from a stream.

Nor am I unique. No human being *could* be left who would be able to carry on *all* the advances that cumulative effort has made possible. We never stand completely alone. We are always, to a certain extent, supported on the shoulders of those who have gone before. Each important advance has had behind it not only the contributions of innumerable individuals but the gradual accumulation of knowledge and ideas by previous generations. We today are part of the present machine age. The things we buy help support it. But its complexities are beyond us.

However, as I climb the slope to the house, I find I am less concerned that I, as the last man on this planet, would be technologically uninformed than I am about how imperfectly I could report on the trees and the wild flowers, the birds and the insects, the rocks and the star-filled sky. In a thousand ways I would be deficient in my knowledge of the natural world. How scanty an account even the wisest among us can give of this earth on which all our days are spent.

N O V E M B E R   2 0 .   I hear a lone whitethroat singing beside the lane when I go down to the road for the morning mail. And when I open the letters that I bring home, I find one concerns another of these same singers. A retired biology teacher in Wilmette, Illinois, tells of banding a male white-throated sparrow in the fall of 1948. Eighteen years passed. Then the band was returned to Washington. The bird had been picked up dead in New Jersey. This is a remarkable record on two counts—because of such longevity in a small bird in the wild and also, in view of the north-and-south migration of the whitethroats, that it should be banded in northern Illinois and later found 700 miles to the east in New Jersey. A still rich field for research is the east-and-west swing of north-and-south migrants.

Some years ago, in *The Auk,* the journal of the American Ornithologists' Union, I encountered a dramatic instance of the kind. Not far from Palmyra, in upstate New York, on August 25, 1969, a numbered band was placed on the leg of an eastern mourning dove. Slightly more than a year later, in September 1970, the same bird was shot in the Shasta Valley of Siskiyou County in northern California. How, in that time, had it gone from north of the Finger Lakes, near the southern shore of Lake Ontario, all the way to California and the far northern county that adjoins Oregon?

It is known from banding ducks that these birds may move for surprisingly long distances east and west between one southward movement and the next. Events in their southern wintering areas, where waterfowl from east and west may mingle together, provide an explanation. When spring approaches and the waterfowl pair off, the males tend to follow the females north. In this manner, some of the ducks may end their northward flight in spring in a different part of the country from that in which their southward flight began.

In a somewhat similar way the continentwide swing of the Palmyra mourning dove may be explained. Although such birds do not pair off on their wintering grounds, flocks from the east and flocks from the west may on occasion intermingle. In the northward movement of spring, the eastern dove may have been swept along in a western flock that carried it up the Pacific Coast almost as far as Oregon.

N O V E M B E R 2 1 . It is at the farthest reach of our walk today that we find ourselves in the midst of an extensive stand of the handsome striped or spotted wintergreen, *Chimaphila maculata*. We count more than a hundred of the striking plants, each with its slender pointed leaves ribbed with white. The green of those leaves is as rich in November as it was in July and August. And so it will continue all through the winter months. Whenever we brush the snow aside we will find buried beneath it this color of summer days.

By the time we come home the air has grown milder. Sunshine pours down from a cloudless sky. Chipmunks race along the walls. In this time of uncertain, unsettled weather, this is one of those days that is commented on and appreciated. It is too short-lived to be considered an Indian Summer. Perhaps we might call it a Chipmunk Summer.

N O V E M B E R 2 2 . To follow a winding stream to its source, to trace it to its small beginnings—beside how many brooks and creeks and rivers has that been my desire! On this lowering morning, when the thermometer stands only a few degrees above freezing and a chill mizzling mist of a rain is falling, I set out to track to its origin the flow of Hampton Brook.

Usually in our contact with a watercourse we resemble someone who dips into the middle of a book, reads a chapter or two, and is left wondering what came before and what follows after. At Trail Wood we have become familiar with most of the chapters in the story of this largest of our streams. In all seasons we have wound with its wandering serpentine through our woods among the knolls and ridges left by the glaciers. We have watched it end its flow at Hemlock Glen, where it joins Little River hardly more than a quarter of a mile from our southeastern boundary. Sitting beside the Brook Crossing, in the Far North Woods, we have seen its clear water, sliding over sand and gravel, appear from a stone viaduct under the old railroad right-of-way. During most of the two miles of its length, Hampton Brook flows within our acres.

Upstream beyond the viaduct, when I set out this morning, the story of this watercourse is a closed book. We know its source lies somewhere in the Natchaug State Forest that stretches away toward the west. But that is all.

I slide down the other side of the elevated right-of-way through a stand of slender gray birches and pick up the stream at the opposite end of the viaduct. It has just completed its descent of a steep, boulder-choked ravine carved out by the freshets of spring. All down its length the declivity is clad in moss and ferns with clusters of low shrubs among the great rocks and fallen trees crisscrossing above this trough on the hillside. They lie around me as I commence my cautious ascent.

As I advance, the brook grows more ill-defined. Part of its water flows underground, part in rivulets—threads and skeins intertwining among the rocks, continually joining and parting and joining again. The hush is profound. Everything—moss and ferns and fallen trees—glistens with fine droplets deposited by the drift of the rain.

When I climb out at the head of the ravine, the land opens up before me. Boggy ground, studded with clumps of sedges, mottled with the dull sheen of dark pools of standing water, dense with growth and supporting a few widely scattered trees, extends across an acre or two. I recognize this marshy tract in the forest. I have been here before, arriving on a dirt road one spring just after the ice had melted. Then the wood frogs were mating. Then the air was filled with the clamor of their clacking din. Now the silence is unbroken as I look around the scene again. This stretch of wetland, only a few hundred yards above the old railroad right-of-way, represents the place of beginning for Hampton Brook. Here I read the initial sentences in the first chapter in its story.

As I descend the choked and boulder-strewn ravine again, the atmosphere of lonely wildness grows stronger. No human sound, except the slow slog of my feet or the small crack of a stepped-on twig, breaks the somber stillness. I see everything in a gray and watery light. It enhances the sense of mystery I feel in these veiled and shadowy surroundings. On this day of rain, under these lowering skies, how remote from the world of man, how far from its activity, appear the beginnings of our woodland stream!

N O V E M B E R   2 3 .   With many-branched stems interlacing into mats of low vegetation between the wheel tracks of the lane outside our kitchen door; the goosegrass, the birdweed, the doorweed, the wireweed, the common knotgrass, *Polygonum aviculare*, has passed through its various stages and has come to the end of its yearly cycle. Now it is blanched and dry, spreading in yellowish smudges along the center of the lane. And always on these days of advancing autumn, it is the center of attention for the sparrows. The seeds of this prostrate member of the buckwheat family appear most attractive of all to these small birds.

In the chill wind that is whipping around the house, this morning, we watch the tree sparrows and the juncoes working over the mats, picking out the tiny seeds that are distributed in the leaf axils along the stems. Suddenly the place where they are feeding becomes the most fascinating spot at Trail Wood.

For a new bird has joined the group. It is unlike any bird either of us has ever seen before. At first we suspect it may be some tropical species escaped from captivity. It is the same size as the juncoes, but its head and neck are pure white, as though it is wearing a snowy hood. Its underparts are also white. But its back, beyond the hood, is blue-gray, slaty-hued.

This beautiful stranger joins in the hunt for seeds. It is accepted by the other birds. Its movements, its activity in no way set it apart from the tree sparrows and juncoes. Only its shining hood distinguishes it. Watching it through our field glasses as it feeds in the shunshine, we detect a faint darker smudging across the white at the top of its breast, the almost invisible evidence of the normal plumage of a junco. The conspicuous stranger, we decide, is a partial albino individual of the familiar *Junco hyemalis* of our winter days.

Almost every year we have partial albinos in one form or another. I remember a house sparrow with a pure white tail, a blackbird with white wings, a starling speckled with spots of white. And earlier this fall we had the dark "Newfoundland" robin with white shorebird streakings in its wings.

All the rest of this day, the junco with its eye-catching hood of white remains, feeding among the other birds along the lane. The chances are infinitely remote that ever again we will encounter this same abnormal pattern of plumage in another junco. On this day in fall, nowhere else on earth, probably, is there a counterpart of this small, strikingly beautiful bird that has alighted here. Our visitor may well be that rarest of the rare, a little bird unique.

NOVEMBER 24. Our Indian Summer has come at last. Usually it arrives sometime during the third week of November. This year the downward tendency of the thermometer has continued until this final week of the month. But today, under a warm sun and with the breeze from the south, the mercury takes an upward swing. Before noon it has reached sixty degrees Fahrenheit in what will probably be the warmest day we will know until spring.

There is always added poignancy inherent in the transitory nature, the fleeting existence, of this swiftly passing time of autumn reprieve. It is that "last brief resurrection of summer," as a writer of the early nineteenth century put it," "a resurrection that has no root in the past, nor steady hold upon the future, like the lambent and fitful gleams from an expiring lamp." While our enjoyment of it is most intense, we feel it slipping away.

So today Nellie and I are out on the trails. We make the most of all the genial minutes. In the out-of-doors history repeats itself on every autumn day—and on every summer and winter and spring day as well. Walking about, stopping at old familiar places, we note how in a hundred small ways events of the past have become events of the present. We observe the tiny repetitions of the year.

And as we come home across the fields where, in more places than one, a woodchuck now lies curled into a ball in a hollow in the earth deep beneath the grass, we talk of all each sleeping marmot will miss that we would be loath to miss in the days to come—the fairyland scenes of soft snow mantling the branches of trees and bushes, the delicate beauty of the frost, the glitter of icicles in the moonlight, the play of pastel colors over the fields of snow as the sunset dies. Such things the sleepers will never see.

We continue on trying to imagine ourselves in the place of the woodchuck lost in a sleep so long and so profound it is close to death. What does the woodchuck know? Is its slumber like a black velvet curtain, a long darkness, a blank, a void of feeling and awareness completely? Or do visions of summer clover and the bitter relish of dandelion leaves, dreams of sunshine and dew occasionally light up its months of darkness? No human can ever be entirely sure. Only the woodchuck knows.

N O V E M B E R   2 5 . On this second day of milder weather, I notice a long, straight line of raised earth descending the slope of Firefly Meadow. It is the burrow of a mole, but a strikingly different burrow from the wandering tunnels I usually encounter. I pace it off and find that for 162 feet it follows a course as straight as though it had been laid out by a line stretched taut between two stakes. Then it curves around a half-buried boulder only to straighten out once more and continue descending toward the lower ground. Digging blind, the mole, for some unfathomable reason, has traversed the meadow from top to bottom as though following a compass course.

I stand beside the thin line of the raised meadow turf pondering on the mystery. This is something new and I am at a loss to account for it. In these final days before the ground freezes and excavating through the soil just below the surface ends, moles are usually active. But why the long straight line? Why not the usual meandering progress? I can think of no satisfactory explanation. The winter of the mole is at hand; its customary habits are beginning to change. But this consideration represents only a vague and unsatisfactory inference, not a solution to the mystery.

For all moles, including the excavator of this straight tunnel, these are the days of a migratory urge. However, their migration is a minimigration, a vertical migration measured in feet. It carries them deeper into the ground, below the frost line, in advance of the hard freezes of winter. There beneath the ice and snow and zero winds of the world aboveground, the hidden life of the tunnel maker is still touched with aspects of the unknown. Apparently it does not hibernate or store up quantities of food for winter. Following its deeper tunnels, it continues its search for such prey as earthworms and grubs. No doubt its metabolism is slowed down and its food requirements are lowered, because supplying its summer appetite would be impossible in winter. In those warmer months, a mole may consume half its weight in food each day.

Whatever the unseen hidden winter life of moles is like, it has succeeded in keeping the species alive. For these tunnelers date back to the Pleistocene. Among North American mammals, only the opossum has a more ancient lineage.

NOVEMBER 26. I have been trying to remember. I can close my eyes and see gray squirrels in August days of humid, exhausting heat walking in slow motion along the tops of stone walls. Their feet appear dragging as they advance. But have I ever seen a red squirrel walking on the ground? That is what I have been trying to remember. I have observed them run, race, scamper, climb at high speed, leap, and leap again. But even at rest, a red squirrel gives the impression of being tense, ready to explode into instant action. Its spring is always tightly wound.

These reflections run through my mind this morning as I follow the trail to Broad Beech Crossing in the last of the warmth of our departing Indian Summer. I have been having a conversation of sorts with a red squirrel. Near Wild Apple Glade, it hailed me. I had entered its territory. In our area, overrun by gray squirrels, the smaller red squirrels are few in number. But their vigilant sense of proprietorship is undaunted. When nuts are ripening during many autumns, we see a red squirrel come from the woods to our largest hickory tree. In an awe-inspiring consumption of nervous energy, it races up and down the trunk, out on the branches, along the top of the stone wall below, driving away raiding gray squirrels from a tree it considers its own.

My repartee in my dialogue with the Wild Apple Glade red squirrel has been calm. Its tirade has continued in a sputtering denunciation, complaint, accusation, condemnation, reproach, challenge, censure, threat, malediction, and imprecation. The sounds, now a monologue, follow me when I move on along the trail.

I look back. I see the squirrel launching itself from limb to limb, still sputtering, still peering down, following me with its eyes as well as its voice. Although I cannot see them, the lenses of those eyes are yellow-tinted. It has been suggested that, just as yellow sunglasses aid our vision on bright days, this tinting of the lenses of the red squirrel's eyes may provide clearer vision when in its leaps its aims for twigs and small branches seen against the brightness of the sky.

N O V E M B E R   2 7 .   In this bright, crisp midafternoon, with the roast turkey—light meat and dark—the dressing, the gravy, the sweet potatoes, the cranberry sauce, the pumpkin pie of our Thanksgiving dinner behind us—or rather with us—Nellie and I start out on a kind of shakedown cruise along a circuit of our woodland trails.

Twice—once in the open pasture, once in the woods—we come upon flowers blooming so late in the year. Both are yellow. The first is a compact disk topping a short, hollow stem, one of the last of the dandelions nestled at the base of a sheltering grass clump. The other—the final flower of the woods—blooms several feet in the air. With its four slender, ribbonlike petals curled and twisted like shavings, it clings in small tousled clusters to the twigs of several of the witch hazel bushes beside our path.

That path leads us into the woods and across the upper flow of Fern Brook and over the green carpet of the ground pine of Ground Pine Crossing to the Lost Spring below Old Cabin Hill. A century or more ago, this spring was enclosed in a rectangle of rocks by the inhabitants of the pioneer cabin at the top of the hill. When we found it, during an early winter at Trail Wood, it was hidden beneath a tangle of fallen trees, which we dragged aside. Now it is mainly the deer that drink at the spring, pawing it out from time to time to increase its flow.

After the spring is left behind and we have climbed to the Old Colonial Road, we break off twigs from a black birch tree, as we sometimes do in our walks, chewing them to enjoy the wintergreen flavor they contain. It is curious how this same flavor is found in the white berries of the moxie vine, the red berries and green leaves of the checkerberry, and the twigs of the black birch. Where else in nature is the taste of a parsnip except in the parsnip, the taste of a red raspberry except in the red raspberry? Yet this flavor of wintergreen is encountered in species remote from each other, in trees thirty feet tall as well as in small and ground-hugging plants.

It is late afternoon when we come out of the dusk of the darkening woods into the brighter light of the meadows. High above us, delicate feathery clouds, ice crystals in the upper sky, at least 25,000 feet above the earth, have formed on a grand scale graceful patterns such as frost crystals produce on a windowpane. The frosted windowpanes of the upper sky!

Halting often, advancing slowly, gazing upward as we trail home across the meadow, we watch these insubstantial cirrus clouds transformed by the sunset into flame feathers thousands of miles long. We see their reddish glow sweeping across the heavens, shifting from tint to tint. Then, in the swiftly coming end of this late-November day, we see the colors pale and ebb away. By the time we reach the pasture bars, the high clouds have lost the last of their brilliant tints. They have changed into mere brushstrokes of wood-ash gray.

NOVEMBER 28. On the road to the South Woods today we stand looking down at a tree that has crashed across the trail since the last time we came this way. For both of us, our minds are busy with remembrances of the same past event.

It occurred in the spring of 1970. We were in the heart of the Big Thicket, that magnificent and abused remnant of the original wilderness at the eastern edge of Texas. *Audubon* had commissioned me to visit the area and gather material for a feature on the battle to preserve portions of it under the National Park Service. Nellie and I had flown over it. We had traveled down back roads through it. We had tramped into areas of rare, wild beauty and had seen the wreckage of destruction across wide sections of it. Toward the end of our stay, we floated downstream on the current of the Neches River for the better part of a day with Geraldine Watson of Silsbee. Familiar with the rivers and swamps, knowledgeable and devoted to the preservation of this irreplaceable wilderness area, she proved an ideal guide.

We were eleven miles in from the nearest hard-top road when we reached Timber Slough and launched our twelve-foot, flat-bottomed boat, squared off at both ends and a little narrower at the bow than at the stern. Nellie sat in the middle, Geraldine Watson sat in front, paddling at intervals on one side or the other to steer us around snags and fallen trees, and I, with another paddle for use in emergencies, sat in the rear. For a stream so serpentine, the current of the Neches is surprisingly swift. It swung us in great arcs around the curves where it had gouged out a deeper channel and undercut the bank on the outside of every turn. At the same time the slower water on the inside of each of the bends had deposited a shining white sandbar, a feature of the stream that had led the Indians to call it the Snow River. The path of the Neches is through the heart of an almost untouched wilderness, an area so remote and uninhabited you can drift for three days downstream without encountering a community of any kind.

On both sides of us along this stream the lowland jungle of the Big Thicket pressed close, huge muscadine grapevines throwing their coils over the tree branches, loblolly pines, with thick, plated bark, some a hundred feet high, towering over us, wild wisteria tumbling in descending cascades of purple. Sometimes we swung to shore to gather handfuls of the fruit of the mayhaw hanging on bushes like tiny red apples. On this spring day we were on a river of bird song. The gold of prothonotary warblers flashed in and out among the green of the bushes. Once, to our intense surprise, we saw a pale-green luna moth flutter across the stream in the sunshine and alight in a clump of willows. How remote seemed this wilderness beauty from the slashed areas, the wandering streams bulldozed into straight ditches, the vast stands of regimented, quick-growing evergreens that we had encountered where man had been at work.

Sometime in the afternoon we were swept into a wide turn, carried by the swifter race of the current close to a vertical bank that rose half a dozen feet above us along the outer edge of the curve. High water, after rains the week before, had gnawed into all these banks, undercutting the roots of many of the riverside trees. We had passed several lying outstretched in the water.

I was bending down, making a quick entry in a small spiral-ring notebook on my knee, when I heard Geraldine Watson call: "Help! A tree is falling on me!"

I looked up. A high sweet gum, perhaps fifty or sixty feet tall, was swinging outward, descending like a railroad-crossing gate in front of us. Undermined by the high water, its roots were letting go just when the current was hurrying us toward it. The racing water below us was at least twenty feet deep. The area was uninhabited, rarely visited. Even if we escaped serious injury and could cling to the tree, days could pass before there would be any chance of rescue.

With all my might, I backpaddled, slowing us down, while Geralding Watson swung the head of the craft out into the stream. But the current was too swift, the distance in which to maneuver too short. We were hurried, drifting sidewise, toward the falling tree. Had it followed its original course, its branches if not its trunk would undoubtedly have sunk the boat and, in all probability, drowned us. What saved us, by a narrow margin, was the fact that the gnawing of the water had dug a little more deeply among the roots on the upstream side. They gave way first, while the roots on the downstream side held slightly longer. This swung the tree so it fell not straight out but at a slight angle slanting away from us. Even so, it crashed into the water so close the spray that shot into the air when it struck the river sent a rain of drops over us. With both of us paddling, we were able to skirt the fallen branches and drift away downstream.

Never before in the out-of-doors, and never since, have we had so near a miss, so close a brush with complete disaster. This is the event of the past that fills our minds on this November morning, as we stand looking down at this other prostrate tree lying across our path.

NOVEMBER 29. Cold has turned the key. Ice locks in the pond.

Ever since our short Indian Summer ended, dawns have lighted a fine fringing of frozen water along the edges. This lacework of crystals has widend or narrowed according to the drop or ascent of the mercury. Yesterday this band of forming ice was no more than a foot wide, a shining rim framing in the dark pond water. But when we went to bed last night the mercury was plummeting. This morning it records thirteen degrees—almost twenty degrees below the freezing mark. I walk down the slope to the pond. Ice, in a thin but unbroken sheet, extends from shore to shore. One of those sudden, dramatic landmark events of the year has come in the night.

For a pond, the four great changes of the twelve months are the spring and autumn overturn of its water and the coming and departure of the ice. Of the four, the arrival of the ice comes with the greatest abruptness. On this morning, the smooth, transparent sheet forms a wide window through which I look into the unmoving, encased water below.

Standing on Summerhouse Rock, I see a back swimmer apparently frozen in the underside of the ice. Then it kicks with its extended sculling legs and I watch it advance with little spurts beneath this clear lid that has been clamped over its world in the night. A short time later a pollywog swims straight up, bumps its nose on the transparent ice, turns and swims straight down again. For all the aquatic creatures inhabiting the pond, new conditions bringing new habits have come with a rush. Soon the larger fish will seek out the deeper hollows of the pondbed. There, largely with lost appetites, they will fast for long periods during the coldest times of winter.

Slowly I advance along the path encircling the pond edge, reflecting on all the changes this one night of cold has brought to our acre of water. Its endlessly altering surface now is stilled. Gone are the cat's-paw sweeps made by the gusting winds, the sparkle of sunshine on little waves, the mirror of the open water that for so long has reflected the images of the sun and moon and stars and a thousand shapes around its edges. With the coming of the ice, much of the life seems to go out of the pond.

Throughout the greater part of the winter—perhaps because the pond lies in a depression and is protected by woods—snow, darkening the water below, spreads in an even, level mantle of white. Superimposed on the hard lid of the ice is the soft lid of the snow.

N O V E M B E R 3 0 .   Several hours after the earlier dusk of this last November day, we watch the rising of the moon. We see it lift, disengage itself from the maze of the silhouetted treetops, then mount, in slow motion, up the seeming curve of the cloudless sky. This November moon, only a few days past full, the beaver moon of the Indians, is pale, cold silver. Its shining disk is almost as round, almost as large as that of the harvest moon of earlier fall. But gone is the warmth, gone the golden light that lengthened the days for pioneers working in their fields.

Our way lit by this colder moonlight, we walk in the night. We see its pale rays glinting on the new ice of the pond. We see its silver light spreading across the frost mantle of the meadows. Everything there is glowing and silvered—double silver, the silver moonlight on the silver frost. We breathe in the air, cold and pure. We stop now and then to look up at the stars, paler stars in the moonlit sky. But when we stop beneath our hickory trees and catch the light of the stars among their branches, they shine with their old brilliance. Here, where our eyes are shaded from the dominant light of the moon, we see the glitter of pin-point lights enmeshed in the treetops, the stars and the contorted, leafless limbs intermingling.

This will be part of our nocturnal walks until foliage returns in spring. This will be part of the winter beauty we will enjoy. The glory of the autumn leaves is gone. But now, through the months of winter, we will have trees filled with stars. One magic goes, another magic comes; the magic of autumn, the magic of winter. The year is the great magician.

DECEMBER 1. John Burroughs once wrote that in our climate December is the month when nature finally shuts up house and locks the door. On this first day of the last month we seem already settled down in winter, although the season, officially, is still three weeks away. The growth of summer, the changes of fall are both behind us. Ice storm and blizzard, snow and bitter wind lie ahead. We dream of spring, of a warm and easy time. But before that comes there intervene the winter months.

I watch the juncoes chase one another with bright little cries as they feed along the lane. I see them dart about in the fullness of health. Once more, as so often in the past, I reflect on how for them there is no foreboding, no sense of insecurity, no sense of impending change, harsh and sudden. Instinct prepares them for coming events. They face the winter and the rigors of winter with the confidence of inborn abilities. Man, more removed from nature, acting less by instinct, more by calculation, plans mentally to meet the problems of the season of cold. He carries the burden of the future as well as the present, a burden the juncoes do not know.

Standing in the cold wind at the top of Firefly Meadow, I look down at the pond, locked in its winter ice. I look at the bare trees, their branches waving in the gusts. I look at the dead slope of the pastureland. How sad for us would be December if we had no foreknowledge of spring.

D E C E M B E R   2 .   Suspicious as a crow!

We watch this often-noticed characteristic reassert itself this morning in a little open space toward the brook. Our Thanksgiving turkey—a rather large one—has reached the point of diminishing returns. About ten o'clock, I put the framework, with its fragments of adhering meat, out for the crows to pick clean. Within five minutes, three of the birds, glossy-black plumage glinting in the sun, are circling the yard. They alight in the highest treetops along the brook, swaying on the slender branches, peering down at the food below.

This careful reconnaissance, accompanied by intermittent cawing, continues for several minutes. Then, one after the other, they slant down and alight in the open space. According to my wristwatch, ten more minutes pass while they walk around the carcass of the turkey, examining it from all sides, always keeping their distance, coming nearer, moving farther away, going off for several minutes at a time only to be drawn back again. All appear fearful of a trap.

The first to pick at the skeleton edges cautiously ahead. It advances a slow step at a time, head far outstretched. On its first approach, it leaps back before actually touching the meat on the bones. A long pause. Then the same crow comes in again, this time sidewise, stretching out its foot like a man edging out on thin ice. Finally it darts in and with its heavy bill snatches a bit of meat. Its backward leap this time is an airborne bound. It carries it several feet to the rear.

Almost half an hour goes by before all the crows are satisfied there are no hidden dangers lurking in the turkey remains. Convinced at last, they crowd around, put their feet on the skeleton, holding it down while they snatch up and gulp down mouthful after mouthful of food. But rarely does a minute go by without one of the birds lifting its head and sweeping its glance around, alert for danger. It is, I notice, the timidest of the three crows that eats most steadily of all once the birds have dispelled their inherent distrust of this meal provided by man.

DECEMBER 3.   Where the dead December grass swirls like a stilled pirouette beside a small clump of wild cherry sprouts, our attention is captured by something silvery gray as we cross the Starfield this afternoon. It is only three or four feet away from the path. Veering aside, we discover the object is heart-shaped and fashioned from sheets of thin, insect-produced pulpwood paper. So low has this nest of a large colony of white-faced hornets been built that its tip touches the ground. All summer long, while it has been largely hidden by the grass, we have passed it—as we had passed the underground nest of the yellow jackets—without realizing it was there. Months on end have elapsed while we and the hornets have gone our separate ways in peace.

More than half a year ago, on some spring day, a fertilized queen wasp, emerging from hibernation, had selected one of the smallest of the sprouts—only two and a half feet long and hardly more than a quarter of an inch in diameter—as the support for her nest and the colony it would contain. As layer after layer and cell after cell were added to the nest, and members of the colony increased, the weight bent the wild cherry wand farther and farther down. This is the lowest white-faced hornet's nest we have ever seen at Trail Wood. To the credulous ones who still believe the often exploded superstition that the height of a hornet's nest above the ground foretells the depth of the winter snow, this Starfield nest would augur the king of open winters.

Gone now are all the hundreds of large black and white insects that came and went on summer days. Young queens, bearing within their bodies the future of other colonies, are now hidden away in crannies and beneath trash where, deep in a state of winter dormancy, they will spend the months until spring. All the other members of the once-teeming throng of white-faced wasps have been slain by cold.

We break free the deserted nest and carefully dismantle the

tough paper shell to examine the horizontal layers of cells within. All the cells except one are empty. That one contains a young wasp, perfectly formed, its wings ready for use, its antennae for smelling, its feet and claws and mandibles ready for their various employments in life. Its egg had hatched, its body had formed—too late. Just when its growth was complete, when it was ready to emerge from the cell, its life was ended—before it began—by freezing cold. We stare at it, its form so unmarred by death. In its unlived life we glimpse one of nature's elemental tragedies.

For in nature's scheme of things, it is the unsprouted seed and the unhatched egg rather than the frog run over on the highway or the songbird picked from the air by the hawk that represents disaster. Life takes its chances. But life that never has a start, that never puts its inborn capacities to work, that never fertilizes a flower or disposes of carrion or holds down the population of some species likely to multiply too rapidly, that never reproduces its own kind—this in nature's plan is the real, the essential catastrophe.

D E C E M B E R 4 .   Across all the tangles of dry, dead weeds scattered over the open fields, the breeze plucks and strums among the stems. It elicits small, almost inaudible sounds. I lean close and catch little rattlings and tickings and scrapings and flutterings where the arid seed heads and seedpods shake and clash together.

Many seeds already have fallen to the ground. Others have ridden away on parachutes of silk or have clung with the hooked hairs of their little burs to the fur or clothing of passersby. But many plants, like the rough-fruited cinquefoil, will continue to shake out their seeds, scattering them in the successive storms of winter. Compressed and dry, in varying shades and shapes, the tapestry of colors of next year's flowers now, in December, lie waiting in the seeds.

In the northern states, New England, in early times, was a great port of entry for plants as well as for people. The seeds arrived in bedding. They emigrated to America attached to the wool of sheep. They came in ballast. They crossed the sea in the hay and grain brought to feed livestock during the long voyage from the Old World to the New. They were intentionally introduced to provide garden flowers and herbs and familiar sources of food. So established have these alien plants become, so common and accepted are they today, that when we look across the rainbow fields of early summer, it is difficult to decide whether flowers that are native or flowers that were introduced predominate.

At the edge of a small town in central Michigan one summer day, I walked about in a rambling frame building examining shelves crammed with small glass vials. Thousands of them, sitting side by side, were packed with rare seeds from faraway parts of the world. Some had been gathered by missionaries in remote sections of Africa, some by travelers to the interior of Asia, some by professional collectors wandering in South America, the South Seas, and Malaysia. The building was the headquarters of a company specializing in new and exotic plants. The owner poured the contents of vials on a tabletop. Each time the rare seeds came tumbling out like a cascade of delicately sculptured miniature gems. That summer day of seeds—in one of those sudden waves of memory—returns in all its details as I stand on this December morning amid the dry plants of this pasture field.

Among all those thousands and thousands of seeds that had journeyed so far to reach the Michigan town, probably none had had so strange an adventure as had the collection of African seeds that once arrived in Uppsala, Sweden, addressed to Carl Linnaeus. They had been collected by an Italian naturalist named Donati. Before he could embark on a vessel for Europe, all his possessions, including the seeds, were stolen. The thief, finding the addressed packet among the stolen goods, sent it on to the Swedish botanist. Except for this sequence of events the seeds would have been lost. For the vessel on which Donati sailed for home was wrecked and all its passengers were drowned and all its cargo disappeared.

DECEMBER 5. "Do cats eat bats? Do cats eat bats?" That was the question Alice kept repeating to herself as she fell down and down the rabbit hole into Wonderland. It runs through my mind as I pause beside the frozen pond on this cold and windy December afternoon. *Do* cats eat bats? Well, I have known cats to *catch* bats. One in Maine was an expert. But I cannot recall ever seeing one eating the bat it caught.

Standing here, with the earflaps of my cap pulled down and the collar of my jacket turned up against the slashing of the wind, I wonder what brought cats and bats into my mind on such a day as this. I finally conclude that it is probably the fact that it was on this same spot, near the leaning apple tree, that Nellie and I, on an autumn evening, had watched four migrating bats, one after the other, follow a straight-line course from the northeast to the southwest across the pond. How far, I wonder, did their seasonal journey carry them? And most important of all, how had they guided their course, navigating in some mysterious way toward their distant goal?

The long riddle of the bats' ability to avoid objects in the dark and locate the nocturnal aerial insects on which they feed was solved by the classic disclosure of the natural sonar the animals employ. But within the strangely shaped little heads of these flying mammals there reside other abilities that still baffle explanation. Chief among them is this enigma of how, unaided by vision, they pursue their unmarked paths over great distances through the trackless air.

One of the most striking demonstrations of this navigational ability that I can recall occurred one summer off the coast of Maine. Allan D. Cruickshank, who told me the story, was leading a party of bird observers on a visit to a rocky seabird island some miles from shore. Their boat was caught in a dense fog that lasted for several hours. During this delay, the passengers noticed a bat that had gone to sleep under the superstructure of the boat. It awoke and fluttered around and around the vessel. Each time it tried to return to its hiding place, the noise or movement of the people on board frightened it. Finally it turned away and—although it had no assistance from its high-pitched squeaks above the sea—it disappeared, taking a beeline course for the coast. Within its little head, what still-unfathomed compass guided it? How did it orient itself in traveling through those miles of fog?

Someday science may supply answers to those questions as it has to the centuries-old mystery of how bats avoid objects in the dark. But today they represent a riddle as deeply shrouded as the explanation of the ability of certain seabirds to find their way through the densest fog. I recall sitting, one evening, in the long living room of a house beside Puget Sound while the famous authority on Alaskan wildlife, Frank Dufresne, related an experience of his on a boat off the Aleutian Islands. Fog had reduced visibility to a circle around the ship that had a radius of less than fifty yards. Out of the fog on one side of the circle burst a flock of small seabirds. Winging straight and fast, they passed over the vessel and disappeared into the fog on the other side. Flying blind, they were on course, heading as though following the pointing of a compass needle toward the seabird cliffs of the nearest land.

D E C E M B E R 6 .   It is nearing the end of this still December day. Returning from the Shagbark Hickory Trail, Nellie and I pause at the top of the hill overlooking the western end of the pond. Everything below—the extended ice, the clustered cattails, the distant bridge, the tans and russets of the dead vegetation clothing the descent from the north—all are bathed in a delicately tinted light. A thousand times we have stood where we are standing. A thousand times we have looked down on this same scene. But now we are—literally—"seeing it in a new light." And so it is throughout our miles of trails. After years of daily association, we are subject to surprises still.

In the house at Trail Wood, we have become so accustomed to the sound of the large wall clock in the living room that most of the time—our minds busy with other things—we fail to notice its chiming of the hours and half hours. This is of small importance. But how great would be the importance to us if the day should come when the songs of the birds, the opening of the wild flowers, the passing of the butterflies should become similarly so familiar we no longer would experience awareness of them. Many a countryman sees little because he feels he has seen all. But there is always more and more and more. Life is never long enough to exhaust the multiformity of new experiences in the out-of-doors. There is never an end to seeing, to hearing, to feeling. There is never an end to the small adventures of seeing something new in what we have come to know so well.

DECEMBER 7. For the past five minutes, we have been watching a cold-tailed squirrel.

Hopping across the light snow that fell in the night, unrolling their distinctive chainlike trail of tracks behind them, five gray squirrels have come out of the woods. All except one individual have the large, bushy, silvery tails that help these animals balance themselves in their leaps and that long ago gave them the nickname of "bannertail." But, perhaps because of some bodily deficiency during this particular season, the fur on the tail of this one squirrel is thin and sparse. It offers scant protection from the winds of December. It has the same threadbare appearance we associate with the tails of gray squirrels in August.

With the first cold of late autumn each year, we notice how rapidly the thicker winter pelts of the foxes and skunks and raccoons are produced. The growth of the denser hair seems achieved in the space of a few days. Thinking of the fur of these fur bearers nudges a memory into life in my mind. It concerns a curious belief held by a man living in southern Texas.

Once toward sunset when we were riding through the Welder wildlife refuge near Sinton, Texas, with Clarence Cottam, the director, an airliner droned overhead. All about us the coyotes began to howl. Always, Dr. Cottam told us, this sound toward evening produces the same effect. A similar chorus rises in the hills around a small Texas town when the six o'clock whistle blows. I ask Nellie as we leave the cold-tailed squirrel behind if she recalls the explanation given by one old-timer in the region for the howling of the coyotes. The man had explained, in all earnestness, that it is the fur on the coyote's body that causes the reaction. Loud sounds stir up the hairs and heat up the fur and make the animals uncomfortable. And so, he said, they howl.

DECEMBER 8. Wandering among the juniper clumps on Juniper Hill in the bright sunshine of this cold December morning, I hear the whistling of a cardinal. The clear musical sound carries from the trees along the brook. It comes over the lane, over the pasture, over the pond to where I stand. It reminds me of how, in the years since we have been in Hampton, the cardinal, the mockingbird, and the tufted titmouse—all birds associated with the South—have increased in this region.

The first cardinal to visit us in winter, I recall, was extremely shy and easily alarmed. The newcomer was the odd bird, out of its territory. It flew up frequently. It gleaned around the edges of the feeding areas. It gave way to other birds, especially the aggressive bluejays. Even now, we notice a characteristic of our cardinals is that they are early and late feeders. In winter, as soon as the day breaks, we see them coming for scattered seed almost as early as the tree sparrows. They tend to make their harvest of food before the more dominant birds arrive. And in the fading light after the sun has set, when the larger birds are gone, the cardinals are active. When frightened away, they tend to remain away for a longer time than many of the other species.

In the last few years, cardinals have been nesting at Trail Wood. Once they raised a brood in the tangle of the pillar rose close to the kitchen door. We see the males, at mating time, pick up and pass to the females plump sunflower seeds. We see the young brought to the terrace and introduced to the source of food found in seeds falling from our hanging feeders. In winter, the red of the redbirds amid the snow and the clear, carrying whistle of their repeated calling brings color and cheer to days of wind and storm. For so long associated with southern states, the voice of the cardinal is now one of the dominant sounds produced by the songbirds in our northern winter.

D E C E M B E R 9 .   Fear in the wild is fear as fear should be. This I have seen demonstrated twice in recent days.

Just now, above the deep ravine that falls away to the east from Witch Hazel Hill, Nellie and I have rounded a turn in the Old Woods Road. We halt, arrested by a loud and sudden snort among the trees ahead. An antlered buck, taken by surprise, bounds away in alarm. We see its white flag rising and falling, shining out among the oaks and maples and hickories as in great soaring leaps it puts distance between us. A hundred years ago, in these same woods, we might have encountered this same frightened reaction of the white-tailed deer. But the fear of the fleeing animal is confined to the time of immediate peril, to the proximity of danger. In the distance, we see its leaps become shorter, more relaxed. It slows down. It stops. It turns and peers back from among the trees. Its fright is replaced by a curiosity that in the hunting season is sometimes fatal.

A couple of days ago, while crossing the meadow, I looked toward the house at the sudden screaming of the jays. A Cooper's hawk had invaded the yard and was teetering on one of the slender upper branches of a hickory tree. The screech of the bluejays scattered the feeding small birds. A gray squirrel, in the terrace grass, sat up and looked around, and a moment later, as the screaming continued, it streaked for the apple tree, shot up the trunk, and raced far out on one of the limbs. There it froze, bright-eyed and alert to danger. Minutes passed. Then there was a flash of movement at the top of the hickory tree. Hurtling down from its lofty perch, the hawk zeroed in on the motionless squirrel. As though paralyzed by terror, the little animal flattened itself and hugged the limb. In its pass, the hawk shot by, swung in a curve over the yard, and disappeared toward the North Woods. Less than two minutes passed before the gray squirrel, its fear apparently forgotten, was back nosing among the grass clumps and the small birds were back at their feeding.

Terror in nature is real terror, not imagined terror. As far as we can tell, for most wild creatures, fear rises and falls swiftly. It comes in the presence of danger, spurs the creature to employ to the maximum all its speed and skill to survive, and then, the danger past, it swiftly ebbs. When the danger is gone, the fear is gone. There appears to be no residue of neurotic brooding, no prolonged and paralyzing fright. It is the actuality of danger, the immediate happening, not the imagined possible future event, that alarm and terrify. There is no evidence that danger, when it is past, continues to haunt the mind as is frequently the case with human beings. Wild fear is an aid to preserving life, an aid that comes when it is needed and largely disappears when it is no longer required. It is fear of great intensity but also fear confined to the time of danger.

DECEMBER 10. I part a clump of dry goldenrod beside the trail. Within I find a tan-colored, slightly pear-shaped ball about the size of a plump Concord grape. It appears to be part of the dead plants around it. But I note how threads of silk anchor it in place among the dry stems. The envelope of the ball is smooth and tough, designed to weather all the gales of winter.

Invisible within this globe are packed hundreds of honey-colored eggs, the eggs of that weaver of the largest of our orb webs, *Argiope aurantia,* the golden garden spider with its striking markings of black and yellow.

Prepared in the sunshine at the end of summer, this tan ball, with its store of close-massed eggs, has outlived its maker, has outlived all the summer's generation of orb weavers. Through it the species will be carried over into another year. Through it, untaught skills will be transmitted, skills that will produce the intricately constructed silken snares of the next summer.

I let the parted goldenrod fall back into place. For a moment I stand picturing in my mind's eye all the possible spider webs inherent in that mass of eggs—orb webs spangled with dewdrops at dawn, billowing in the wind, vibrating with the exertions of their makers while they enmesh captured insects in swathings of silk. At the same time I reflect on how few of all the teeming horde of spiderlings that will hatch and pour from the ball when spring arrives will live to practice their inherited skills or use the silk their bodies are capable of producing. And finally, as I walk on, a recollection takes shape in my mind of one of the most unusual uses man has made of the silk of the orb-weaving spiders.

There lived in Louisiana, a century and more ago, a wealthy planter named Charles Durand. To reach his mansion, visitors drove for more than a mile down an avenue bordered by immense live oaks. About 1870, when his two daughters both became engaged, Durand planned a never-to-be-forgotten double wedding at his plantation home. For weeks before the event, workmen were out combing the fields, collecting orb-weaving spiders. Then, a day or two in advance of the wedding, a multitude of these arachnids were released among the lower twigs of the trees to spin their delicate webs and thus provide silken decorations extending the whole length of the avenue.

When the guests reached the house, they found other spider webs, specially treated, glittering with metallic brilliance among the branches of one wide-spreading oak under which the double wedding was to take place. This spectacular effect had been achieved early that morning. Climbing ladders with bellows, workmen had sprayed the sticky silk of the spider webs with gold and silver dust.

DECEMBER 11. What am I doing at this moment of midnight, this instant when the tenth of December ends and the eleventh of December begins? I am standing in the snow, on a night when the temperature is twelve degrees below freezing, facing toward the north. Beside me, solidly planted on its tripod, my camera also faces north. Its shutter is open and in an hour-long exposure its film is recording the circling lines of light traced by the stars of the northern sky in their slow, imperceptible wheeling advance around the hub of Polaris, the apparently fixed North Star—in reality a double star with a brightness greater than our sun's but dimmed by distance.

This has been the night of the annual party and final meeting of the year for the Hampton Bird Club. When Nellie and I come home, driving up the lane and into the garage under the still brilliance of the sky, it is long after eleven o'clock. Before we go indoors, we gaze about us. We see all the familiar scene around us—the house, the lane, the fields, the pond, the woods—in a magic light, the illumination coming from a sky ablaze with stars. Rarely have we known starlight so bright, the heavens so studded with points of brilliance. This is an exceptional hour, a time to take advantage of; so, instead of going to bed, I bundle up in my heaviest clothes, put on woollen socks and leather boots, and, carrying my camera equipment with hands encased in fleece-lined mittens, emerge again under the star-filled sky.

While the wheeling points of brightness are tracing their images on the film, I, moving about or standing alone in the brilliance of the night, roam with my eyes over the vast panorama of stars and constellations and planets that arches overhead. On this crystalline night, these heavenly bodies, infinitely far away, infinitely remote from each other, seem to merge together, to crowd closer all across the sky.

At last, this hour of stars at an end, I pack up my camera and carry it, and its record of movement in the December sky, back to the house. I look up one final time before I go inside. Astronomy, dealing with the far, far away—the most remote objects in man's universe—was the first of all the sciences. The earliest civilizations sought to solve the mysteries of outer space. The first navigators, the astrologers of old, the poets of long ago, all looked up at these same stars. And now—in days when men have walked on the moon and spacecraft have probed as far as Mars and beyond—they still remain aloof and enigmatic.

DECEMBER 12. All across the snowy fields today, small birds are investigating the projecting tips of dry plants, hunting for seeds. Around each clump of goldenrod and Saint-John's-wort and rough-fruited cinquefoil I see the surface of the snow imprinted with a maze of little three-toed tracks.

Often I stop in my walk this morning to watch these gleaners of the winter seeds—here to observe how the goldfinches are working at the heads of the nettles; there to note, beside the wall, purple finches feeding on the seeds of the ragweed; here to see juncoes, with their white outer tailfeathers catching the sun, fluttering up to pluck seeds from dry catnip plants; there to observe how the tree sparrows advance in little hops around the base of a stand of cinquefoil, pecking at the snow, getting seeds scattered by the latest wind.

All these birds are ones that throughout the day come and go getting the food we have put out. The juncoes and tree sparrows dine on our birdseed, the chickadees on our suet and sunflower seeds, the goldfinches on our thistle seed. But, as I observe today, they all go to the fields to vary their diet. Similarly the hairy and downy woodpeckers, which seem to gorge themselves from morning till night at the suet bags we have hung in the apple tree, hammer away on trees in the nearby woods, adding a grub here and a grub there to the suet that forms their staple diet. The white-breasted nuthatches come and go, getting and secreting sunflower seeds. But we also see them probing among the cracks in the bark of all our trees, investigating crannies for overwintering insects and spiders and pupae, adding variety to the ample source of food they find concentrated at our feeders.

Toward noon, when I climb the trail past my log cabin, I come upon a trio of chickadees. Calling back and forth, they are working among the deep-red clusters of the massed dried fruits of the dwarf sumacs. I watch them thrusting their little bills among the close-packed fruits. So far as I can observe, it is tiny creatures secreted there, rather than the fuzzy fruits themselves, that the trio is gathering. To other winter birds, it is the sumac fruits themselves that are important. On those occasions when we have had overwintering robins at Trail Wood, such red clusters on the sumacs appear to provide their favorite food.

Retracing my steps across the fields in coming home, I am amused by a new activity of the juncoes and tree sparrows. I find them investigating the footprints I have left in the snow. Weed seeds, carried by the wind, have been trapped in the pitfalls of my tracks. All along my backtrail I see the small birds disappearing and reappearing among these depressions where they find the seeds collected.

DECEMBER 13. A bug in a box comes riding up the lane in one of my coat pockets this morning. It is part of the day's mail, the mail that so often brings surprises. As I slip the small container in my pocket, I glance at it. It is postmarked "Sherman, Texas." But what is inside remains a mystery until I unwrap and open the package on my desk. Out come a folded note, a piece of persimmon bark, and a dead insect. The latter is between an inch and an inch and a half long, dark and tinged with bronze. Beneath its head I notice a slender, three-jointed beak folded tightly against its body. Above its back rises a curious projection shaped like part of a cogwheel or the comb of a rooster. I recognize the insect as a wheel bug, *Arilus cristatus*, the largest of the *Reduviidae*, those widespread, predatory insects known as assassin bugs. Mainly they prey on caterpillars and other soft-bodied insects.

Then I unfold the note. It relates the curious circumstances under which the insect was encountered. It appeared on a window screen where a praying mantis, an armed insect twice its size and weight, was already clinging. The mantis, with its heavy, spiked forelegs lifted, ready to imprison its victim as in the jaws of a trap, began deliberately stalking the newcomer. The wheelbug showed no alarm. It appeared instead to be stalking the mantis.

Like two gladiators, each wary, each equipped with a deadly weapon, the predatory insects circled slowly, seeking a favorable position. Suddenly, without warning, the mantis lashed out with its forelegs, aiming just behind the head of the assassin bug. But the spiked jaws of its trap slid off the armor of the prothorax without obtaining a grip. While my correspondent watched through the screen from a distance of no more than a foot or two, the mantis retreated and its smaller opponent resumed its stalking. Again the two predators maneuvered closer. Again the mantis struck, this time attacking from the front. Again its spiked forelegs failed to clamp in a solid grip. Then, for some time, the insects rested, as though measuring each other's menace. Soon the slow circling for position commenced again.

The end of the duel came suddenly. And surprisingly, the smaller insect triumphed. There was a movement forward, a quick thrust of the hollow lance of the jointed proboscis, and the needle-sharp tip plunged into the body of the mantis. There it remained embedded while the assassin bug pumped in secretions that produced a deadening effect on its prey. Then deliberately, through its hollow sucking beak, it commenced draining away the nourishing fluids from the larger body of its victim.

Later, my correspondent reported, a cluster of eggs appeared on a branch on a nearby persimmon tree. I find them attached to the small fragment of bark. Were these, she wanted to know, the eggs of the victor in the duel on the window screen? I examine them with a hand lens. Each egg is shaped like a slender jar with a narrow neck and flaring top. And the cluster of eggs on the persimmon bark is hexagonal. Both the six-sided form of the mass and the jarlike form of the individual eggs are characteristic of the wheel bug that lies on my desk, an insect that, 1500 miles from Trail Wood, had been encountered under such dramatic circumstances.

DECEMBER 14. Three thoughts on a cold walk in the woods in December.

(1) On a return to old familiar scenes, it is remarkable how remembered trees seem to step forward to meet you in a landscape. To local inhabitants they are almost unconsidered because they are so familiar. But to the returning traveler, they are landmarks of special importance. We look over the old trees as we do our dear friends and acquaintances to note the changes that the years have brought.

(2) Forty years ago, the English historian G. M. Trevelyan wrote: "Two things are characteristic of this age. . . . The conscious appreciation of natural beauty and the rapidity with which natural beauty is being destroyed." In the intervening years, both characteristics have been accentuated. Man may end by succeed-

ing in unraveling most of the secrets of nature about the time he succeeds in destroying nature. In this final quarter of the twentieth century a pervading sense of uncertainty is in the air. When Omar Khayyám wrote the lines: "When You and I behind the Veil are past,/ Oh, but the long, long while the World shall last . . . ," the idea seemed self-evident. But now, with the hydrogen bomb and other forces of destruction, for the first time in history we are not quite so sure.

(3) In all our travels, zigzagging more than 100,000 miles through the seasons in America, we never found any boring country. When we were in England, making our journey through *Springtime in Britain,* people warned us that we would find the far northern tip of Scotland, around John o' Groats, a particularly uninteresting area. We found it fascinating. For the naturalist who does not depend on people and people-made things for his enjoyment, all kinds of country have their charm. When the out-of-doors seems boring, it is due not to lack of interest inherent in the place, but to a lack of recognition in the beholder. There is no place without its own special attraction to one who looks with understanding and with care.

As I am setting down these three remembered thoughts, sitting at my desk after warming myself before the fireplace, Nellie comes in to recall the beauty of the great trees we saw in the rain forest of the Olympic Peninsula. She concludes: "I would travel all the way to the West Coast just to see the Sitka spruce again."

A minute later she is back, adding: "And, if I lived in the West, I would travel all the way to the East Coast just to see a sugar maple in the fall."

She is washing the dishes and she is ready to go anywhere except back to the kitchen.

D E C E M B E R   1 5 .   A mile or two to the east of Trail Wood, beyond the valley of the Little River, at the end of the Windy Hill Road, friends of ours, Edmund and Dorothy Parks, have cut trails that wind among their thickets and woods. When they are out on these trails during the hunting season, they take along a cowbell, ringing it at intervals to warn anyone who has wandered onto their land with a gun that they are in the woods.

This morning, when I meet them at the post office, they recall a connection between their cowbell and their chickadees. As our woodpeckers come flying when they hear us pound suet into the tree, their chickadees come flitting through the woods when the clanging of the bell begins. They associate the sound with our friends and they associate our friends with food. For a long time, in the out-of-doors, human beings have been studying the actions and reactions of birds and animals and the sounds they make. For just as long a time birds and animals have been observing human beings. They, too, study *our* actions and reactions and the significance of the sounds *we* make.

DECEMBER 16. No bluejay sounds an alarm. No small bird pays any heed. No sign of recognition greets the robin-sized black and white and gray stranger that alights in our apple tree. At first glance, it resembles a chunky mockingbird. For a time all is still. Below the tree, juncoes and tree sparrows continue their feeding.

Then a bluejay shrieks. Then the small birds scatter. With no warning in advance, the stranger has plummeted downward from a tree limb. Without a sound—as though wielding an ax—it has delivered a deadly blow with its bill to the back of the skull of one of the feeding juncoes. Now it stands over its dead or dying victim, where it lies on its side without a quiver. The swift and silent attacker is that predatory butcher-bird, the northern shrike.

Nesting as far north as the evergreen forests extend into Canada, it invades New England only when winter scarcity of food drives it south. Then its cold-weather diet consists mainly of mice and small birds. In summer it adds grasshoppers and other insects. Among the birds it preys on most commonly in winter, when it comes south, are the juncoes, tree sparrows, and house sparrows. But on occasion a northern shrike will attack birds as large as robins, hairy woodpeckers, cardinals, and mourning doves. In a surprising number of instances its victims seem unaware of its menace until it is too late.

We watch the shrike, seeing it magnified through our field glasses. It twists its head, owlwise, while its gaze, fierce and intent, makes a swift survey of its surroundings. Its bill, we notice, has a light lower mandible and a darker upper one that ends in a hawklike hook at the tip. At times one of these shrikes will grasp a victim and, holding it by the back of the neck in its bill, shake it like a terrier shaking a rat. In this instance I see the rapacious songbird—and it is a songbird with a surprisingly beautiful melody on its far northern nesting ground—seize, in a single swift movement, the neck of the junco and dart into the air, carrying its burden with ease as it flies across the pasture and into the woods. There, if it is hungry, it will consume its prey at once. If not, it will hang it up for future use, as is the wont of its kind, wedging it into the crotch of a bush or tree or impaling it on a thorn.

The speed with which the shrike flies, transporting its burden dangling from its bill, is a revelation. Its movements, it seems to us, are as swift or swifter than those of a sharp-shinned hawk. The velocity with which it dives on a victim and its split-second shifting of direction in flight are reflected in the fact that there are even records of chickadees falling victim to the northern shrike.

One ironical footnote to the story of these boreal predators concerns the Boston Common. In the early days of the house sparrow, not long after it was introduced from England—when it was classed as a protected bird before its proliferating numbers made it a widely disliked problem species—its growth in population attracted northern shrikes to Boston. In a single winter, fifty were shot by men employed to patrol the Common and kill the native shrikes in order to protect the house sparrows introduced from abroad.

DECEMBER 17. Everywhere we walk on our Trail Wood land, Indians have walked before us. The valley of the Little River cradled the famous Nipmuck Trail that Red Men had been following for centuries before the first line of published history was set down in New England. Overhanging ledges of rock a little lower down the valley are blackened by the smoke of Indian campfires. And in the village garden plots still yield occasional arrowheads.

So far, at Trail Wood, we have never discovered an Indian artifact. The only artifacts we have encountered have descended from far later dwellers in the region. Once I dug up the bowl of a clay pipe stamped with a shamrock and the words "Freedom for Ireland." At other times we have uncovered various relics of Colonial ways of farming.

The largest of these is a "lye stone." It came to light under the hickory trees. Gray, formed of metamorphic Hebron schist, about four feet across and three inches thick, it bears on its upper surface a circular groove two and a half feet in diameter. From one side of this circle a straight groove about three inches long runs to the edge of the stone. It was some time before we learned the true identity of what we had discovered.

In earlier days, this grooved stone had been employed in obtaining lye for making soap. Within the circle, a hogshead, with both ends knocked out, was upended and filled with wood ashes from the fireplace. Water, poured in the top, percolating downward, leached out the lye. The stone was set at a slightly tilted angle on a raised platform so the liquid was caught and carried around the circular groove and then down the straight groove to drip into a receptacle below. This Colonial lye or leach stone now rests on a pedestal of flat fieldstones I cemented together in my Insect Garden. It forms the base for a sundial that occupies the center of the circular groove.

Toward the end of this morning I make a fresh discovery of the kind. Several days ago the snow melted and everywhere the ground lies bare. Just before I reach the bridge in going down the lane, I notice that a dry and brown mass of feathery hay-scented ferns, that on summer days rose almost as high as the wall, has been trampled by some animal in the night. Peeping above the flattened mass is the top of a gray boulder about the size of a plump, rounded watermelon. I have never looked closely at it before. Now I see a yard or more of rusted chain trailing from its top. When I bend above it, I find the links emerge from a hole bored in the rock, a hole that appears to have been filled with molten lead, anchoring the chain in place. What part did the boulder and the chain play in the farming of this land in former times?

I return to the house and begin calling up friends in search of an answer. One suggests the rounded stone might have been attached to the end of a wide gate as a counterbalance. Another thinks that because it is close to the brook it could have been used for tethering a team of horses—after they were watered at the stream—while the farmer ate his lunch. A third has the idea it could have been employed for tethering a bull; a fourth that it was a counterweight on a well sweep. So the suggestions come in. Night finds me with the riddle unresolved. I am still collecting guesses.

DECEMBER 18. It is eight thirty in the morning. Nellie and I are following the path over the meadow to Ground Pine Crossing. Beneath its low ceiling of darkened sky, the morning is hushed and expectant. The thermometer stands exactly at the freezing mark. The snow begins before we reach the woods.

But it is no ordinary snow. Catching whatever illumination there is, startling in their whiteness, large feathery masses of snow, many soft flakes clustering together, descend slowly around us. They drift and hang. They twirl in indolent spirals. Rather widely spaced, twenty or thirty feet apart, these buoyant little islands in the air gradually sink lower, reluctant in their slow descent.

We stand, minute after minute, watching them idle downward. For nearly a quarter of an hour, we wander in the open field among these descending clumps or accumulations of flakes that ride the air like thistledown formed of ice. During our Squaw Winter, a month ago, I saw goose-feather snow. But these are far larger clusters. We debate the peculiar meteorological conditions that produced them. Then the masses grow smaller. They give way to single flakes. A change has taken place in the sky. The descent of the snow assumes its old familiar character and we continue on surrounded by the thin white shifting curtains of the silent storm.

There is another witness to this unusual snowfall—a red squirrel in a tree at the edge of the woods. It shows no recognition of anything abnormal in the super goose feathers that drift down around it. We see it streaking along the branches and making surefooted leaps among the twigs while the buoyant snow clusters pass it by.

A red squirrel is surefooted but the length of its leaps fall considerably short of those of the larger gray squirrel. This fact led to an amusing event in a treetop one autumn on a farm not far from Trail Wood. A red squirrel had elected himself defender of a hickory tree. Each day it found itself outwitted by one gray squirrel using the same stratagem. With the red animal in close pursuit, the gray animal would head for a limb that divided into a wide fork. Then as its pursuer neared it on one branch of the fork, it would hurl itself across the gap to the other branch. The leap was too far for the red squirrel. It would go tearing back to the crotch and race up the other branch. Again, as it drew near, the object of its attack would leap back to the other side and the same sequence of events would be repeated. Thus by making easy leaps while its little pursuer tired himself by running fruitlessly back and forth, up and down both sides of the fork, the gray squirrel avoided the furious attacks and remained undislodged among the hickory nuts of the treetop.

DECEMBER 19.   Each time we enter the North Woods along the Fern Brook Trail these days we find our way speckled with the striped gray hulls of sunflower seeds. The chickadees that come and go hour after hour at our feeders have buzzed away in beeline flight, carrying their seeds in their bills to the privacy of the woods. Here, holding them in place with one foot on a branch, hammering them with their bills, they split them open, extract the plump meats, and let the two halves of the light hulls drift spinning down through the air. Certain trees appear favorites. Under them, as winter advances, the trail becomes littered with a carpet of fallen hulls.

Watching birds at the feeders where the chickadees get their food, we observe the varying ways in which they handle the sunflower seeds. The bluejays hammer open single seeds or, at other times, bolt down a dozen or more, storing them in their throats and flying off to some secret place to cache them away for future use. The evening grosbeaks perch as long as competition permits, running the seeds through their heavy ivory-hued bills, cracking them open, dropping the hulls and consuming the meats with machinelike regularity. Purple finches also tend to remain and continue their feeding as long as possible. But the chickadees are different. They streak in, snatch a seed, shoot away, crack the seed open on a tree limb, then sprint back again. From dawn until well after sunset on these short December days, we always see chickadees coming and going around our feeders.

Just outside one of the windows of the library in the village a few years ago, Eunice Fuller, the librarian, saw two different feeding methods collide head-on. An overwintering goldfinch monopolized the single opening in a hanging feeder filled with sunflower seeds. It continued searching for bits of broken seeds while an impatient chickadee darted, called, fluttered around the feeder, and swooped close to the perching finch. Finally just as the goldfinch lifted its head with a sunflower seed in its bill the chickadee shot down and seized the other end of the seed. The finch held on. For several seconds the two strained in a tug-of-war. A little larger, a little stronger, the goldfinch won the contest. But its rival, the chickadee, in an important way, also won. For the finch took wing. The instant the perch became vacant, the chickadee flitted in, snatched a seed, darted away again—its goal attained.

DECEMBER 20. My mind is teeming with memories of the Great North Woods of Maine when I come back from Kenyon Road this morning. What we have seen in the August sunshine of summer vacations deep in the forest between Jackman and the Canadian line, at wild Crocker Lake, below Slidedown Mountain and along Sandy Stream, returns, on this dull December day, in lightning flashes of recollection.

For I bring with me a large flat package that I find hanging on our mailbox. It bears a Jackman postmark. Every few steps as I advance, I stop and lift the package to my nose. It is surrounded by the redolence of balsam boughs woven into a Christmas wreath. Long-time friends of ours, Harry and Ann Sopp, for years have sent us such a homemade wreath fashioned from boughs gathered near Moose River. Like a breath of the endless forests of northern Maine, it arrives about this time each December.

Even more than sights, even more than sounds, scents awaken memories of the past. Each deeply inhaled breath triggers a new panorama of remembrances. There was the time Nellie and I, farther south, lived for a week, sleeping on balsam boughs, in a trapper's cabin far back in the forest. There was the night of the full moon, one August, when we paddled a canoe down the "moonglade," following the shining path of mellow light that stretched across the dark lake water. There were days of exploring "Deadwater," that stretch of lowland wilderness inhabited by beavers. There was the long-abandoned lumber road that ran on and on, climbing through the forest toward the Mooseback Mountains and Bootlegger's Notch. There was once a dinner with lumberjacks at a lumber camp in the forest. And there was a climb to the vast jumble of lichen-covered rocks where the whole side of Slidedown Mountain broke loose and crashed down one spring, leaving a sheer cliff where ravens nest on narrow ledges. Again there was a patch of glistening mud at the edge of a beaver pond, a patch no more than a dozen feet across, where we found hoofprints and pawprints recording the passing of forest animals. In that small space we saw the tracks of a deer, a moose, a coyote, a beaver, a raccoon, and a bear.

When I reach the house, I hang the package in the entry, leaving until tomorrow the installing of the wreath on the front door. Throughout the afternoon, in going and coming, each time I pass that way I catch the rich forest fragrance of balsam fir. And my mind, for still another time, is transported in a rush far to the north, across three state boundaries, to a more wild and primitive scene.

DECEMBER 21.   The birds are bigger birds this morning. Wherever I see them they are fluffed up, insulating themselves against the cold. When I step from the kitchen door, feeling the nip of the cold in the first breath I breathe in, I glance at the thermometer. The upper end of the mercury touches the zero mark. Clad in sweaters and a heavy woollen cap, encased in a greatcoat and mittens lined with fleece, I, like the birds, have increased my bulk, have expanded my plumage. So I wander about on this day of zero weather in search of whatever I can see. Roaming thus, I have the sensation of inhabiting a snug little portable house—my skull—and looking out through its two windows—my eyes.

Each time I blow out my breath, it drifts away like a small cloud of steam. We see such condensed vapor produced by horses and cows and the larger dogs. How about smaller animals? How about birds? I try to remember, as I walk along, whether I have ever seen the breathing of the birds on winter days create similar, but tinier, puffs of white vapor. With their higher body temperatures, its creation seems likely. But I can recall only one clear remembrance of the kind. At the end of our autumn journey, on a particularly cold morning, we stopped at Merritt Lake, the famed waterfowl sanctuary at Oakland, California. Watching the pintail and mallard feeding close to shore, we remarked on how each time the mallards lifted their heads and quacked loudly tiny puffs of vapor drifted away. But at Trail Wood, although we have watched hundreds of small birds feeding on days when the thermometer has fallen even as low as twenty degrees below zero, we have never seen anything of the kind. The explanation, no doubt, is that the quantity of moist air expelled by the lungs of these birds is insufficient to make an observable amount of vapor.

THE
WALKS
OF
WINTER

DECEMBER 22. Winter begins with sunshine. Winter begins with wind. Winter begins with the calling of a bluejay, the cawing of a crow, the whistling of a cardinal. Winter begins with a shining lacework of ice edging the running water of the brook, with the sweet fragrance of hickory smoke in the air.

On this day that inaugurates winter dawn comes late and dusk comes early. After shouldering logs and carrying them to the huge fireplace in the living room, I return to stand for a time beside the bars taking stock of the snow-covered fields around. "Four seasons," I remember John Keats said, "fill the measure of the year." On this day we have crossed an invisible boundary, entering the fourth and last of this quartet into which the twelve months of the year are gathered.

As always at such a time, the time of quitting fall and entering winter, I marvel at how swiftly the days have passed since June, how rapid the pace of our journey through that half of the year; and now how long, slow, laborious appears the time that stretches through the winter months on this other journey from the shortest toward the longest days. It is as though the globe has speeded up for half its yearly course and now is slowing down.

I look up as a small band of goldfinches pass overhead in

rocking-horse flight. Then my gaze returns to wander over the white fields, the dark woods, the gray stone walls, the brown weed tops thrust above the snow. Summer diversifies; winter simplifies. The multitudinous tints we know in summer have been reduced to variations of browns, whites, blacks, greens and grays, to colors we can count on the fingers of one hand. In the out-of-doors, changes in the hues around us form a measure of seasonal time. They reflect the slide from season into season in each ensuing year.

DECEMBER 23. The lone bluebird. A bluebird of December. It flies above me over snow-covered fields as I trudge home in the early sunset of this shorter afternoon. Its soft warble recalls the April skies. Throughout the winter each year a few of these gentle-voiced singers drift about our Hampton region. This lone bird that I see and hear for several days has reappeared at Trail Wood.

I scuff along the paths, watching the colors fading rapidly from the sky, the swift prelude to the coming of the winter night. The two—sunset and night—are, these days, separated by a shorter twilight, abrupt in its departure. On the other side of the year are those lingering, gradually fading dusks that bring to an end the latter days of June.

The reflected tintings ebb from the snow as the glowing tintings fade from the sky. Looking across the graying fields, I see our white farmhouse under the bare hickory trees. It is sinking steadily into deeper darkness. Light already shines from the windows of the room where logs are blazing in the fireplace. Outside the kitchen door the thin column of mercury sinks lower in the thermometer. At such a moment as this—as on some lonely beach facing the vastness of a darkening sea at the end of a summer day—the coming of night takes on an added solemnity. Everything trivial is stripped away.

Once, on a trip through Texas, after days in wild areas, I was browsing through the library of our Rockport friend, Doris Winship, when I came upon those wonderfully sonorous lines taken from "On Night" in *The Book of Common Prayer*. They embody the rhythms that Winston Churchill echoed in some of his most effective wartime speeches. I copied them down and have thought of them many times. They run through my mind as I plod steadily homeward.

"O Lord, support us all through the day long, until the shadows lengthen and evening comes, and the busy world is hushed, and the fever of life is over, and our work is done. Then in Thy mercy grant us a safe lodging, and a holy rest, and peace at last."

D E C E M B E R 2 4 .   The shortest day in the year has come and gone. Imperceptibly now, each day will be longer. But still early sunset, early dusk, early dark, late sunrise, late dawn are part of this month of shorter days. In no other one of the twelve are the hours of daylight so few.

The rising of the sun on this last morning before Christmas comes in a special way. I look down the slope toward the brook and see brilliant stabs and flashes of light scattered over the weeds and bushes. Walking among them, I find they are strewn with plates of frost, many tilted at exactly the right angle to reflect the rays of the eastern sun. They are dense among the dead flower heads of the goldenrod, along the outer branches of the plum tangle, down the rough rails of the rustic fence.

And after breakfast, when I start across the pasture, these jewels of the morning spangle the drifts with their own particular form of winter beauty. I scoop up a handful of snow and examine it with my small pocket lens. On the surface, magnified by the glass, spikes and ferns and crystals of frozen moisture merge together into plates of ice that catch the sun.

In my slow passage across this wide field of bespangled snow, I encounter two sets of far different tracks. One, with footprints round and in a straight line, records the wandering progress of a hunting fox. The other appears to come from nowhere and to disappear into nowhere. It is made by a crow that alighted from the air, dragged its feet for a couple of yards across the snow, then took off again. On either side of the final impression of its feet, I study the sharply defined imprint of the wingtip feathers. They record the first powerful downbeat that made the crow airborne again.

Looking at these markings left imprinted on the drifts by the bird's feet and wings, I am reminded of another crow that grew up as a pet on a farm in the valley. When visitors arrived, it always hailed them with a loud and cheery "Hello!" As it grew older, it

was, as are most wild creatures that are made pets, pulled in two
directions—toward the tame life to which it had become accustomed and toward the wild life to which it was instinctively attracted. Each time a crow winged by overhead, it called a greeting.
Once, in autumn, when crows were gathering together and a small
flock alighted, cawing, in a nearby treetop, it flew to the roof of the
house in great excitement. Over and over, louder and louder, it
called as long as the wild birds remained. But the crows paid no attention to it. For what it called was not the cawing greeting of their
kind but "Hello! Hello! Hello!"

This day that began with the cold special beauty of the spangled
snow ends with the warm special beauty of our fireplace fire on
Christmas Eve. And the long living room in which we sit becomes
scented in succession by two special fragrances. For as part of our
celebration of the occasion tonight we add to the blazing fire mementos of two far-off places. The first brings us the faint sweet
scent of peat smoke. It is given off by a small block of peat brought
home from Ireland by a friend of ours. The second contributes a
fragrance of the dry country of the American Southwest. It is produced by a handful of mesquite chips that have reached our Trail
Wood fireplace after traveling 2,000 miles from the southern coast
of Texas.

DECEMBER 25.   Christmas Day in the morning!

It begins with birds, yellow and black and white and blue, a
score of jays and nearly half a hundred evening grosbeaks, perched
in the apple tree by the terrace. They are scattered among the
branches like living ornaments on a decorated Christmas tree.

Christmas Day in the midmorning!

When I descend the slope to the western end of the pond, I
come upon a network of tracks where the wild apple tree arches
over Veery Lane. They all are heart-shaped. Unseen during the
hours of darkness, deer have come from the woods in search of the
small frozen fruit that lies buried beneath the snow. The deer,
coming so close, were the guests unknown of our Christmas Eve.

Christmas Day at noon!

As we are eating our leisurely holiday meal before the fireplace,
I look out at the feeding birds and notice tiny, gemlike flakes sifting
down, shining and turning in a mistlike drift of snow. By the time

our dinner is over, this almost imperceptible descent has ceased. All the time the sun continues shining.

Christmas Day in the afternoon!

From the highest point of the Starfield, Nellie and I notice how far we can see into the woods to the west, how the snow-clad hills and ravines are now clearly visible among the bare trees. Winter unveils the topography of the land that summer screens with its foliage. We look down on the ice-covered pond and speculate on the unseen life inhabiting the dim, unending twilight of winter beneath it. As we have done so many times since we came to Trail Wood, this afternoon we make a Christmas census of plants still green. In sheltered places we discover ground pine and shining club moss and haircap moss, the striped leaves of the pipsissewa and the ribbon leaves of the wood sedge. Green are the needles of the white pine and the juniper and hemlock and cedar. But the deepest, richest green of all on this Christmas day is the green of Christmas ferns amid the snow. For a hundred yards along the eastward-facing slope that climbs steeply above the trail beyond Hyla Pond, we walk below a tilted carpet dense with their massed fronds, dark and glossy-hued.

Christmas Day in the evening!

After dark has fallen, Nellie and I return to the Starfield. It is before the rising of the moon and the cold shimmer of the stars sweeps across the whole arch of the sky. It is always, for us, a deeply stirring, strangely spiritual experience to stand in silence beneath so vast a star-filled sky. After a time we turn back along the meadow path. By starlight we find our way home at the end of this day, this special day, at Trail Wood.

D E C E M B E R 2 6 .   Four muskrats huddle together in a dark cluster at the edge of Summerhouse Rock. The warmth of the sun absorbed by the rock has melted an opening, a temporary aquatic door to the pond, through which the quartet can come and go. At intervals I see one of the water rats leave the mass, take on individual form, slip into the pond. A few moments later, it reappears and climbs back on the rock, a cattail root in its mouth. For a long time after their feeding is over the four animals bask in the sun. On this mild day, they may well be making their last appearance above the ice until spring arrives.

During the night the weather has moderated. When I step out before breakfast this morning, snow fog is adding its silvering to the air. Before ten o'clock, the thermometer has risen to forty degrees. For the wild creatures, as well as for me, the air is filled with a sense of spring. Downy woodpeckers are drumming on dead limbs; bluejays are yodeling. The rising temperature, after days of cold, brings to us all the deceptive impression that winter—just beginning—is nearing its end.

In many places along the course of my walk this morning, I see the changes brought by this melting day. Along the warmer southern side of one stone wall, where the snow has settled until it is only an inch or so deep, a maze of tunnels branch and intersect like veins in the circulation system of an animal. These burrowings, just above the ground like mole tunnels in the snow, are the work of meadow mice.

Where on a recent night a deer has followed an aimless course over the pasture, crossing and recrossing this path of mine, I notice how the melting of the snow around the sides of the tracks has increased their size. Some have expanded until the hoofprints appear as large as the tracks of a man.

It is while I am bending over them that I find myself forgetting the deer and becoming engrossed in the travels of a small black spider. Brought out by the unusual warmth of the day, it is running hurriedly over the snow. Several times I see it come to the chasm of a deer track, yawning like the Grand Canyon before it. Each time, instead of circling around it, the little spider forges straight ahead—down the precipice of one side, across the floor of the expanded track and up the opposite side in an almost vertical climb to reach the surface of the snow again. I can only guess at the purpose of its journey. What unknown food is it searching for? What unknown hiding place in the meadow is it seeking on this more genial winter day?

DECEMBER 27. Again the weather has changed in the night. Yesterday's warmth is gone. This morning the temperature is below freezing; the skies are lidded; snow slants on a tilted path out of the north. When I set out to follow the brook to the waterfall, I feel encased in a moving shell of white. Each time I pause, I hear the faint, fluttering sound of the flakes striking and sifting down among the tangles of the bare branches of the bushes along the way.

On the terrace, I noticed in passing, all the small birds feeding on seeds scattered on the snow are on edge. They fly up into the apple tree in sudden alarms. It is on such dark days, amid the veiling curtains of snow, that those bird hawks of winter—the Cooper's and sharp-shinned—are most likely to appear. They flash out of the storm to cut out one individual among the many. Behind me, as I follow the brook, I hear, muffled by the insulating effect of the snow-filled air, the sporadic screaming of the jays in some real or fancied alarm.

Later, when I come plodding back and reach the little gate at the far corner of the yard, the screaming breaks out anew. Now it is loud, now it carries a new pitch of excitement and alarm. Shadowy and indistinct, the streamlined form of a hawk bursts out of the screening snow. The bird sees me and cuts away, a small dark shape dangling from its talons. The jays are still darting from limb to limb in the apple tree when I reach it. They still scream their hawk alarm. And, dispersed among the branches, cling the silent smaller birds, the tree sparrows and the juncoes. Beneath the tree I find little drops of blood and tiny dark feathers on the snow. Once more a junco has been the victim. With us, these snowbirds from the north are the ones we see most often caught by hawks.

I have hardly stamped the snow from my boots, come in, and shut the door when, looking out, I see all the small birds scattered over the ground again. They are nervous, alert, feeding rapidly. But they are following—even in this hour of the hawk—the main business of their December days: the obtaining, in sufficient quantity, the food they need to survive in winter cold.

DECEMBER 28. News of the world of men I read as I eat breakfast. News of the world of nature I read afterward, walking over the fields of snow. The coming of daylight, these winter mornings, reveals the secrets of the night chronicled on the pages of the drifts and published in the language of footprint and wing mark.

Even before I leave the yard, I read the record of the activity of a deer that crossed the pasture and scarred the snow beneath our pear tree, hunting for fallen fruit. Nearby, where there is a gap in our north wall, a lacework of mouse tracks links the ends of the wall, chronicling how the small rodents have raced back and forth between the protection of the rocks. Crossing these tracks are other tracks, round and catlike, the trail of a fox that has caught the scent of mice.

Along the top of a wall in the woods, I see the tracks of a ruffed grouse suddenly stop. Overhead, the snow is disturbed along the branches of a wild apple tree. Sometime since daylight, the bird has launched itself upward to feed on the winter buds of the twig tips. Again, two parallel lines, narrow grooves about a foot and a half apart, extend for more than a yard across the surface of a drift. They end where footprints shaped like the silhouette of a flying goose begin. I read this simple story written on the snow. A grouse has come in with deeply downbent wings. Just before it has touched, the long primaries of its wing tips have traced the parallel lines. Then the bird, no longer airborne, was supported by those outspread plates along its toes that form its winter snowshoes.

Wing-tip markings of an even larger bird have left their written message on the snow at the edge of the woods not far from the entrance to the Ground Pine Crossing Trail. Here an owl ended its silent, deadly swoop through the dark. Another event, which I am at a loss to explain, I find recorded in the snow surrounding a large bayberry bush near Juniper Hill. Cottontail rabbit tracks run around and around the clump. They have trampled a circular path nearly a foot wide. It suggests a rabbit racetrack.

The last of these items I find published on the snow was written by the hoofs of another deer. I see where it has emerged from the woods into the meadow and returned again in several places. Always, I am certain it is the same deer. For along the line of the heart-shaped indentations there is a series of short, straight, narrow troughs in the surface of the snow. The animal—either shot or injured—is dragging one foot as it advances.

These are the stories printed by mouse and fox and deer and grouse and owl and rabbit that I read on my walk this morning. They represent but a few accounts in the large volume of news published by the extent of the Trail Wood snow. All across our 130 acres the movement of feet and wings has chronicled other events—large and small, trivial or mortal—that have taken place mostly unseen during the hours of the night.

DECEMBER 29. The outstanding event of this day comes at its end. It comes after dark. It comes with the rising of the moon. Nellie and I clad in our warmest clothes, set forth along the snow-covered path beside Hampton Brook. We are on our way to see a frozen waterfall by moonlight.

All the scene around us—so familiar, so clear, so brilliant in sunshine—now lies in another world. Everything is softened, grown more vague, mellowed and silvered by the moon's rays. Shadows of the trunks and branches of the brookside trees stretch across the snow of the path we follow; but they are shadows often light gray instead of black. Looking across the stream, toward the east, we see the white slope of Monument Pasture with its skep-shaped mound of stones, now clad in snow, rising like a white haycock at its summit.

The night is serene and hushed. Our advance is slow, our pauses many. But as we progress, the sound of the waterfall grows louder. When we reach it, we see the narrowed flow of dark water plunging in the midst of glowing stalactites and stalagmites of ice. All the rocks of the pool below are sheathed in shining frozen spray. Masses of foam, congealed, white on black, tumble with the tumbling water, then whirl away on the current. Everything—the masses of ice, the coating of the rocks, the frozen foam glow or glint or sparkle—bathed in moonlight. From our viewpoint we see the magic of the scene heightened by backlighting, the rays imparting to the clear ice a luminescence that penetrates through, rather than around, the smooth irregular forms.

Coming home through the stillness of this silvered night, with another year so near its end, we see our white cottage rising against the stars, flanked by the bare hickory trees, set amid the snow-clad fields. Our conviction is certain. This is what we have always longed for—just such a house, in just such a setting, seen in the midst of just such a life as we live at Trail Wood.

DECEMBER 30. All day long a bleak wind out of the northwest rakes across the fields and pounds the trees and tosses the feeding birds about. The commonest, the juncoes and the tree sparrows, hug the ground and seek the lee of every object that provides the slightest shelter. Their wings dart out to one side or the other as gust follows gust; like outthrust arms, they steady them as they are rocked by the wind.

In midmorning, as has been their custom in recent days, a dozen bobwhite quail come to the plum tangle. They feed for a time on the cracked corn Nellie has scattered over the snow— plump little birds, with heads lifted high, moving across the drifts in their smooth ball-bearing runs. They appear to flow along with none of the up-and-down jolting apparent when many creatures run. Their hunger satisfied, they seek the sheltered side of the tangle where they are reached by the winter sunshine, the only source of warmth on this day of falling mercury. We observe with delight how carefully they arrange themselves into a starfish-shaped cluster, their short, dark little tails massed at the center, their striped heads pointing outward. Thus they relax, prepared to shoot in all directions if confronted by sudden danger.

Later, after the quail have drifted away, I put the remains of another turkey, this time our Christmas turkey, in the lee of the north wall. Watching through our glasses, we note the time it takes the various wind-buffeted birds to discover this new source of food. Within one minute, a starling is pecking at the carcass. Within two minutes a bluejay has joined in the feast. And less than ten minutes elapse before two crows materialize from somewhere beyond Hampton Brook. Pitching on the rough surf of the gusts, they drop down and alight to monopolize the food. Our yard at Trail Wood must be a place triple-starred on the unprinted maps these glossy black birds carry within their heads.

As the hours pass, the thermometer keeps sinking. When evening falls on this next to the last night of this twelfth month of the year, we see the mercury still plunging. Before we go to bed, it has left the zero mark behind and is continuing its descent.

DECEMBER 31.   Ten degrees below zero. And the wind is rising. On this last dawn of the calendar year, our temperature, according to the morning weather roundup, is exactly the same as it is in Anchorage, Alaska. Both here at Trail Wood and in Anchorage, the thermometers stand at forty-two degrees below freezing.

In the first light of this morning, I see the little sparrows huddled close to the terrace door, where the bulk of the house stands between them and the numbing gale still booming out of the northwest. All morning, all afternoon, all evening the freezing wind dominates the passing hours.

Throughout the day, I am in and out a dozen times, spreading seed for the small birds, scattering cracked corn on the driveway for the mourning doves and in the plum tangle for the quail, filling sunflower-seed feeders for the evening grosbeaks. Braving the wind, giving their bright, slurring notes, the overwintering gold-finches come to the swinging bags of thistle seeds. When I pound suet into openings on the lee side of the apple tree, chickadees and nuthatches dart down to join the downy and hairy woodpeckers in snatching small fragments embedded in the crannies. In the early afternoon, we look down at the wild plum tangle and see the maze of the bare branches on its southern, protected side burdened with the massed bodies of more than a hundred small resting birds. Here, after feeding, they find the warmth of the sun and relief from the siege of the icy blasts.

Each time I come back from a foray into this Alaskan weather, I stand gratefully in front of the blazing logs of oak and hickory in the large fireplace. Before the coming of the short winter dusk, I carry in more logs to pile on the hearth. From them will come firelight and warmth, the low fluttering music of their flames, the sharp crackling emphasis of the burning wood—simple things that will be companions of our country New Year's Eve.

After the birds are gone, after the dark has fallen, I add new logs to the fireplace. All is bright and warm within. Outside the thermometer is slipping downward again. The wind is snuffing at our windows, hooting in the chimney, flailing among the bare branches of the hickory trees. In this way, on this night of violence, of wind and stabbing cold, this night when the roughest side of nature is uppermost, the year, the calendar year, rounds out its circle and comes to its end.

JANUARY 1. The beginning of another day; the beginning of another year. The wind no longer rages but the silent cold continues on. The thermometer outside the kitchen door registers zero when, exhaling shining little clouds of instantaneously condensing moisture, I set out down the lane and return with the morning paper. In the windy darkness between yesterday and today, the only season that is not confined to a single year—winter—has bridged the gap from the end of one twelve months to the beginning of another twelve months. From now on, until the latter part of March, a new calendar will mark the days of our walks abroad. The paper I bring back is the only one we have ever seen carrying the new year's date.

J A N U A R Y   2 .   On winter walks, as on this January morning, I sometimes stop to look across the open fields and let my gaze sweep from tree to tree along the edge of the leafless woods. Seen from this same spot in summer, trunks and branches and twigs are indistinct or invisible, curtained behind the green billows of foliage. But now in midwinter, with the leaves stripped away, the gaunt skeletons of the trees stand revealed with all their differences apparent. I can observe at leisure each distinguishing and individual assemblage of parts.

Along the line where woods and fields meet, I recognize the silvery branches and dark trunks of the red maples, the crooked limbs of the wild black cherries, the solid construction of the black and red oaks, the stiff-fingered twigs of the white ashes, the leaning trunks of the gray birches. The framework of each tree, lost in its clothing of green in summer, takes on new individuality through curve of limb or tilt of trunk or maze of topmost twigs.

All these trees that I see as my eyes travel down the edge of the field are species that have been familiar to this region since before the time of the first European settlers. Those earlier trees provided the pioneers with shade and lumber, firewood and material for a thousand and one home-crafted aids ranging from ax handles to bobsleds. But two of the largest and most valued trees of those days are now gone from Trail Wood. These are the American elm and the chestnut. Wiped out by the Dutch elm disease and the chestnut blight, their weathered remnants remain to form stark landmarks in our fields and woods.

Gone also from this region is that tree of emotional associations known to earlier generations as "It's Tree." It was once the custom on farms in the area to plant a sapling in the year in which a baby was born. It usually was some large-growing, long-lived species that would stand out in the landscape such as an elm or sugar maple.

Now this charming custom of an older time has disappeared. I know of only one person, Helen Mathews, a friend of ours now past ninety, who can recall an "It's Tree" planted for her when she was young. Not far from it there was a similar, far older tree that had been planted for her father.

JANUARY 3. In the gray light of the earliest dawn during all these winter days, mourning doves circle the yard, alight in the bare trees, then flutter down to the cracked corn and seeds we scatter on the packed snow of the lane outside our entry door. They descend in twos and threes, then a dozen at a time, until the mass of grayish bodies fuses into a shifting, unstable carpet over several square yards of ground. As the light grows stronger, I count the feeding birds. They are continually in motion. I count swiftly, and as I count a story I heard as a child on my grandfather's dune-country farm in Indiana flashes across the back of my mind—the story of a small boy sent out to count the pigs who came back saying he had counted them all except one, the little spotted one that ran around so fast he couldn't count it. My tally of the doves shows more than 150 are feeding in the yard in the daybreak of this third of January.

A dozen or more have descended on the opposite side of the house. The gale winds of the last December storm scoured out bare spots around the larger trees. In one such snowless area encircling the base of the apple tree below the terrace, we have concentrated seed for the smaller birds. The doves have discovered it and I see them ranged around the ring of seeds, their heads inward, their pointed tails projecting outward all around the circle. Looking down the slope, the effect is like viewing the crown of the Statue of Liberty from above.

As soon as its feet touch the ground, each new dove, either at the apple tree or on the snow of the lane, immediately falls to snatching up food with rapid, pistonlike up-and-down movements of its head. On a winter morning, among the push of the mourning doves, there is no stepping aside, no sharing the banquet. Each individual tries to get all the food it can. It always seems to me, when I observe the doves feeding at such a time, that they cram themselves with added haste as the mass grows denser. This may well be true.

In that superb summary of our present knowledge of avian ways, *The Life of Birds*, Joel Carl Welty tells of a German scientist who worked with domestic hens. When he placed a single hen beside a large pile of grain, he found it would eat until its hunger was satisfied and then stop. But when he introduced a second hen, the original bird would go on eating. By counting the kernels it consumed in all cases, he discovered that under the stimulus of the presence of a competitor it snatched up thirty-four percent more grain. And when the rivalry was heightened by the introduction of a third hen, the intake of the original bird rose even farther. It then stuffed itself with fifty-three percent more grain than had originally been required to appease its hunger. So it appears likely that for our Trail Wood doves, as for the hens of the German scientist, competition stimulates consumption, the wants of its neighbors enlarge the wants of the individual.

JANUARY 4. When we put out bird food these winter days we are as impartial as nature. The varied species compete, as they compete in the wild, for seeds scattered on the stone wall under the hickories and along the lane, for suet hanging in mesh bags from the apple tree, for food filling our different feeders. Bluejays share the sunflower seeds with chickadees and evening grosbecks. Cowbirds feed among the tree sparrows. In these bitter winds of January, there are no "good" or "bad," "beneficial" or "harmful" birds for us. There are only hungry birds.

Aside from the gray squirrels that we see coming in from the woods, bounding across the glitter of the ice crust, the birds, in their varied shapes and colors, are the most evident form of warm-blooded life today. And they will remain so all during the rest of the weeks of cold. The "dead of winter"—how much more dead it would be each year without the birds!

JANUARY 5. Snow-covered trails lead me to the summit of Old Cabin Hill. Where I have stood amid the green of so many summer days, I stand amid the white of this January afternoon, looking down at the rude rectangle of stones that once supported a dwelling. Now vines mound over one end and the trunk of a good-sized oak lifts within the rectangle, perhaps where a chimney once rose. In the deeper soil and more abundant rainfall of this region, rooted growth unremittingly overwhelms all. Our trails fill in; the edges of our fields are transformed into sproutlands; the woods eternally come creeping back.

But in my mind there are pictures of drier, harsher regions we have visited, desertland and tundra, where growth is long drawn out and scars heal slowly. We have seen the tracks of covered wagons still traced across arid stretches of the Oregon Trail more than a century after their westward progress ended. We have seen a patch of cleared ground on a desert hillside standing out distinctly in the distance. It had remained without visible change while three generations passed.

The impact of human visitors on such easily damaged environments is now a cause of increasing concern. During my lifetime, and especially during these latter years, I have watched attitudes toward nature change, an appreciation of wild areas grow, a surge of interest build up in experiencing such things, firsthand, as the wilderness, the remoter mountains, the untamed rivers. Contact with primitive nature has become more prized in a time of urban pressure. All this is good—immensely good. The value of wildness is coming into its own. Yet it is becoming increasingly apparent that out of this expansion of interest a new danger is rising. The more wildness is appreciated, the more it is experienced, the more people visit it, the more it is in danger of losing its character of wildness.

Half a million people in a single year now descend the Colorado River into the Grand Canyon, tracing on rubber rafts the route Major John Wesley Powell followed with so much hardship and peril only a little more than a century ago. Even if no litter is left, no intentional changes produced, just the impact of so many human beings where virtually none had been before is altering the environment. Park officials are already troubled over the effect of visitors to the remote and untouched beauty of the Brooks Range region of interior Alaska. Its tundra conditions can sustain relatively few visitors without lasting damage.

Here at Trail Wood our tracks are soon obliterated. But in many places, places more remote, places especially attractive to those who love the wilderness, a major threat is coming from the ones who enjoy the wildness most. The sudden growth of interest in primitive nature, the swift multiplication of travelers to remote areas, has produced the paradox of a need to protect the wilderness from those most deeply concerned with its protection.

JANUARY 6. A sharp-shinned hawk that has been haunting the yard wastes some of its time this morning trying to catch a chickadee.

A few days ago, I saw this same predator come streaking in, speeding low across the yard toward the terrace apple tree. It had almost reached it when it tilted up into a rocketing climb, burst over the crown of the tree, and snatched a sparrow from its perch on one of the topmost limbs. But in the tiny chickadee it encounters its match. The movements of this midget are bewilderingly swift. At times its changes in direction among the branches of the apple tree are so rapid it gives the impression of going in several directions at once. Watching it, I find it is confusing me as well as the hawk. And all the while, like a great cheering section, nearly twenty bluejays keep up the cacophony of their screaming.

In time the hawk turns away and in the movement admits defeat. Then, as though in a rage, it begins harrying the screaming jays. It is a small-sized sharpshin, probably a male, not much larger than a bluejay itself. Over and over it dashes into the apple tree or pursues a jay into the forsythia bushes or the Wild Plum Tangle. It makes swift passes at every jay that leaves its cover.

But this is a morning of frustration for the hawk. Its tormentors grow bolder. The blue and white birds begin playing a dangerous game with its talons. They fly jeering back and forth across the yard. Twice the hawk almost has a jay in its grasp, but each time the fleeing bird twists away. And all the while the taunting chorus continues. The final indignity comes when the hawk alights and rests for a time on one of the posts of the rail fence above my Insect Garden. A daring bluejay swoops past and tries to peck it on the head as it goes by. The sharpshin ducks, then shoots in pursuit. But the jay plunges into the sanctuary of the wild plum maze. So the cat-and-mouse game goes on. It continues sporadically for the better part of the morning. When the hawk finally speeds away toward the woods beyond Azalea Shore, it leaves the yard without a chickadee and without a jay.

JANUARY 7. That silent artist that produces the beauty of frost and snowflake—the cold—has sketched two small masterpieces on muddy puddles in the wheel tracks of our lane. During melting hours yesterday, water collected in low spots at the bridge. Then, before this water drained away, in gusty weather as the night closed in, a quick plunge of the mercury transformed the surface of each puddle into a sheet of ribbed and patterned ice.

This morning, Nellie and I find these twin shining plates, extending for a yard or more along the wheel tracks, engraved from end to end with intricate designs. The sidelighting of the low rays of the climbing sun bring out the texture of the ornamented ice— the sweeping curves of parallel lines, the sudden flourishes, the sharp angles and triangles, the lacy bands that swing like combers foaming toward a shore. All is varied. No two square inches exhibit precisely the same designs.

In places the wavering, close-packed lines suggest a contour map of hill-and-valley country. Straight lines bordered by a lacework of finer lines resemble the shafts and barbs of feathers. One circling swirl with a central eye immediately brings to mind the pattern found on the tailfeathers of a peacock. As our eyes roam over the ever-changing scene of webs and lines and gliding curves, we have the sensation of flying over strange and interesting country; over terraced hillsides and rivers bordered on either side by bluffs, over outspread farms where long straight roads cut through fields furrowed by contour plowing.

We walk on stimulated by the fertility of nature reflected in this beauty recorded on sheets of puddle ice. We debate how all the varied markings came into being, what were the factors involved in their production. Stresses in the ice and the gusts of the uncertain winds that blew at the time the water congealed, we conclude, were the chief aids in creating this ephemeral form of art—seen by only two pairs of eyes, enjoyed by only two brains in all the world, coming into being and disappearing forever.

JANUARY 8. Along the trails through the woods today, my tracks record where I have turned aside in the snow to examine the form and feel of winter buds. A few I cannot name. I pick them to take home for the pleasure of indoor identification. But as I see them now, their names unknown, they seem no less beautiful, no less attractive. We can respond to the beauty of a colored cliff without ever knowing whether it is formed of granite or sandstone. We can enjoy the pure hue of a wild flower without ever hearing it referred to as *Oenothera biennis* or even as the evening primrose. Our enjoyment of nature is not based primarily on instant recognition.

There is something else besides technical knowledge that takes precedence. It is an intensely felt relationship—a relationship compounded of a sense of wonder, a response to beauty, an undying curiosity. This may or may not be associated with an immediate recognition of the identity and an awareness of the accepted scientific name of the things observed. Gaining that knowledge is a lifelong pursuit. Its attainment represents an added dimension in our nature contacts. But it is the second step. First glance for beauty and interest; second glance for knowledge. It is rarely the man with all the scientific names at the tip of his tongue that has impressed me as enjoying the richest relationship with nature. I have come to believe that it is something else—a simpler, more primitive, more deeply affecting response—that is the common denominator in those to whom their hours out-of-doors have meant the most.

JANUARY 9. I go out in driving snow. I come back in stinging sleet. The shift takes place while I am sweeping off the capping rocks of the stone wall under the hickory trees. In this protected place the piles of birdseed I pour on the flat tables of rock will remain accessible long after other seed is buried by the storm. In turning away, my head lowered against the driving particles of ice, I notice, down the lane near the bridge, a small cluster of dark birds on the white snow. They appear to be feeding. But no seed has ever been put there. I slog down the descent and see them flit into the bushes. When I reach the spot I find holes, two or three inches across, dug out by the scratching feet of the sparrows. Each of these openings in the snow cover has laid bare sand and fine gravel that has been washed down the slope. The birds have found no food here. But they have found something that without which, for them, ample food is useless.

Without grit in their gizzards to aid in the crushing and grinding of hard seeds, the songbirds of our winter months could not survive. Experiments have shown that seed-eating birds deprived of grit lose weight and eventually die. Some years ago in Great

Britain, the decline of the famous red grouse of Scotland led to an exhaustive study of its habits. The conclusion was that in many places the accumulation of organic matter had buried the natural grit too deeply for the grouse to reach. The investigating commission proposed, as one remedy for the declining population of the game bird, that artificial grit be scattered in the areas it inhabited.

JANUARY 10. In one drawer of the old desk in my writing cabin beyond the pond there is—among various mementos of our travels through the seasons—a polished greenish stone about half the size of my hand. Tens of millions of years ago, when the Badlands of South Dakota were covered with lush tropical swamps, this stone had been gulped down by a dinosaur. In time the prehistoric monster died and all that remained after the decomposition of its huge body was a small mound of smooth gastroliths, or stomach stones, the oversized grit that had helped it digest its food. Birds evolved from reptiles and it was from those remote ancestors that these avian creatures inherited their gizzards. But whereas the giants of the Mesozoic had gizzards only, the birds have developed a second stomach that functions in a different way. Their muscular stomach, the gizzard, grinds up seeds, nuts, and harder foods; their second, or glandular, stomach—in which enzymes are secreted for the digestion of protein foods—takes care of softer food such as insects. Fluid secreted within the gizzard hardens into horny ridges or plates which serve as millstones in grinding up the food. This work is aided by the abrasive action of pieces of grit swallowed by the bird. In the course of time, the grit wears away and must be replaced. It was this that the cluster of small songbirds had been doing by the bridge in yesterday's storm.

The French scientist and naturalist, René de Réaumur, two centuries ago reported the amazing results he obtained in experiments with the strength and grinding ability of the gizzard of a domestic turkey. He found that a small tube of sheet iron, which could be dented only when subjected to a pressure of eighty pounds, was flattened and partially rolled up after only twenty-four hours in the gizzard of this bird. Another scientist, in a more recent test, showed that in the space of four hours a turkey's gizzard can grind up as many as twenty-four English walnuts within their shells. Again it has been demonstrated that such a gizzard is able to crush nuts so hard their shells resist pressures of from 124 to 336 pounds before they crack. The grinding stomachs of these birds—and both the wild and domestic birds have the same scientific name, *Meleagris gallopavo*—contain about forty-five grams of grit.

JANUARY 11. From the top of Juniper Hill, the high point with the farthest view at Trail Wood, I see the land sloping into the valley of Little River, then lifting again to the ridge beyond. On this winter morning, slopes and valley, woods and fields are clothed in soft, immaculate, new-fallen snow. After the gentle descent of the flakes in the night, the day has come still and cold. But in my woollen socks and high-topped boots, my heavy shirt, and goosedown jacket, I view in comfort the winter loveliness spreading away before me.

In this windless hour, every bush and tree and clump of juniper is mantled in clinging snow. Cedar trees rise, slender figures robed in white. Sumac clumps lift in zigzags their snow-laden branches. Oaks and maples tower above the undergrowth, each limb extended in double image, the black line of the wood supporting and contrasting with the white line of the snow. As far as my vision extends—to the tiniest trees of the distance—the scene is one of still and airy beauty. Such moments as this compensate for all the bitter winds of our northern winters.

On a summer day, an interviewer once came to Trail Wood from New York to get material for a magazine feature. Discussing winter, he expressed as his personal opinion that the only really beautiful feature of the season is a mountain slope in Switzerland dotted with the colorful costumes of skiers and with a large and warm hotel waiting in the valley below. That I, too, no doubt would appreciate. But here in this scene that I appear to have all to myself—for I see no other sign of life across the white panorama below—I am in the midst of another, an older type of beauty, one of infinite diversity. It—as well as the summer flowers and the autumn foliage—is an indispensable element in rounding out the year.

JANUARY 12. The gray goshawk is back in the hickory tree. For several years about this time in winter, one of these swift accipiters from the forests of the north has appeared at Trail Wood. Each time it has remained for several days, then passed on to other hunting grounds. It is reasonable to assume that we are seeing the same bird in successive winters. More than was previously realized, banding has revealed the frequency with which wild birds return to the same wintering areas, just as they do to the same breeding areas when they come north in spring.

Through our field glasses, Nellie and I spend a long time watching this large and handsome bird of prey. From its coal-black crown and the striking white stripe above its eye, down its blue-gray back and laterally streaked silver-gray underparts, to its long, nearly square tail, it is a bird streamlined for velocity. We see it rocking on a topmost branch, turning its head, taking stock with its brilliant orange eyes of all the scene around it and below.

We notice the hush that has fallen over our surroundings. No bluejay calls. The quail of the plum tangle huddle in silence. The mourning doves, which have fluttered down on previous days in flocks sometimes exceeding 150 birds, have melted away to hardly a dozen. The speed and alertness of the goshawk sends waves of fear running through wild creatures such as none of our other winged predators produce.

When we see it drop from its high perch in a long accelerating swandive, it cuts through the air like a rapier. We feel the thrill of its plunge. We follow every movement of this bird—so swift, so graceful, so beautiful in flight. Mainly it preys on larger birds, on grouse and quail, robins and flickers, bluejays and mourning doves. It also snatches up squirrels and rabbits. No other hawk is more audacious in raiding poultry yards. There is even a record of a farmer who had just chopped off the head of a chicken and thrown the flopping bird on the ground when he saw a flash of gray as a goshawk shot down, sank its talons into the headless bird, and flew away.

In 1929, because of damage done to poultry and game birds by these winter visitors from the north, the state of Pennsylvania offered a bounty of five dollars for every goshawk destroyed. The results were like those obtained by other similar misguided programs of destruction. During the first year, bounty seekers sent in 372 hawks. Of that number only 76 were goshawks. The other 296 included species that are on the whole beneficial to farmers and many that are harmless both to game birds and poultry.

JANUARY 13. I walk out under a low and heavy sky, a sky that threatens snow—but no snow falls. Everything is still and subdued in the heart of the woods. I encounter no animals, only the tracks of animals in the snow: the footprints of a fox, the markings left by squirrels, the wandering tracks of a cottontail rabbit among the underbrush, and, along the bank of Hampton Brook, the spread-toed print of a mink, a rather rare sight at Trail Wood.

Coming home about half past three in the afternoon I am descending through the sproutland that borders the South Woods near Seven Springs Swamp, when I surprise a dark robin. It is one of those last of the migrants that each year move down from the deeply shaded coniferous forests of northern Maine and the provinces of eastern Canada. It remains on the ground as I halt and stand watching it. Then as I advance it hops rapidly away across the snow. One wing, I notice, droops slightly. But in spite of this handicap, the bird appears well and strong.

Twice, as I follow it slowly, it makes short flights, once into a juniper clump, the other time to the branch of a small oak where dry, tan leaves still cling to the twigs. When in the air, the robin leans or tips toward the injured wing. It appears to extend that wing less far and to beat it a little more rapidly, as though fluttering. But it remains airborne. Unable to make sustained flights, it had brought its migration to a halt on this hillside. But this handicapped bird, if it can continue to escape from fox and hawk, has a chance to survive until spring. For on this slope it is surrounded by both cover and ample food—the blue fruit of juniper bushes, the frosty-gray fruit of bayberries, the massed red fruit of sumacs.

JANUARY 14. Soon after daybreak, the gray squirrels appear from the woods. Natural food is growing scarcer. Buried nuts lie frozen in the ground and blanketed with snow. The wild squirrels of the North Woods have discovered emergency rations in the cracked corn we scatter for the mourning doves and blue-jays and bobwhite quail. By nine o'clock twenty or more are feeding in our yard.

Seven or eight of them are our "regular" squirrels, local residents, the ones that have been coming to the terrace since late summer and early fall. They have become acquainted with us and with our ways. They feed just outside the kitchen door, keep their distance when we appear, and are rarely frightened into taking flight. And when they are, they retreat only as far as the nearest tree.

But the other, the larger, group is made up of strangers, of wilder squirrels that have been drawn from farther away, deeper in the woods. They are more tense, more shy. They feed only on the corn scattered farthest from the house. If I swing back the inner kitchen door and become visible through the glass of the outer storm door, these wilder animals become uneasy. But the squirrels of the terrace remain unalarmed even if I tap on the glass or wave my arms. And if I step outdoors, the farther squirrels streak in a gray pack toward the distant woods. They never stop until they are safe among the branches of the treetops. Already they have beaten a wide band of tracks across the snow. It is a long time before they begin drifting back to the yard, one by one or in little groups advancing cautiously.

Just before I leave the house for a winter walk this morning, the radio is reporting fair weather for the day. But I notice all the squirrels—strangers and regulars alike—are feeding voraciously, with ravenous appetites, snatching up food the way birds do before a winter storm. As events prove, on this day and in this locality, the animals know more about the weather than the weather bureau does. For I come home from my walk a little before noon in falling snow. The storm continues the rest of the day. The flakes are still swirling around the house when darkness falls.

JANUARY 15. Tensely, minute after minute, Nellie and I have been watching a life-and-death confrontation in the sky. Near the edge of the North Woods, a master of attack and a master of evasion are pitted against each other. At first glance the contest seems hopelessly one-sided. The attacker is that avian projectile the goshawk of our hickory tree. The evader is that apparently slow and cumbersome flier the crow. But the skill of one is parried by the skill of the other. The velocity of the gray hawk is offset by the almost instantaneous tilts and sidesteps and shifting evasions of the slower crow. It is like swordplay in the sky.

In every attack the goshawk reverses the stoop of the peregrine falcon. Instead of thunderbolting down and striking its prey from above as the peregrine does, it rips through the air in a long power dive toward the earth either to the rear or to one side of the flying crow. Then in a nerve-tingling reversal of direction, when it seems on the point of crashing headlong into the ground, it skyrockets upward in a towering zoom, zeroing in on the bull's-eye of its target from below. Always the hawk appears aimed like a rifle bullet. But always, as though miracle is followed by miracle, the slower-flying bird flips to one side and the goshawk goes flashing by, its talons empty.

A second crow, which might have fled the scene of danger, moves screaming from tree to tree as though trying to attract the attention of the hawk. But it never takes to the air while the predator is near. All during the running attack upon its companion it caws incessantly, but from within the protection of a maze of branches. It recalls to our minds Homer's description of the Greeks going into battle: "And shouted loud the *hindmost* throng."

At intervals the harried bird alights in one of the treetops. This brings the duel to a temporary halt. It is only when the crow is in the air that the hawk attacks. But always the struggle is resumed again. The crow rests for a few minutes, then, by taking to the air, exposes itself to attack once more. It gives the impression of enjoying the lunge and parry of this dangerous game, of exulting in its own prowess in evasive aerobatics.

A score of times we are sure the black bird will be snatched from the sky. But, on each occasion, the goshawk reaches its target only to find it gone and empty air where it had been. As the plunge of the hawk, the tilt of the crow, the cawing of the noncombatant bird continue on, Nellie and I lose track of time. The crow is still unscathed when I look at my wristwatch. I find the aerial duel has been continuing for nearly a quarter of an hour. And when it ends, when the goshawk scuds away, it is—incredibly—the crow that is triumphant.

JANUARY 16. There once lived in this region an eccentric farmer, a bachelor who dwelt by himself in a remote farmhouse. His single claim to distinction was a whimsical diversion that occupied him during much of his lifetime. Whenever he struck a wooden match to light his pipe or his lantern or to start a fire in his fireplace, he carefully blew out the flame and saved the unburned stick and added it to a pile accumulating in a back shed. The great ambition of his life, he confided to his friends, was to construct a mound of matchsticks high enough to reach the ceiling. Year after year he watched the pile grow larger. When he died, although he had failed to reach his goal, the matchstick mound had risen until it was almost as high as his head.

I wonder why, on this winter morning as I stand beside a clump of juniper, I remember his singular obsession. Then I discover the link in a wide band of lacework woven by tiny footprints in the snow. I lean down and note how the little pawprints wander and crisscross, fan out and contract, interweave and disappear as they trace the running progress of white-footed mice foraging for seeds under the cover of darkness. The gathering and transporting and storing up of food by these small and timid creatures occupy their lives. They, like the man with his matchsticks, are hoarders. During all the years of his collecting, the farmer had rivaled the gleaning mice in the persistence, if not the practical aim, of their activity. And often, when life has ended for a provident mouse, I have found the pile of laboriously assembled seeds—the edible wealth of its lifetime—remaining behind, secreted in crannies, among stored cordwood, within piles of gunny sacks within a shed. The fate of man and mouse were parallel. The matchsticks for the man, the seed stores for the mouse—both remained behind, memorials to the single-mindedness of their endeavour.

JANUARY 17. A world encased in ice, a glittering crystal world, extends around me. During the night, fine, freezing rain drifted down, and on this mid-January morning the rising sun highlights gray birches burdened with ice and bending low, snowdrifts armor-plated with glaring crusts, stone walls sheathed in smooth, transparent shells.

When I start across the fields for a closer view of the ice-filled woods, my progress is lurching and knee-punishing. Here the crust supports me, there—suddenly and without warning—it lets me through. I soon turn back. It is on such a day as this that deer are particularly vulnerable when run by dogs. Their small, sharp hooves break through while the padded feet of the lighter dogs do not.

Although my walk this morning is short, I find there is much of interest close to home. In several places I stand watching the varied ways in which alighting birds come in contact with the slippery crust. A bluejay swoops in, legs outthrust. Its feet touch the ice. They shoot out from under it and it goes sliding along on one side. An evening grosback skids to a stop, stands motionless for a moment, takes a step, and one leg slips out from under it sidewise. The lively little juncoes arrive in fast, straight-ahead landings. They toboggan forward a foot or two before they come to rest. The tree sparrows, I notice, more often than not flit down and end their approach with a different tactic. They make a sharp turn at the last instant, braking their speed before they swing back and make slower contact with the ice. Their slide over the crust carries them only a few inches. Mourning doves flutter down in an almost vertical descent. Even so, I see several alighting on a patch of tilted crust slip and fall and go skidding along like rowboats sliding on their keels. So short are their legs they hardly seem to fall at all when they lose their footing on the ice.

One last amusing occurrence catches my eye before I leave this slippery world and come indoors to warm myself before blazing logs in the fireplace. I am puzzled by the actions of a bluejay whacking its bill on the surface of the crust. As I approach, it flies away. When I reach the spot, I find it has been trying to obtain a sunflower seed clearly visible but securely locked within a layer of hard, transparent ice.

JANUARY 18. I remember once walking at sunset across a wide sea meadow on the south shore of Long Island. Far out on the flat land, caught among the dense, wiry masses of the swirling cordgrass, I came upon a rainsoaked copy of *The New York Times Magazine*. Dated some years before, it had blown from a pile of trash dumped at the edge of the meadow. When I looked down at it, I saw it lay open at a double spread of photographs—pictures of my own that I had taken depicting the life of the honeybee. A chance encounter? A falling into place of a vast kaleidoscope of time and space? A happen-so without significance? It may have been all of these. But it is also one of those happen-sos that leave in our minds a vague sense of disquietude, a sense of puzzlement, a feeling we are face to face with some powerful, illogical, unseen force that runs through our daily lives.

A friend of mine once drove to work, stopping for gasoline along the way. Just as he crossed an intersection, later on, another car, running through a red light from a side street, crashed into him. No one was seriously hurt. But for a long time afterward he puzzled over all the varied factors—the minute he lost getting the car started, the seconds he gained when traffic lights were green instead of red, the time that elapsed when the filling station attendant had difficulty making change—all the pluses and minuses of time that brought him to that precise corner at that precise moment of the day.

This afternoon, I remember the magazine on the sea meadow, I recall my friend at the intersection as I stand in the snow where the open fields draw close to the North Woods. My mind has been reenacting an event of the night as my eyes have ranged over varied markings on the snow. I see where a cottontail rabbit has hopped into the open, where a drift is gouged and scattered by its struggles, where the wing tips of a barred owl have left their imprint on the surface of the snow. For the rabbit, too, the shifting kaleidoscope of time and space had brought it to this place and moment of its death. All its wanderings, all the shifting course traced by the flying owl had intersected at this spot on the snow beside our northern woods. Once more, as amid the green of that day in June, I am face to face with the workings of fate.

I wonder, as I walk slowly on, if any Red Man, long before the coming of the Pilgrims, had ever stood in the snow among these same New England hills and pondered on the working of unpredictable chance in the lives of rabbits and men, had ever sought—in the words of *The Golden Bough*, Sir James G. Frazer's monumental history of primitive beliefs—"to fathom the secrets of the universe and to adjust his little life to its awful mysteries."

JANUARY 19. We are out in the woods breaking up fallen branches that project above the snow, gathering sticks to replenish the dwindling pile of fireplace kindling stacked in the entry shed. Often we pause. Often we look around us. Often we talk of the things we see. And so it is, in our shuttling back and forth across the Starfield toward the end of this bright January day, we bring home more than kindling. We bring living memories of the things our eyes have glimpsed and our ears have heard.

Coming back in the sunset, carrying the last of our loads across the snow, we pause to enjoy the moment that impresses us most deeply of all. Our little flock of tree sparrows, perhaps numbering thirty birds, has gathered in the plum tangle. There, at the end of this still day, they are lifting their voices in a wild medley of chiming sounds that suggests, again, the tinkling of a profusion of clear and sparkling icicles.

From now on, during days when the temperature rises, the tree sparrows will become more vocal. Their full song—the sweet repeated melody the males sing on their northern breeding grounds—we will not hear until spring. But these simple notes, in their unpretentious way, add music of moving beauty to these midwinter days. In our future memories of this one particular day in January, the sinking of the sun and the tinkling chorus of the tree sparrows—sight and sound—will be linked together in our minds.

JANUARY 20. Watching day after day the wildlife of our sanctuary farm, we have become increasingly aware of the diversity of the dispositions of the creatures we see. We are reminded of this this morning when a small flock of cedar waxwings, rare at this time of year, alights at the top of our hemlock tree. These birds always give the impression of being amicably inclined, of enjoying the society of their kind. In contrast are those other flocking birds, the bluejays and the crows—bickering, quarreling, prone to explode into cawing or screaming melees at any provocation.

Some birds, by temperament, concentrate on their own business; others, like the house wrens, concern themselves with everybody's business. Some, like the wood thrush and the brown creeper, are reclusive; others, like the catbird and the robin, live and build their nests close to human dwellings.

Those insects with fiery stings, the wasps, vary greatly in their dispositions. The *Polistes* are generally pacific, using their weapons only when they or their nests are threatened. But the yellow jackets are irritable, quick to launch an attack when anyone comes near. And poles apart are the calm of the feeding cottontail and the taut excitability of the chickaree, or red squirrel.

Every farmer knows how individuals among his livestock—his cattle and horses and pigs and sheep and chickens—exhibit a wide range of dispositions. Some are placid, others high-strung; some are submissive, others fractious; some are harmless, others dangerous. This difference in character was taken note of some years ago in the aftermath of an accident on a New England dairy farm. Through a misadventure, a Department of Agriculture truck hit and killed one of the farmer's cows. Together with a letter expressing regret, the Government enclosed a form to fill out to claim damages. The final question, referring to what had been done with the dead animal, read: "Disposition of the cow?" The farmer, with a Yankee twist of humor, wrote: "Kind and gentle."

JANUARY 21. In a sheltered spot close to the edge of the North Woods, the swirling of a gust has partially uncovered the woolly rosette of a mullein plant. I scrape away the snow and feel the thick flannellike leaves. Recalling a time when on a previous winter I made a careful census of the little hibernators in such a plant, I know that between these leaves, as between a stack of woollen blankets, tiny forms of life—minute spiders and beetles and springtails—have found a snug winter retreat. There they will be secure while ice and snow lock in the winter fields.

One element of that security puzzles me as I walk on. So far as I have observed, not one of our birds has shown any awareness of this convenient source of winter food. I have never seen bird tracks in the snow around the overwintering rosettes of these biennial plants. I have never seen the leaves disturbed or torn apart. In the pinch of midwinter, it is surprising that some bird has not uncovered the possibilities offered by all the tiny creatures hibernating between the mullein leaves. How long will it be, I wonder, before some such enterprising and omnivorous bird as the starling makes this discovery?

JANUARY 22. The morning of our birds begins earlier now. The days are noticeably longer. Since the darkest time of the winter solstice, we have gained half an hour of daylight. A month already has been subtracted from winter.

As the dawn strengthens, I stand watching the dusky forms of small feeding birds become more distinct. I see them separate into different species—whitethroats, juncoes, tree sparrows, a song sparrow, and two overwintering field sparrows. Soon they are joined by chickadees, cardinals, tufted titmice, white-breasted nuthatches. The sun clears the ridge beyond the river valley and the air is filled with the calling of bluejays and the trilling of half a hundred evening grosbeaks. Endlessly repeating the "zeee" of their ascending calls, goldfinches swarm over the mesh bags of thistle seeds hanging from the buckthorn bush beside the terrace wall. I make out, feeding among the sparrows on the ground, two or three tiny redpolls, flighty and wild. A little later, an overwintering flicker joins the downy and hairy woodpeckers at the suet bags, seeking a January substitute for ants.

I examine each bird through my field glasses with special care. The reason is linked with one of those rare and beautiful displays sometimes encountered at this season of the year. Along the summit of Juniper Hill, all the oaks, all the hickories, all the feathery birches glitter in a thin sheathing of ice. Mistlike rain, freezing in the night, has clad every twig and branch and trunk in a veneer of frozen moisture. As I watch, the ice begins to fall away, softened by the rays of the sun, descending in a shining cascade to litter the snow.

Where were the birds when the freezing rain veneered the trees? I check them one by one. Nowhere can I detect a trace of moisture frozen on their plumage. How did they escape having their feathers encased as were the twigs? Most spent the night in thick undergrowth or in the shelter of evergreens. All have high body temperatures which may have helped. But there is another explanation. On occasion I have seen birds come flying in through mizzling winter rain when it was freezing on contact with the ground. Yet their wings remained entirely free of ice. The reason, I am sure, is that—like the deicers, the expanding and contracting leading edges on the wings of airplanes—the moveable feathers of the bird break up the developing ice before it can form a coating and thus keep its wings and body unencumbered.

J A N U A R Y 2 3 .   With three-quarters of this first month of the year gone by, this is still a January without a January thaw. The thermometer stands only a few degrees above zero when Nellie and I begin our "moon walk" about nine o'clock this evening. In this still, lustrous night, we half expect to see an owl scudding in spectral silence in the moonlight or a deer stepping from the edge of the dark woods where the open snow begins. When we pause—which is often—the night is so hushed we can hear, from far down the lane, the low murmuring of the brook pressing downstream between its ice-clad banks. Slanting across Firefly Meadow, the moon rays pick out little mounds of snow, accenting them with light on the eastern side, shadow on the western.

We seem walking through an old-time woodcut, a study in black and white. The snow-clad fields, the stark silhouettes of the trees extend away around us. Going down the lane and coming back up the lane, we stop to gaze on the shadows of the hickory trees lengthened over the drifts. All slant toward the west. But they will swing in a slow half-circle as the frosty hours of the night advance, rotating in almost imperceptible movement over the snow until, before dawn, they will slant toward the east. Recording the passing of time by this slow sweep of their shadows, our hickories beside the lane form the great moondial of our winter nights.

J A N U A R Y 2 4 .   The new birds are four in number. The size of robins, their plumage is black, washed with reddish brown over the foreparts of their bodies. They look about them with eyes so pale a yellow they appear almost white. When they call, their voices are unmusical, like the creaking of a rusty hinge. On two counts—plumage hue and voice quality—this quartet of male birds that arrived in the yard soon after daybreak yesterday deserves the name bestowed upon them: rusty blackbirds.

We watch them as they turn from side to side in feeding on scattered seeds. The sun catches their pale eyes and again we notice how unpleasant is the impression such eyes make upon us. We feel repelled by their coldness. We miss the warmth and responsiveness we find in the more deeply, richly colored irises of the other birds.

Among these four rusty blackbirds—a species we have seen only infrequently at this season of the year—one stands out. It belongs to that rare class of individuals that always catches the eye—a bird with a single leg. How it was robbed of its other leg we will never know. It manages well, hopping about, balancing itself by canting over, alighting at a slight angle that throws its weight to the side on which it has support. But it is a little slower in its movements, a little less solidly planted, a little more vulnerable. It is a handicapped bird in the worst time of the year—the winter.

How long it has been maintaining itself in its maimed condition, finding food and evading predators, we are unable to guess. But on this, its second day at Trail Wood, its fight for survival comes to an end. When a Cooper's hawk comes pitching down across the yard from the hickory tree, it is a little delayed in its getaway. Without slackening its speed, the raptor veers slightly, opens wide and closes its talons. It flaps away, the one-legged blackbird dangling below it. Later, under one of the hickories, we find dark feathers scattered about beneath the limb where the hawk has fed.

J A N U A R Y 25 .   All night, all morning, snow drives out of the northwest. It piles up against the walls, chokes the lane, packs into crevices of bark to form varied patterns on the different tree trunks, sculptures sweeping, sinuous drifts that writhe away wherever barways let the wind come through. Throughout the night we have heard the sound of plows clearing the roads. Before it ceases, the storm deposits fourteen inches of snow on the winter landscape around us. Kenyon Road is open but between our house and the road stretches the lane with drifts almost as high as my shoulders. A day or two will pass before we can be plowed out.

Late in the afternoon, I take down from a wall in the entry

A winter path leads down Firefly Meadow, around the end of
the pond, and across the bridge over Stepping Stone Brook.

Where tips of drooping branches dip into rushing Hampton Brook, the ice builds up into gyrating, glittering masses.

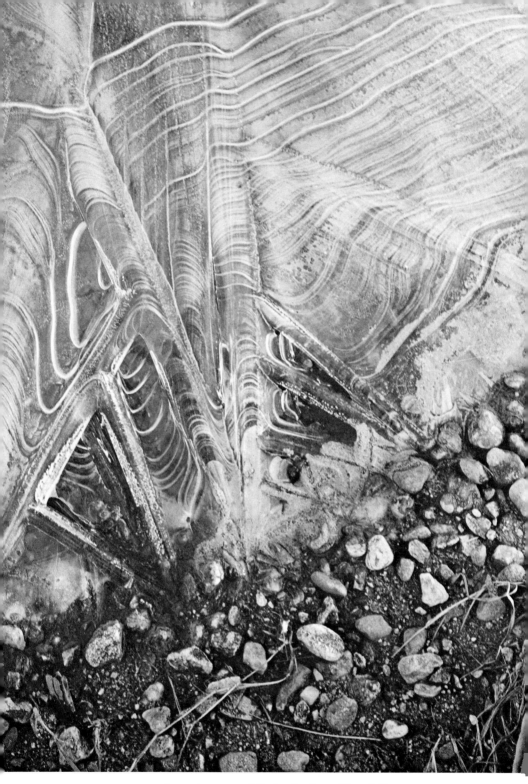

Ice art on a puddle in the lane. The elaborate patterns
were created by wind and cold following a winter thaw.

Small birds have left their tracks in the snow under a
winter source of food, a weed rising above the drifts.

Viewed across the mounded evergreen shrubs of Juniper
Hill, the house on its knoll stands in its winter setting.

Glowing in the backlighting of the setting sun, a dry
goldenrod lifts its burden of soft snow after a storm.

Evening grosbeaks, soon after dawn, harvesting scattered sunflower seed after a heavy fall of snow in the night.

February at Trail Wood. Last year's goldenrod, the pond ice, and the roof of the writing cabin all are white with snow.

shed a pair of snowshoes. They were made three decades ago by the old-time snowshoe maker of Solon, Maine, Charles Holway. The wind is still blowing. When I straighten up from buckling on my webbed shoes, I see the snow smoking along all the walls that run east and west, curling over the tops, racing away in long, downcurving clouds across the fields beyond. My own progress over the drifts produces a steady sequence of miniature blizzards. At each stride, I disturb the surface layer and watch the wind whirl the loosened flakes away.

After several milder seasons, when snowshoes were unneeded, I find I am out of the swing of walking on webbed feet. I am less a snowshoe hare than a snowshoe tortoise as I plod down the slope and over the dam, across the bridge and up the drifted path to my writing cabin. Buried in snow, it resembles a mountain hut in the Swiss Alps. All the juniper clumps around it are mounded over with snow. Wildlife is lying low. I watch a single buffeted crow go beating by, heading directly into the wind.

On a long slant, I come back up Firefly Meadow to follow the lane to Kenyon Road and collect the mail. Steeplebushes all support white steeples along the way, their sturdy stems bearing easily the burden of snow. It is the toughness of these stems that gave to the plants their common name of hardhack. Early settlers found they had to "hack hard" to sever them when mowing hay with scythes.

For the fun of it, as I come back up the lane, I turn aside and make slow-motion wingovers by angling up the drifts to the wall tops and angling down again. When I look toward the house I see blue-tinged smoke streaming away from the fireplace chimney. More than once it occurs to me how fine it is to feel so good in this time of winter storm. With my snowshoes hanging on their nail in the entry shed again, with the snow shaken from my clothes, exhilarated by my walk in rough weather, I come back to the warmth and dancing light of the fireplace. With food in the pantry, oil in the fuel tanks, wood stacked on the fireplace hearth, we are prepared to enjoy this snowbound interlude.

JANUARY 26. The wind still blows and the snow still drifts for several hours this morning before the weather moderates. Out on my snowshoes again, I stand near Nighthawk Hill, looking down where the long gusts funnel between two close-set tussocks, drawing out a tongue of snow, a Lilliputian drift that forms and curves away downwind as I watch. It brings to life a time of special excitement and adventure experienced years ago.

Nellie and I were returning from a reconnaissance in March to the area along the New Brunswick border at the northern tip of Maine, where we planned to end our journey through winter a year or two later. On the morning we turned south, a man eating breakfast in the restaurant at Caribou said: "I hear it is storming downcountry."

By the time we reached Mars Hill, the light of the sky was dimmed by darkening clouds, and just before Houlton, the snow began.

"It is rough below." That was the information we received at the filling station where we had our fuel tank filled. The attendant was right. From then on we were in the raging wind and blinding snow of a Maine blizzard. All the hills, all the woods, all the farms were swallowed up in driving clouds of white. The only other machines we met on the highway were occasional snowplows.

It was near Danforth, on a high ridge far above the dim, wavering outline of Grand Lake off to the east, that we traversed the most memorable stretch of this tumultuous day. The mile-long slant of the land from the lake up to the ridgetop ascended toward us in an unbroken expanse. Nothing impeded the sweep of the gusts that came charging from the northeast. Sheet after sheet, the driven snow rushed toward us, enveloped us in a white smother, and plunged on. Here at the summit of the slope the drifts were deepest. Only a few minutes before, a snowplow had gone through leaving behind a sheer four- or five-foot wall on the side toward the open expanse. The top of this thrown-up snow ran beside us with a serrated or ragged profile.

As each successive gust flung its streaming burden of snow over the crown of the ascent, the Vs or notches along the uneven top of the wall channeled the snow, concentrated it, directing it across the highway in transverse curving drifts that continually took shape and grew before our eyes. The wind shrieked, the snow scudded by, the drifts mounted. All down the recently cleared highway before us we could see a long succession of newly formed drifts swiftly rising, each curved like a white scimitar. Looking back, we were amazed at how our tracks had been obliterated almost in seconds, at how high the new-formed drifts had risen. Barriers of snow closed in before us and behind. The car slowed as we plowed into each successive drift, gained a little momentum in the open spaces between. For an uncertain time, in what we realized afterward was a narrow escape from being stranded, we broke our way through the increasing walls of snow and at last descended to lower, more sheltered land. If we had been only five minutes later, our way would have been blocked and we would have been trapped on the ridgetop in the storm.

On that high land, amid the sheets and streamers of snow, we seemed in the very heart of a vast snowdrift factory. It is this memory that the forming driftlet in our field recalls so vividly this morning—just as it was the time among all the hours of that north-country blizzard that provided our outstanding recollection when the day ended and we had attained Topsfield and Calais and had settled down in safety for the night.

J A N U A R Y 2 7 . The lane is open. The snow is settling, the drifts packing. In various places where I have dug out openings and scattered seeds, the small ground-feeding birds are huddled densely together. I am amused by watching one of the "winter chippies," the little red-capped tree sparrows. It stands in the midst of a small mound of seeds, pausing at intervals in its rapid feeding to scratch vigorously just as though it were among leaves where seeds were few and widely scattered. It demonstrates the endurance of instinctive acts.

In contrast, in recent days another of these small birds has taken to flying up to a feeder almost four feet above the ground. While all the others of its kind are scratching in the age-old way, it is experimenting, trying something new. None of the other ground feeders follows its example. It is the only tree sparrow we have ever seen ascending to such a feeder. This little bird is the one among many, the individual, the pioneer, the explorer blazing a new trail of its own.

J A N U A R Y 2 8 . That smallest of the falcons, a kestrel or sparrow hawk, crouches on the snow near the sundial, its wings outspread and downbent. From beneath, the sharp tip of a black wing jerks upward, continues rising and falling, revealing where a starling struggles in the grip of its talons. It seems most likely it is the sick and sluggish bird we have seen in recent days.

The sparrow hawk bends down its head. Small black feathers, torn from its prey, begin littering the snow. I change my position, come out into the open. And in so doing I learn again the importance of leaving a feeding hawk undisturbed. To frighten it away not only is likely to prolong the suffering of the victim but, because it leaves the hawk with hunger unsatisfied, may contribute to the death of another bird. As soon as I come into sight, even though some distance off, the nervous little raptor takes fright. It leaves the starling, too heavy to transport through the air, and flies to a fencepost. From the black tip of its beak to the white tip of its tail, its length is no greater than that of a robin. But its narrower, more sharply pointed wings have a span five inches greater than the wings that support a redbreast in the air. For some time the hawk remains on its perch, watching me and turning its head from side to side. Then it lifts into the air in quick and buoyant flight and speeds away.

I plow down the garden path through the shrinking drifts. The starling is dead. But it has not died where the hawk had held it. Although mortally injured, with the flesh torn from its back and neck until the vertebrae are fully exposed, it had pushed itself ahead over a space of more than two feet and into the edge of a tangle of weeds. Its end would have been mercifully quicker if the hawk had had its way.

Interfering in the fate of wild creatures, taking sides in nature, is a recurring temptation. Instinctively we tend to sympathize with the victim fighting for its life. As a philosopher once put it: When the lion eats the lamb, we always say "poor lamb"; never "happy lion."

I remember one early-summer day when I discovered a large blacksnake with its head thrust inside a bluebird box where I thought there were baby bluebirds. I killed the snake. Then I found the serpent was getting not nestlings but eggs and the eggs were not those of bluebirds but of a pair of house wrens that had driven the bluebirds away and filled the box with the sticks of their bulky nest. Moreover, a little later, when I checked on the birdhouse to which the bluebirds had moved, I discovered the

wrens, to whose aid I had come, had slipped inside while the occupants were away and had punctured every egg.

So much for another interference. Hereafter when I am
tempted afresh to become an intruder instead of a spectator, to
alter the fate of creatures living normal existences in a normal
web of life, I will hesitate. I will remember those experiences of
the past, such experiences as with the wrens and the bluebirds,
this starling and this hawk.

J A N U A R Y   2 9 .   Twice this winter a melanistic buck has
been sighted disappearing into woodland close to Trail Wood. At
the opposite end of the color scale from the albino deer that a few
years ago spent a winter in the James L. Goodwin State Forest
area, it is reported to be almost coal black. Nellie and I have been
on the alert for a glimpse of this rare animal. But so far, along the
wood paths we have watched in vain. Out in milder weather
today, we spend several minutes trying to convince ourselves that
fresh deer tracks in the melting snow *might* have been made by
the melanistic buck.

Bluejays, in the fields, are going over the softened snow. We
study them for a long time through our field glasses. They pick
here and there among the drifts. What food can they be finding?
The first possibility we think of is that small spiders may be
abroad, spiders like the dark midget we saw last month toiling
down one side and up the other side of a deer track in the snow.
Another suggestion: in this moderating weather, they may be getting springtails, those multitudinous snow fleas of the winter
thaws. But after a moment's reflection, we discard them as a
source of food for the bluejays. Ten thousand would hardly make
a mouthful for a jay.

The farther we go this afternoon, the warmer the day becomes. The gentle breeze has swung to the south. Everywhere
the snow is swiftly melting. Everywhere drifts are shrinking.
Now, in these last days of the month, gales and blizzards, icicles,
and below-zero temperatures seem suddenly, miraculously a
thing of the past. So long delayed this year, that time of reprieve,
of mild and sun-filled weather—the time of the January thaw—is
bringing its temporary changes, sudden and dramatic, to Trail
Wood.

JANUARY 30. The breeze still blows from the south. The sunshine still pours down from a sky clear, pale burnished blue, where fewer than a dozen clouds, far spaced, hang becalmed. As Nellie and I look about us at the beginning of our walk to the Far North Woods this afternoon, we see that—except in hollows and along the edge of the trees—the snow has disappeared from the open fields. Two days ago they lay outspread sheeted in white. Now they extend away gray and tawny with the intermingling hues of wet winter grasses and dead goldenrod.

At innumerable places along our walk today we catch the liquid whisper and murmur of little threads of water, miniature temporary streams flowing over the saturated ground, winding among the grasses and fallen leaves, creeping downward along tortuous ways to reach the nearest brook. In the sudden mildness of the weather, uncounted tons of snow and ice have returned to the water from which they were formed. Hampton Brook is a raging torrent. Our little waterfall has become a Niagara.

Taking advantage of this day of spring in January, we follow the Old Woods Road as far north as our land extends. We skirt Witch Hazel Hill, looking down into the deep valley of the beaver pond. We work through openings in old stone walls. We step from rock to rock across the cataract of the Brook Crossing. We turn aside to visit the triangle of large white oaks that forms a landmark in this northern portion of the woods and continue on to examine the ragged face of a little cliff where our only polypody ferns are rooted. Then we wind with the remnants of the old wagon tracks down the final decline to Griffin Road.

Near this northern boundary line, a lone gray squirrel bounds away downhill along the path ahead of us, its feet silent on the sodden leaves. It is the only animal that we encounter in the woods. But in many places along the flat tops of old walls we see the little kitchen middens, the piles of shell fragments of hickory nuts or discarded pine-cone scales that squirrels—gray and red— have left behind in their winter feeding.

From end to end, we hear no birds in the Far North Woods. It is our yard at home that constitutes the populous bird oasis at Trail Wood. Only when we have reached the Fern Brook Trail and are nearing the Starfield do we encounter the first bird voice. It is the "yank! yank!" of a white-breasted nuthatch. Not far beyond, a downy woodpecker, stimulated by the abnormal warmth, is rat-a-tatting on a dead and hollow limb. Then, as we come out into the open fields again, we catch sight of another bird, a larger bird, a red-tailed hawk quartering over the meadow along the edge of the western woods.

JANUARY 31. More of the wonderful same! More of mild air, more of warm sun, more of gentle winds from the south. More hours filled with minutes like balm for the winter-weary. For a third day this late-January thaw continues.

Again we range widely over trails no longer blocked with snow. We replenish our fireplace kindling among fallen branches no longer buried beneath the drifts. We follow Hampton Brook upstream to listen to the roaring of the waterfall in the floodtide of its meltwater. We climb to the top of Juniper Hill. We circle the pond. We feel liberated with the snow gone, with former restraints to walking suddenly removed.

All the wild creatures are responding, as we are responding, to this taste of spring in winter. We watch gray squirrels digging for buried nuts in the thawed-out ground. We hear, from half a dozen directions, the xylophone solos of downy and hairy woodpeckers hammering on dry, resounding limbs. About noon, a little flock of eight silent male red-winged blackbirds drops down to feed on the carriage stone. Overwintering in the coastal swamps, they have wandered inland in this time of thaw.

Late in the afternoon, when I return from a final foray to the woods, I find the house echoing with the songs of British birds— skylarks and nightingales, English blackbirds and missel thrushes. Nellie is playing one of the bird-song records we brought back from our springtime in Britain. For a time on this day of milder winter weather birds of England sing within and birds of New England call without.

The night falls with the air still mild. Tannish forest moths flutter around the terrace floodlight. But the wind no longer blows entirely from the south. The chill gradually increases. The swing is back to winter. Toward bedtime, at the end of this last of the January days, we walk out to see the blaze of the stars glittering across a sky still untouched by clouds. Coming back along the Starfield path, we pause to listen to that strange, menacing sound—a short bark ending in a snarl. A red fox is hunting in the night.

FEBRUARY 1.   Gone are the redwings. Gone the sunny skies. Gone the long-delayed, doubly appreciated thaw that brought January to a close. Back is winter. Back the stabbing, cutting edge of the cold wind from the north. Back the low-dragging clouds heavy with snow. Gusts tilt and buffet the small birds, fluffed into balls of feathers. During the hours of the night there has come one of those sudden reversals in the weather that are characteristic of the midwinter months.

Now, toward noon, as I am beating my way into the wind across the Starfield, the snow begins. At first it streams out of the north in small hard pellets. I feel the sting as they are hurled against my face. Through slitted eyes, I see them racing by. Within the shelter of the woods, I stop to catch my breath. Where only yesterday we walked in mild and sunny weather, the trails lie in dim, murky light. Overhead the treetops wail in a rising and falling tumult, swept by uneven gusts. As I plod on deeper into the interior of the woods, I am accompanied by a kind of faint shimmer where the reduced light catches the snow filtering down along the dark tree trunks. During short-lived lulls, I pick out the low sifting or sibilant hissing sound of the flakes striking twigs and small objects around me. Not once in my hour in the woods do I hear the voice of a living creature. Everything has taken shelter. Everything is lying low.

In time I follow the example of the wildlife. I swing back toward the warm sanctuary of home. Even before I reach the Fern Brook crossing, I notice the character of the snowfall has altered. The flakes are larger, more fluffy, more numerous. Outside the woods, I find the air dense with the softer snow. From the vantage point of Nighthawk Hill, the highest ground of the Starfield, I look about me. Already the snow had thrown a thin veil of white over fields that were bare when I entered the woods.

FEBRUARY 2.   As Landlord of the Woodchucks, I am out making my rounds like an old-time English squire visiting his tenants. In midafternoon on this second of February—Groundhog

Day—I am checking the burrows for any sign of activity. The weather is clear and cold. Amid this snow-covered landscape, the hours pass in sunshine and shadow. In visiting the remembered marmot holes in the fields and along the edges of the woods, I am not concerned with that old folk belief that if on the second day of the month the groundhog sees its shadow there will be six more weeks of winter weather. In these hours of appearing and disappearing sunshine, whether the woodchuck sees its shadow will depend most of all when it appears and looks around. Shadow or no shadow, Nellie and I are prepared for six more weeks of cold. What I am interested in is evidence of a midwinter break in the long sleep of the woodchucks.

I move from hole to hole looking for any disturbance of the newly fallen snow around the entrance. Only once do I find what I am looking for. But there the tracks entering and leaving the opening are unmistakably those of a cottontail rabbit. All our groundhogs have slept on, have continued in the unawareness of hibernation through this Groundhog Day.

However, if the woodchucks fail to contribute to the interest of my walk, the house sparrows do provide a time of diversion at its end. When I started out nearly half a hundred were feeding with other small birds on seed scattered on the snow. When I come back I see them again. Now they are collected at the far corner of the yard close to a bluebird box that has been left on its post throughout the winter. Like tree swallows on telephone wires during the autumn migration, they are ranged side by side in dense rows along the barbed wires of a nearby fence.

All are chirping in great excitement. One after the other, little groups of four or five flutter up to the box. They hover at the round entrance hole or alight on the box. Then another band rises and takes their place, repeating the puzzling performance. None of the house sparrows, so far as I can see, enters the hole. After all the birds have flown away, I examine the box carefully. There is nothing inside that has caused the excitement. Perhaps the sparrows have engaged in some mass response to an early mating and nesting urge. But I cannot be sure. What I have witnessed takes its place among those many encounters in the out-of-doors that leave us unable to explain the cause of what we have seen.

FEBRUARY 3. Who can tell or guess or comprehend exactly the response within an individual? What sounding line can measure the depth of an emotion? Following this snow-covered trail winding northward from Witch Hazel Hill this morning, I remember a young woman of about twenty who once accompanied us along this same path in that wonderful time of year when the wild flowers of spring are everywhere. From beginning to end she hardly uttered a word. We were not sure whether she had appreciated anything or enjoyed anything along the way. But a year or two later I met her again. I found she had remembered everything she had seen. In silence, she had been stirred deeply, had responded intensely, had come away with memories that had endured.

The one who exclaims the most is not necessarily the one who appreciates the best. The one who stands silent, impassive, unchanging—except perhaps for a special lighting in the eye—may be the one affected most profoundly, the one on whom the scene is making an impression to be remembered through life. The one who is touched not at all and the reserved person who feels most deeply of all may appear the same from an external viewpoint. It is often the case that we must know people for some time in order to gain a true understanding of their capacity for appreciation or the depth and intensity of their emotional response.

FEBRUARY 4. It all starts with the scream of a jay. Nellie and I look down the slope to the decaying apple tree whose branches form the arch through which we see the pond from my study window. The calling bird, screeching in excitement, hops from limb to limb. On each successive perch, it cocks its head far to one side to peer into a hollow in the tree trunk. Then it shrieks anew. From all directions other jays come flying. We see them hastening across Firefly Meadow, winging their way over the pond, hurrying from the North Woods over the Starfield. All are screaming as they fly.

Within minutes the tree swarms with jays. The uproar rises to a crescendo. All the birds hop about, all peer continually down into the hollow. Within that cavity one spring wood ducks built their nest. Another time, gray squirrels made it their home. This winter, under the roots of the tree, a woodchuck, oblivious to all the hubhub above its hibernaculum, is lost in its winter sleep. What has caused the excitement? What have the bluejays discovered within the shadowed cavern of the old tree trunk?

The only thing Nellie and I can think of is that an owl may have taken shelter there. We walk to the tree. The bluejays scatter to neighboring trees, there to continue their screaming. I find a stick and whack the trunk of the tree. We half expect to see the face of an owl appear at the opening. Nothing happens. While Nellie remains on watch, I go to the garage and return with an aluminum ladder on my shoulder. Ascending it, I reach the mouth of the cavity. Like the bluejays I cock my head and peer intently within. I find I am cutting off most of the light and can see but little. Nellie returns to the house and comes back with a flashlight. Directing its beam inside, I examine every inch of the interior of the hollow. It is completely vacant. There is no sign of any living thing.

By now the excitement has died down. The bluejays have scattered. I am left, as I was left by the house sparrows at the nesting box two days ago, puzzled by what I have observed. What alarmed the first jay? Did it see some shadow in the cavity, some entirely imaginary enemy, that set it screaming? Or was it merely trying to generate a little excitement on a dull day with a false alarm? At any rate, it has brought not only all the bluejays of the area but Nellie and me as well to inspect the interior of this old apple tree.

FEBRUARY 5.   Daybreak comes slowly this morning. Its transition from dense gray to lighter gray is gradual and protracted. Dimly I make out a cluster of moving, dusky shapes spread out over the tilted carriage stone. Nearly a hundred mourning doves are pressed together feeding on the cracked corn we scattered there last evening. Their numbers have increased again since the goshawk left.

In the midst of the doves I glimpse a form that is larger and darker. It, too, is indistinct in the murky light. At first I take it for a grouse feeding among the doves. Then it lifts its head. It is an unbirdlike head with two ears, long and upthrust. Surrounded by close-packed mourning doves, our corn-fed cottontail is breakfasting on the scattered grain.

Without doubt, this is the same rabbit that has found shelter in the woodchuck hole under the wall beside the lane. A little later, when I go down the slope for the paper, I glance at the hole and at the tracks around its entrance. I turn aside to examine the snow more closely. It records the story of an encounter in the night. Ringing the entrance rabbit tracks and larger pawprints intermingle. The snow is plowed and trampled. Sometime during the hours of darkness, a hunting fox sighted or scented the cottontail out in the open. It had shot away in close pursuit. The two, one just behind the other, arrived at the woodchuck hole. The rabbit hurled itself underground and vanished. The fox plowed long grooves in the snow as it slid in breaking its headlong charge. Then I see where it circled around and around the entrance, the smell of its lost quarry strong in its nostrils, where it turned aside, swung back again, finally trotted away down the lane.

In this story written in the snow, this flight in mortal peril in the dark, and in this peaceful feeding among the doves in the slow breaking of the dawn, I observe two aspects, two opposite poles of experience, in the life of our foraging cottontail.

F E B R U A R Y 6 .   To see something new, to observe something we never encountered before, to make a fresh discovery—these are always possibilities as day after day throughout the year we wander over our Trail Wood land. Close to the bridge along the lane, about midmorning today, I come upon a little revelation of the kind.

As I approach the spot, I see a number of tree sparrows fly up from the snow on the north side of the lane. Around the base of several plants where they have been feeding, I find a maze of little tracks. And scattered among them on the snow are what appear to be minute black seeds. Hundreds of times in the open fields I have come upon similar scenes—the snow imprinted densely with sparrow tracks and seeds or seed hulls strewn about under some goldenrod or rough-fruited cinquefoil or Saint-John's-wort thrust above the snow. But this is different. In this case what appear at first glance to be seeds are not seeds. They are instead the spore cases shaken down from the fertile fronds of dry sensitive ferns.

Never before have I observed or heard about birds feeding on the reproductive bodies of ferns. Undoubtedly the dustlike spores would be rich in nourishment. But in that exhaustive text on the subject, *American Wildlife and Plants*, by Martin, Zim, and Nelson, only the seeds of herbaceous plants are listed in the diet of tree sparrows. There is no mention of fern spores. Nor is there any reference to this source of food in Arthur Cleveland Bent's monumental series on the life histories of North American birds.

When I examine closely the plants where the birds have been feeding, I find the upright fertile fronds, supporting the spore cases massed like rows of beads at their tops, are nearly a foot and a half high. By landing among them and pecking at the capsules, the small sparrows have sent the food showering down on the snow below. I break open some of these tiny containers. Dark chocolate-colored dust streams out—living dust in particles so tiny they can be distributed by the wind. It is in this almost microscopic food that the birds have found a source of nourishment on this midwinter morning.

F E B R U A R Y 7 .   I meet a lost meadow mouse among the snowdrifts this morning. And I help it find its way.

In wind sweeping down from the Arctic, I have been tramping along the edge of Mulberry Meadow, keeping company with one of its boundary walls. During the night a brief mizzling fall of freezing rain encased the drifts in a thin shell of ice. Ahead of me, scattered over the rise and fall of the snow, I recognize small fragments of windblown bark. In one hollow, protected from the wind, I see what I take for a piece of gray bark shift its position in a twitch or jerk as I draw near. I look away. I look back and it is in another hollow. The piece of bark has legs. I steal a slow, stealthy step nearer. Again a retreat of the fragment. This time I see clearly the short tail, the plump body, the dark gray fur, the small bright eyes of a meadow mouse.

Apparently, during the night, the animal had stolen from one of the many tunnels under the drifts to wander about on a foraging expedition. Perhaps the shell produced by the freezing rain had blocked its return. Perhaps it had lost its way. Now, for a few moments, we remain unmoving, eying each other. I am remembering, as I stand here, another lost mouse of a different species that I once encountered under somewhat similar circumstances.

I first saw this mouse huddled at the side of Kenyon Road as I was extracting mail from our rural delivery box after another night of freezing rain. Highway plows had cut through the drifts before the rain, leaving on either side vertical walls of snow, now clad in ice. The cowering animal, a beautiful little white-footed mouse with large nearly translucent ears, appeared to shrink within itself. It remained pressed against the base of one of the perpendicular cliffs of ice-covered snow, giving the impression it was too chilled to move. There was no chance of its burrowing through the ice armor and finding protection within the snow. With a swoop of my gloved hand, I scooped it up and thrust it into the warmth of one of the pockets of my greatcoat. After it had ridden part way down the lane, I broke through the crust of a drift and dropped it into the soft snow. There, finding sanctuary, the revived whitefoot burrowed quickly out of sight.

Something of the kind happens now. With a little more difficulty and after several misses, I capture the lost meadow mouse. Holding it in the leather glove encasing my hand, I carry it to the boundary wall. Here, on the southern side, where the wind-driven rain in the night had leapfrogged over the wall, I discover a narrow strip of snow unglazed with ice. Mouse history repeats itself. As the whitefoot had done, the meadow mouse immediately commences burrowing downward. So swiftly does it descend through the soft drift that it appears to melt into the snow. But the round hole left by its tunneling remains—one among the thousands of such burrows produced by the winter mice, burrows that wind and intersect unseen beneath the snow.

FEBRUARY 8. For a second time this winter the goshawk appears in the hickory tree. I see it rip down, flicking in and out of snow fog curling up from drifts on this day of warmer weather. I glimpse it riding its tip-top perch on the loftiest branch of the highest hickory. Fluffy white feathers that flare out like short pants around its upper legs ruffle and flutter in the wind. Their moving, shining whiteness catches the eye. It draws attention to the spot where the motionless hawk is perching.

A dozen bluejays, silent and unmoving, cling within the maze of branches of the terrace apple tree. The goshawk launches itself into a downward plunge, abruptly checks its speed, and alights at the end of one of the apple limbs. It is no more than a couple of yards from the nearest jay. A tremorlike movement runs across the interior of the tree. But there is no hopping about, no confusing crisscrossing movements such as greet that milder menace, a sharp-shinned hawk. Every bird remains anchored in place. Not a bluejay breaks cover. The goshawk keeps its intent glare centered like a searchlight on the perching birds. They stare back without a movement of their heads. Minute after minute this war of nerves drags on. Then the hawk snaps open its wings and scuds away. The bluejays relax. They begin hopping from branch to branch. Their time of terror has run its course.

Later on this same day I witness another instance of that delicate balance that exists between the hawk and its prey. I see this visitor from the north cross the yard in an arrowing advance that ends in a skyrocket climb to its favorite perch at the summit of the hickory. On the way up, it passes a lone mourning dove pulling itself tightly together, huddling motionless on one of the lower limbs. The dove continues immobile, as still and apparently lifeless as a knot on the branch on which it rests. Surely the piercing eyes of the goshawk have seen it. Its perching place is in a relatively open area. A return plunge with talons ready and the bird of prey could easily pluck it from the limb.

But this, apparently, would be at odds with the goshawk's psychology. In their relations with their prey, predators often appear to obey certain rules, as the knights of the Middle Ages observed the codes of chivalry. In the far north, goshawks have been seen remaining unmoving for long periods until a ptarmigan they were watching took wing. As soon as the bird was airborne, they plunged and picked it from the air. In this instance, the goshawk waits for the dove to fly. But the dove does not fly. Thus the smaller, defenseless bird escapes. There apparently are limits to the predations of every predator. Skating on the thin ice of these limits, its prey often insures its own survival.

FEBRUARY 9. I watch a red squirrel racing from treetop to treetop, streaking to the end of a slender limb, launching itself into space to reach another almost twiglike limb in another tree. It is so confident, so agile, so surefooted, so perfectly balanced. I plod on through the snow experiencing the same emotions that have come so often in the past on encountering, in the wild creatures around me, abilities I can never know—that initial touch of envy, that enduring longing to live other lives in addition to my own. One of my earliest desires—remembered from long, long ago—was for some Aladdin's lamp, some magic potion that would let me live those other lives, let me know exactly what it would be like to be a red squirrel in a treetop, a mole tunneling through the still, damp earth, a snowy tree cricket filling the night with mellow music, a swallow skimming over the meadow.

FEBRUARY 10. A covey of bobwhite quail. A cottontail rabbit. A bluejay. Each in its own way, they provide moments of diversion in the course of my winter walk today.

During the night, several inches of light snow fell and drifted into the hollows. As I am following the edge of Pussy Willow Corner, I come upon the quail, nearly a dozen. Only half alarmed, they scatter at my approach, running belly deep through the snow, each advancing bird surrounded and accompanied by a whirling cloud of shining flakes.

The rabbit I encounter near the path that ascends the slope of Monument Pasture. It is hopping slowly, its feet sinking deeply into the downy mass. It almost disappears with every jump. Lacking the slight crust over which it can speed in long, effortless leaps, it labors in the elusive footing until it reaches the end of the drift. Beyond, its feet sink only a few inches. Unhindered now, it bounds away up the slope and over the hilltop.

It is later in the day when I watch the bluejay discover surprise and adventure in the deeper drift piled up against the wall beneath the hickory trees. Rounding the house, the jay dips down and alights on the surface of the snow. For weeks it has been accustomed to coming down on more packed and solid drifts. This time it plunges out of sight, disappearing in the fluffy mass. There is a screech of alarm, a great fluttering, a cloud of scattered snow, and the startled bird bursts up and into the air again.

I come in remembering another bird—a crow—I once observed under the apple tree below the terrace. It came swooping down into the yard over high drifts left by a nightlong storm.

Near the base of the apple tree another crow picked at a piece of suet. The newcomer ended its descent with a quick twist and pass at the feeding bird. The latter leaped into the air, sidestepping the attack, and the diving crow—like a schoolboy pursuing another boy and tumbling into piled-up snow when the first boy dodges—plowed into a drift. Snow spurted up as though in an explosion. Then flapping black wings flailed the drift and the crow, like the bluejay, popped up, lifted free, was airborne once more.

FEBRUARY 11. The sunshine slanting from a different angle, the wind sweeping aside curtaining ferns in summer, the chance movement that catches our eye—all at one time or another lead us to little nooks and alcoves and miniature gardens, tiny magic places along our way.

We have turned aside from our regular path up Juniper Hill this afternoon and are taking a shortcut among the sprawling clumps when we reach a barrier formed by half a dozen prostrate junipers interlaced together. I am pushing my way through the heart of this maze when I look down and see at my feet a small elfin landscape filled with dwarfling plants, a garden of miniature colorful lichens. It extends over the earth and onto the dead wood at the center of one of the clumps. Here, crowded together, shielded from the snow, hidden from the world, are scores of those two *Cladonia* lichens, *pyxidata* and *cristatella*, the green pixie cups and the scarlet British soldiers.

Lifting tiny funnels or goblets, the pixie cups tilt this way and that. Each is an inch or less in height. Intermingling with them are masses of the red-crested lichens that for generations have been named, because of their color, after the scarlet coats formerly worn by soldiers of the British army. They are about the same size as the pixie cups. But the summit of each resembles a brilliant dab of red sealing wax. Mingled together, these diminutive primitive plants extend their living mosaic over the ground and decaying wood, hidden by the tangle of the juniper boughs. We spend some time enjoying this sight of rare and hidden beauty that, like so many others of its kind, we have come upon by chance.

FEBRUARY 12. The trail of the forlorn opossum—the animal with threadbare fur and frostbitten ears and tail we see looking for suet scraps under the apple tree on zero nights—shows where it has floundered away across the yard. Where does it live? Where will that trail end? What shelter from cold and wind does it find when its nocturnal searching for food is over?

I pull on high-topped leather boots and slip into heavy clothing and begin following the ragged trough it has plowed across the snow. It leads me through the gate by the bluebird box, amid blackberry canes, under a seedling apple tree. Keeping to the path beside the brook, it nears the waterfall. Then suddenly it veers to the right. I see where the animal has plunged down the steep bank of the stream. Here the trail ends. It ends beside a decaying stump at the mouth of an abandoned woodchuck hole. Deep in its interior the opossum—a creature that seems unfit to survive through a winter's day but that as a species has endured for aeons—is curled up, sleeping the bitter hours away, finding in its sanctuary all the rudimentary comfort it knows in this bleak and frozen period of the year.

FEBRUARY 13. There lived in Hampton up to a few years ago a man who used to recall that his father, as a very old man, often spoke of "a larger kind of quail" he used to hunt as a boy, a bird he never saw anymore. He remembered it was barred and had a head he thought looked like a quail's head except that at times large "golden patches" were visible on the sides of its head and neck. Nellie and I were baffled until we looked up the extinct heath hen, that eastern relative of the prairie chicken, once so abundant here. The remembered field marks tallied.

Standing again near the waterfall today, we recall the tragic story of this game bird. Its mounting slaughter continued until the only place in America it was found was on the Massachusetts island of Martha's Vineyard. There, in 1916, disaster overwhelmed these survivors. Fire swept through the thickets of shrub oak, berry bushes, and pitch pine at the time when the birds were nesting. It destroyed all the eggs and most of the brooding adults. The last heath hen in the world was sighted on the island in the autumn of 1931.

Standing here beside this ice-covered brook—a stream that for how many centuries nobody knows has been part of the drainage system of this land—I remember those lost birds, never to be seen anywhere, anytime again. It is entirely possible that to some of them the flow of water along this same stream's rocky bed was part of their lives long ago just as it is part of ours today.

When we return to the house, I take from the mantle above the fireplace in my study a fragment of weathered bone. It once functioned as part of the skeleton of a great auk, a bird of the ocean that met a similar fate. The fragment was picked up years ago by an acquaintance of mine on lonely, storm-swept Funk Island, off the coast of Newfoundland. This remote dot of land, bare rock at one end, turf and deep humus at the other, provided one of the final breeding grounds for this now extinct seabird.

Perhaps as early as the twelfth century, Viking explorers—and certainly in the sixteenth century, in May 1534, Jacques Cartier—visited this rocky outpost and found it swarming with large black and white flightless "penguins." The birds, unaccustomed to men, showed little fear. Their slaughter was easy. "God made the innocencie of so poor a creature to become such an admirable instrument for the sustentation of man" is the way one seventeenth-century writer in England expressed the prevailing attitude toward the great auks.

They were killed for food. They were butchered to provide bait for the fishing fleets on the Grand Banks. They were slaughtered for their feathers—which were removed by immersing the birds in scalding water. The fat-filled bodies of the discarded auks were used as fuel for boiling the water. This systematic destruction reached its inevitable end. The last day on which a great auk viewed the world its kind had known for so long was June 3, 1844. On a skerry, or rocky islet, off the southwest point of Iceland, a landing party on that day discovered two birds and at once gave chase. "They showed," reports an early record of the event, "not the slightest disposition to repel the invaders, but immediately ran along under the high cliff, their heads erect, their little wings somewhat extended. They uttered no cry of alarm and moved, with their short steps, about as quickly as a man could walk." Both, almost at once, were overtaken and captured. Their viscera are now preserved in the Royal University Museum at Copenhagen, Denmark. When they died, a thread of life snapped, never to reappear on earth.

FEBRUARY 14. All through the night, whenever we awake, we hear the hooting of the wind around our house on the knoll. We hear it wail and scream and pound against the shutters and then go roaring away through the bare branches of the hickory trees. Day comes, a blizzard day without a daybreak, with only a gradual seeping of light through the storm. We look out on an indistinct world that is filled with violent motion, on lashing trees vaguely seen behind the curtaining snow, on clouds of flakes hurled out of the northeast and flung scudding in ghostly waves over the drifts. Everywhere we look we see a white scene of turmoil. This is the great winter storm of the year. It rages throughout the day.

When we switch on the radio for the weater report we learn some of the gusts have reached eighty miles an hour. This is, the announcer notes, one of the major storms of the decade. If you have to meet an emergency, Nellie says, it helps to know it is a historic one. We notice how hungry we are in the midst of the storm. We notice how the tumult has raised our spirits, given us a sense of elation and excitement. To those in good health, a blizzard is an adventure. It is to those who are ill that the possibility of being snowbound looms as a menace.

At intervals, as the hours go by, I struggle through the storm, buffeted by the wind and half blinded by the snow, to scatter more seed on places scoured out by the gusts in the lee of tree trunks and at the corners of the house. Once, as I am coming in, I hear in a momentary lull the "per-chic-ory" of a small band of hardy goldfinches invisible in the storm above me. A moment later they swarm down to join the ground feeders, searching for the smallest seeds. But most birds lie low. No crows appear. Twice a downy woodpecker materializes from the storm to ride the swinging suet bag. For most of the wildlife this is a time of belt-tightening, of endurance, of conserving energy, of sitting out the gale.

Several times I plow through snow up to my knees to the middle shed and return with logs for the fireplace. Across the open expanse of the pastures immense banners of snow stream from the northeast to the southwest in a continual procession as gust follows gust. The edge of the North Woods, beyond the Starfield, is completely blotted out. But I can hear the vast surf of the wind tumbling among the treetops. Each time I come back, pushed along by the drive of the air, I reach the kitchen door with the log I am carrying already white with a coating of snow.

Inside, with the flicker and warmth of the flames, listening to the ever-varying clamor of the wind, aware of the little tremors attending the long gusts that shake the house, we feel snug and battened down, able to ride out the blizzard as former dwellers here rode out the gales of 170 years. We are like the rabbit in its burrow, the raccoon in its hollow tree—snowbound but protected by a home.

Soon after three thirty in the afternoon the light begins to fade. The blasts, undiminished, are still raking along our walls and battering against our windowpanes. Floundering through the woods, gaining speed across the meadows, hour after hour the wind continues to scour and sculpture the rising drifts that slant up and over the walls and lengthen across the lane. Already more than fifteen inches of snow have fallen. So the premature dusk blots out the dim light of the day. Our second night of full-tongued wind and eclipsing snow begins.

FEBRUARY 15. Calm after violence. Stillness after tumult. The sky is clear. The sun shines on a world of immaculate white. This is the change wrought by the hours of darkness. The gale blew itself out in the night and we take stock of our surroundings—the curving beauty of the drifts, the snow packed three feet deep against the kitchen door, the angling ridges, in places chest high, that cross the lane. We are back in snowshoe weather.

Nellie and I begin digging out. We excavate paths to the bird feeders. We shovel our way to the shed where the fireplace wood is stored. We open up areas amid the drifts and scatter seed on the ground. Within minutes these openings are carpeted with feeding birds. Among them in the wake of the blizzard we see two rather rare species at Trail Wood—hardy midgets from the north—redpolls and pine siskins. Moving like striped mice in feathers among the larger sparrows, twenty of the little siskins flash the light yellow patches of thir wings and rumps and repeat the ascending "shreee" of their calls.

More than three times as large is the flock of redpolls that has arrived after the storm. We count sixty-four feeding together. Because we usually see these midgets only when they are fluffed up on the coldest days of winter, Nellie and I decide we have a distorted mental picture of them as especially plump little birds. We watch them feeding with the sun behind them. They pick at the scattered seed, their heads bobbing rapidly up and down, their topknots flashing brilliant ruby red each time the low rays of the just-risen sun strike them. This winking or flaring runs in successive waves across the close-packed birds. We watch them entranced. We are well aware that we are witnessing one of the transcendent moments of beauty this winter will bring—the outspread, new-fallen snow lighted and tinted in the sunrise, the grayish little birds from the north clustered amid it, the twinkling coruscation of ruby spots like small lights flashing on and off all across the flock.

FEBRUARY 16. Out on snowshoes again, sliding over the drifts, leaving a winding ribbon of packed-down snow behind me. It is in these surroundings that I see the crows, the sharp-shinned hawk, and the bluejay. They have come together at one place among the drifts, drawn there, each in its own way, as a consequence of the storm.

The bluejay we have noticed for almost a fortnight, a sick bird growing weaker and thinner and more inactive as the days have passed. The strain of the storm and the day without food have brought it close to its end. But, as we have seen so many times, a dying bird is often able to fly even in its last expenditure of energy. It is crossing the yard from the terrace when the sharp-shin comes skimming over the north wall on set wings and snatches it from the air in midflight. Two crows, riding topmost branches in one of the brookside trees, look down on hawk and jay as they tumble to the snow, the hawk shielding its emaciated victim with outspread wings. Much of the normal food of the crows has been buried under the drifts by the storm, sharpening their appetites. While one of the birds remains as a lookout in the treetop, the other dives in a long downward swoop and alights on the snow near the hawk.

I see the black bird with slow, measured tread circle the crouching predator. I see it draw nearer, move farther away, as it traces ring after ring around the sharpshin and its prey. The hawk follows its every movement with its intense stare, turning its head as far as it will go, then, with a shift that is almost a flick of movement, swinging its gaze around to pick up the crow for the rest of its unhurried, distracting circuit. Each time the crow turns away, the hawk bends swiftly and plucks a beakful of blue feathers that scatter over the snow. For the dying bird this more sudden end is less a calamity than a deliverance.

Often we have seen crows pursue bluejays, seeking to force them to drop pieces of bread. But never before have I been present when a crow has tried to rob a hawk of its prey. Perhaps five minutes go by while this deliberate war of nerves continues. Then, as the circling crow turns aside again, the sharpshin suddenly lifts its burden into the air. The bluejay no doubt is a lighter load than a healthy, well-fed jay would be. Moving swiftly, gaining speed, heading for the North Woods, the hawk flies low above the same stone wall over which, so short a time before, it had skimmed to make its strike. Both the circling crow and its treetop companion set out in laboring pursuit. But they are still well behind when their quarry—with its quarry—vanishes among the trees.

FEBRUARY 17. Today Nellie wins our annual February race. She is the first to see the silver catkins of a pussy willow. Each year we look forward to this discovery which she reports when she comes back from walking down the plowed-out lane. Once more it is Pussy Willow Corner—that small triangular patch of lowland beside the entrance on Kenyon Road—that has produced the first of the shining catkins.

We walk back between the walls of blizzard snow rising on either side of the lane, vertical walls left by the plow as it cut its way through the drifts. At the far side of the triangle, a bushlike willow is decorated with a score or more of the furry aments. We see them catching the light of the sun. They are the first installment of tens of thousands of others that will expand as the spring draws nearer.

In the course of time, each of these silver-furred catkins will change from silver to gold as it expands into a close-packed mass of tiny flowers—flowers without petals but rich in yellow pollen. This pollen provides New England honeybees with some of the earliest food gathered for their larvae. Later in the season, the air around all our pussy willow trees seems to vibrate with the humming of innumerable insects harvesting the nourishing powder covering the catkins in the later stages of their development.

No other pollen I know of is earlier unless it is that produced by the small flowers on the fleshy, knoblike globe within the spathe of the skunk cabbage. The first spathe of the skunk cabbage, the first catkin of the pussy willow, the first "Okaleee" of redwings returning north—these are signs of spring we eagerly look forward to as the winter begins to wane.

FEBRUARY 18.   We have named the two gray squirrels "Bright" and "Early." They come from the woods today in the half-light of the slow-breaking dawn. Always these two are the first to arrive, the first to begin nosing about over the snow in search of sunflower seeds fallen from the bird feeders on the terrace. At this time of year, in their silver fur, all gray squirrels look much alike. We cannot be certain which is "Bright" and which is "Early." We are unable to decide whether they are two males, two females, or one male and one female. From their looks alone, we cannot say with certainty we are seeing the same animals day after day. But there are other means of identifying wild creatures than by their individual markings. They can be recognized through the field marks of their habits as well. On this basis, we are positive the two gray squirrels we see in these winter dawns are the same. But as they both have the same habits, we only know that one is "Bright" and one is "Early"—but which is which we cannot tell.

FEBRUARY 19.   Gray squirrels continue to make headlines at Trail Wood.

This morning I watch one burrowing in the snow first in one place, then in another, digging down in search of buried sunflower seeds. It disappears, then pops up again. Each time before it makes a fresh dive under the snow it swings its head in a bright-eyed survey of its surroundings. On one of these explorations, the animal remains out of sight longer than usual. Down from a tree swoops a bluejay. Bent on profiting from the labors of the squirrel, it lands lightly in the snow close by.

For a moment my attention is distracted. When I look back, squirrel and jay are thrashing about on the snow. When its head had emerged, the squirrel had found the bird almost beside it. It had leaped before the bluejay could take wing. I watch the ensuing battle—the silent squirrel twisting and turning as it uses claws and teeth; the bluejay flapping violently, lashing up clouds of surface snow, its heavy bill whacking, its voice raised in screams of fear and outrage. The struggle is short and violent. The jay breaks free. Still screaming, it bolts away, fleeing toward the brookside trees.

Standing almost in this same spot one afternoon toward the end of another winter, I watched a skunk come wandering across the yard in its nearsighted, apparently aimless search for food. It was a year of abnormal gray squirrel abundance and seventeen of the animals hunted for dropped sunflower seeds over the terrace. The skunk zigzagged closer, then mounted the slope. What, I wondered, would these tree-climbing animals do? Would they recognize the potential of this new arrival? Would they see the danger warning in its black and white fur? They did both.

Within a minute all seventeen squirrels were gone. At first I thought they had climbed the nearest apple tree. But when I looked, I saw it held only a single squirrel. And as I watched, it came racing down the trunk on the side away from the skunk, leaped far out from the tree, and went bounding off on a beeline toward the north wall and the woods beyond. I had made no move that had alarmed the squirrels. No other animal had appeared that might have frightened them. Apparently the seventeen squirrels unanimously had answered my query about their reaction to the presence of a skunk when it is encountered on the ground.

FEBRUARY 20. On the other side of the year, in the month of July, I once stood beside Walden Pond with Howard Zahnizer, the father of the Wilderness Bill and one of the most admired and respected of his generation of conservationists. I have always remembered what he said. Commenting on earlier days, he recalled that then you were fighting the plunderers, the ruthless robber barons, and you could attack with no holds barred. But more and more the blacks and whites have turned to gray. Rarely now is it the all good against the all bad. You find yourself fighting the *good* for something *better*. This is a trend that has expanded since that day beside Thoreau's pond. There is, and probably there always will be, the battle against public loss for private gain. It is a battle that is continually beginning again. It is a battle in which there are no final victories, in which the participants never know the ultimate end of the story. In some instances—as in the historic struggle between conservationists and the chemical companies selling DDT—no giving ground, no conciliation is possible. But in most cases, as blacks and whites have changed to grays, the possibility, the area of mutual compromise and understanding have taken on increasing importance.

FEBRUARY 21. One month until spring! But spring seems all around us as Nellie and I follow the narrow path beside Hampton Brook on this day of temporary thaw. Our booted feet splash through puddles and slush. Beside us the water of melted snow rushes down the stream. Daily now the sun rides a little higher. Daily its rays are a little more direct.

Coming from near and far, we hear the clear whistling of cardinals and titmice. To right and left and overhead, chickadees repeat their clear, whistled "phoe-beee" notes, the sound that to earlier generations in New England brought assurance—the "spring soon" call of the waning period of winter. And from somewhere near the waterfall, the rolling, far-reaching drumbeat of a woodpecker pounding on a dry limb pulses through the air. It is an exciting sound, a wild, exultant sound filled with the health and endurance of nature. We are hearing a drummer who drums in advance of the parade of spring.

One other bird we hear and see along the way, and in the hearing and seeing discover something new about its singing. From time to time this winter we have listened to a loud "wheedle-wheedle-wheedle" coming from the dense undergrowth downstream along Hampton Brook. It is the voice of a Carolina wren, a southern bird overwintering at Trail Wood. In recent days we have watched it come to the terrace apple tree to feed on suet. Now, as we stand near the wreckage of the hollow brush pile by the stream—the brush pile within which I once watched events around me and which now lies crushed beneath a fallen tree—we hear its song, abnormally loud for a bird of its size, descending from high in one of the ash trees.

Swinging our binoculars in its direction, we sweep along the length of the upper branches and comb among the twigs in search of the singer. Nowhere can we catch sight of the plump, rufous-brown little bird we are seeking. Again the song bursts out. Again our ears tell us it is descending from the same treetop. Once more methodically we plod back and forth along each successive branch. But all are bare—as bare of singing birds as they are of leaves. Then, as the song begins again, my eye is caught by chance by a movement on one of the lower limbs. There is the wren—its bill pointing straight upward—pouring forth its song. Projected thus the sound possesses a ventriloqual quality. It reaches our ears as from a point twenty feet higher than the spot where the song originates.

FEBRUARY 22. New snow covers the frozen slush this morning in another of those abrupt reversals of the February weather. The birds are waiting for me when I shovel out open patches and scatter seed densely across them. I am returning the snow shovel to the entry shed, where hoes and rakes and shovels are ranged in line along the wall, when I notice the handle of a spade. Small, faint markings are still visible on the wood. I remember the day, the twenty-seventh of last June, when those markings were new. Nellie had come in for a drink of water while working in one of her flower gardens, leaving the spade lying in the grass. When she returned, she found a baby cottontail rabbit, only about six inches long, gnawing at the wood, probably in search of salt. In telling me of that discovery, I remember, she had ended with a question: "If you can't trust a baby rabbit, who can you trust?"

FEBRUARY 23. The night of the fox. Where did it go? What did it do? What adventures did it have? About half past nine on this clear winter morning, I set out over the snowy fields on a little adventure of my own. I begin following all the shifts and windings of the fox's trail, reading as I go the story of activity that was hidden by the night but is recorded by pawprints in the snow.

I first pick up the trail where it leaves the crisscrossing tracks on the terrace and descends toward the brook along one of the paths of my Insect Garden. It cirlces the sundial and turns aside where the animal investigated a hole in the old apple tree and thrust its pointed nose into a little cavern among the roots of an ash tree. It takes a shortcut under the spicebushes and crosses the open space where on summer days we sit on our bench beside Hampton Brook. I see where the fox has crossed the stream and halted at a small opening, eight or ten inches across, where the snow and ice are melted and sunshine blinks in bright little flashes on water bubbling up from a diminutive spring. Here the tracks of the fox and the tracks of birds intermingle. Both have paused at the spring to drink.

Up the steep bank and over the stone wall beyond the brook, the fox has ascended with the greatest of ease. I scramble up, slipping and floundering and leaving a wide trail of my own in the snow beside the precise line of its pawprints. At the top, endlessly winding, the tracks unroll across the rise and fall of Monument Pasture. They change direction in response to some sound or scent among the snow-covered grass clumps. Close to the stone mound of the hired man's monument, I see where the more recent tracks of a ruffed grouse have crossed those of the hunting fox.

Around one grass clump, a hundred feet away, the snow is trampled by a series of quick pounces, and here tiny drops of blood record that the fox has caught a meadow mouse. It is not long after this that the trail enters a tangled maze of wild raspberries and clumps of juniper. Here, in a puzzling interlacing of two kinds of footprints, I find tracks upon tracks crisscrossing, dodging, zigzagging with swift reversals of direction. What occurred here under cover of the darkness? The feet of the fox have made part of the prints, the feet of a cottontail rabbit the rest. Although I search carefully, I can find no bloodstains on the snow. The hunter in the night apparently went supperless in this maze near the edge of the pasture.

Guided by sight or scent or sound or perhaps by memories, the fox has continued its quest for food. On its trail, I push through a dense stand of dry goldenrod, cross the frozen lowland of Pussy Willow Corner, accompany for a time little ice-covered Wet Weather Brook, and climb up onto the wide cleared pathway of the lane. I expect the tracks will lead me to the Fox's Door, that rectangular opening in the wall beneath the mulberry tree. But they pass it by. They continue on to the little hollow where the evergreen ferns are massed. Here the wall, rising as high as my head above the depression, presents its greatest obstacle. Yet it is precisely here that the fox mounts up and over the barrier. I see where it has made one bound part way up the bank and then, in an almost vertical leap, a second bound that has carried it to the top of the snow-covered wall. Later I bring a steel tape to the spot and find the straight-up distance between lower and upper pawprints in the second leap is almost exactly four feet.

Circling around and coming back on the Mulberry Meadow side of the wall, I pick up the trail again. For long stretches of its twisting, looping progress, I notice how the fox has thrust its nose into almost every grass clump. It has made sudden stops. It has turned aside in quick detours. Once it circled around and around and then pushed part way into a snow-covered mound of juniper. Twice its leaps into the middle of grass clumps appear to have yielded it other mice.

Reading this story written in tracks in the snow, I descend the slope of the pasture to the brook—in and out among the prickly maze of a barberry tangle, under a wild apple tree, beside a wall, from stone to stone across the stream. Beyond I see the fox's trail ascending a snow-covered rock and a snow-covered log and then—with prints bunched together, recording a series of leaps— mounting a high embankment to disappear on neighbor's land. How many miles farther had its feet carried it before the dawn had come?

I turn away recalling the varying viewpoints of the past on this animal whose noctural adventures I have been following over the fields. John Burroughs referred to the fox variously as a "rogue" and a "villain"—reflecting the general outlook of his day. But Thoreau, at an even earlier time, saw the fox in the modern concept—as an individual endowment of abilities, a unit in nature's interlocking whole. He took animals as they are, not as "good" or "bad" or even as "little brothers" as St. Francis of Assisi did, but as fellow dwellers on the earth, to be observed and respected and understood.

FEBRUARY 24. Landmarks—little landmarks, great landmarks—greet us wherever we go in wandering over these wild acres. In the South Woods there is the hillside of the seven springs and one tiny glade where Indian pipes are tinted red. In the Far North Woods there is a triangle of three white oaks, the glacial boulderfield along the ridge and the lower ground where our only long beech ferns are rooted. In the West Woods there is the weather-beaten monolith of the chestnut stub and the place where the Ground Pine Crossing Trail winds through a fern jungle of silvery spleenwort near the lost spring.

Some of our landmarks are as enduring as the stone walls, some are as transitory as the windflowers. In between—neither as enduring as the walls nor as fleeting as the windflowers—is the botanical landmark we observe on this winter walk when we come to the Brook Crossing. At the foot of its foaming descent, where the cascade shatters on a final rock, a little bay or cover of smooth water spreads out behind protecting stones. It is hardly larger than a platter. Yet within it, ranged side by side, rise three green columns, tapering to spear-point tips. Each is formed of the tightly rolled leaves of the skunk cabbage. Already these plants are far advanced on the road to spring.

What more unlikely candidate for a landmark is there than a skunk cabbage? Yet year after year, in this same identical place, rising from the same roots of these perennial plants, we have seen such pointed columns ascending like a line of miniature silos or grain elevators, a trio that stands out, that catches the eye, that each year provides us with one of the wild signposts that tell us where we are.

FEBRUARY 25. In another tilt of the seesaw of weather ups and downs that characterizes this time of year, we awake to find ourselves in the midst of a sudden, chinooklike, giant-sized thaw. The wind in the night reversed itself and before dawn warmer air flows over us. By seven thirty this morning the thermometer records forty-five degrees. By two thirty this afternoon the temperature has mounted to sixty degrees.

Everywhere we look on our walks today we see the consequences of this headlong change. Snow fog, fanned by the breeze, swirls in eddies and serpentines above the drifts. It silvers the air and hangs in floating lakes of vapor over the lower land. Banks of snow slump and shrink and vanish. The sound of water dripping, water trickling accompanies us wherever we turn. Along stone walls and on the trunks of trees, patches of *parmelia* lichen are transformed from dull to vivid green by the surplus moisture in the air.

Nellie and I roam over one pasture after another. We tread across a sodden carpet of faded grass that has replaced the snow. We watch crows and starlings combing the soggy ground, picking among grass clumps, investigating the remnants of drifts, snapping up bits of food exposed by the melting of the snow. Again we look down on the winding tunnels of meadow mice, seeing them three-sided, "on the half-shell," their roofs removed.

It is when we come to the bridge where the lane crosses Hampton Brook that we encounter the most dramatic effect of this sudden release of water. The stream is tumbling over the rocks in flood stage, patches and islands of foam veering and bobbing, small ice cakes jostling together in the millrace of the current. Where the torrent reaches the opening under the great slab of rock that supports the traffic of the lane, it piles up into a churning, whirling, roaring maelstrom that gnaws into the stream banks and lifts its crest almost to the level of the roadway. Far into the night we hear this uproar of the floodwater that a single day of abnormal warmth has brought forth from the crystalline substance of the silent snow.

FEBRUARY 26. Toward midmorning today, with the air still mild and the level and roar of the high water receding rapidly, a new source of excitement appears under the apple tree near the north wall. Six strangers, visitors we have never seen at Trail Wood before, drop down on the wet ground and begin pecking among the overwintering fallen apples, now shriveled and decaying and brown. All six birds have stout bills and long tails and are about the size of robins. Other details come sharply into focus as we look through our field glasses and tick off the field marks—gray bodies, two wing bars, heads and rumps tinged with an old gold or dull yellow hue. What we are seeing are those largest of the finches, the pine grosbeaks from the mountain forests of upper New England. All are females. None exhibit any tinging of the rosy red of the adult males.

As soon as nesting is over, these grosbeaks become gregarious again. The rest of the year they travel about in flocks. Those nesting in the highest mountains descend in an altitudinal migration and spend the winter at lower elevations. Otherwise the birds are almost nonmigratory. Often, as long as food is plentiful, they hardly move from their breeding areas. Even in the bitterest winters, they remain in the north. Always it is scarcity of food, not the severity of the weather, that drives them southward. Some of the largest invasions on record have occurred during mild and open winters.

As we watch, we see the birds tear apart and toss about the brown lumps of the fallen fruit. Apple seeds form one of their favorite foods in the years when they come south. As they feed, their movements appear deliberate. And as we work closer, advancing cautiously a slow step at a time, we learn something else about them. This is their exceptional tameness.

Nellie, making no abrupt movements, walks almost in their midst. I, with my camera, move within three feet of one of the feeding birds. It looks up, alert but unalarmed. I lean even closer and it takes wing. But it merely flutters up onto one of the lower limbs almost above my head.

A soft, short whistle, a kind of "cheeee" that shows their relationship to the other finches, continually rises from among the feeding grosbeaks. This call has been described as having a little roll in it. The full song of the males, the song of their breeding time, is said to resemble that of the purple finch but to be wilder and sweeter. It is a melody made up of warbles and whistles and trills. It is sometimes loud, sometimes soft, sometimes ventriloqual.

A quarter of an hour goes by while we watch these strangers from the north. Suddenly, as though on a signal, for no special reason that we can determine, they all take off together. In the air their movements are wild and varying. Like other finches, they follow an undulating path when on the wing. But the pine grosbeaks appear to us to fly with higher waves and deeper troughs than other members of their family. So it is, rising and falling continually, that the little band passes out of sight. It is entirely possible that never again will we see such birds at Trail Wood.

FEBRUARY 27. The aisles of the woods extend away damp and misty. Except for the drip of moisture from the twigs, I hear no other sounds than those produced by my own slogging advance. These cease suddenly as I come to an abrupt halt at the edge of a clearing. Ahead of me I glimpse that sight that always lifts the heart on a mild winter day—the movement of purple-brown velvety wings with an edging of straw-colored yellow. A "thaw butterfly," an overwintering mourning cloak, the earliest butterfly of the Trail Wood year, is circling around and around the opening among the trees.

I follow it with my eyes. I watch it dip and flutter and come to rest on a mossy stump. There it spreads its wings and is still. But a moment later it is in the air again. Somewhere among the leafless trees, it may find a trunk where the bark has been bruised and the thaw sap is oozing out. At such an arboreal spring it will drink deep, obtaining its first nourishment since winter settled down.

This insect survivor of blizzard and ice storm, of gale and sub-zero nights, is a hibernator that sleeps less soundly than does the woodchuck. During the thaws of March and some years, as now, in the latter days of February, it awakens and is on the wing. Like the spathes of the skunk cabbage spearing up through the muck, like the shining catkins of the pussy willow, the emergence of the mourning cloak from the hollow log or sheltered cranny that forms its hibernaculum represents one of the mileposts on the road of the season from winter to spring.

All the way from the Arctic Circle to the Gulf of Mexico and from the Atlantic to the Pacific, the mourning cloak ranges over North America. One of the *Nymphalidae*, it is related to the painted lady and red admiral, the viceroy and the great spangled fritillary. But among all its kind it alone spends the winter hibernating as an adult.

What is the secret of its ability to survive the intense cold of the northern winters? Probably it is the same secret that enables black carpenter ants and overwintering female wasps to reach the spring unharmed. This, it has been shown, is the development within their bodies of chemicals that function like antifreeze in an automobile radiator. Research, no doubt, will reveal many other creatures that are protected in a similar way. Already it has been established that it is glycoproteins in the bodies of Antarctic fishes that enable them to survive through the winter even amid the deadly cold of the sea ice in McMurdo Sound.

FEBRUARY 28. This is the sound we longed for, dreamed about, looked forward to in the darkest days of winter. Rising and subsiding, becoming a storm of mingled voices, then ebbing away, it comes from the treetops along the brook. It is an excited sound, a festive, holiday sound. Like the torrents of spring, it is a rushing, liquid sound that here antedates the spring. It is the great chorus of the first of the homecoming flocks of the redwings.

Bare only yesterday, the treetops along the brook today are clothed with blackbirds. A hundred and fifty or more swirled down to alight among the leafless branches before dawn today. All are males. The females will arrive later. We watch the birds, in the richness of their breeding plumage, flying from tree to tree, each alighting with its scarlet epaulets exposed. Their surging energy is contagious. We feel a sense of elation, a wave of optimism. The stolid endurance of the deepest winter drops away.

We listen to the overlapping chorus of "okaleees" or "bob-y-leees" swell and fade and begin again. The interplay of sounds merges into a rolling, trilling clamor. Standing listening, we catch little dropped notes spilling through the chorus. Then the "oka-leeeing" is replaced by sharp metallic calls as all the birds take off in a cloud of black to sweep in curves, to turn and turn again, and then swirl down once more to the tops of the brookside trees.

It is usually near the end of this shortest month of the year or in the earliest days of March that the redwings come back to Trail Wood. The intermingled tumult that comes down from the treetops seems compounded of relief at the end of a long journey, of ecstasy in reaching an age-old breeding ground, of health at a peak, and life lived intensely. We are swept along by the excitement pulsing through these hundred and more smaller bodies. This wild musical clamor of the first of the returning redwings reechoes in our minds long after it is left behind. For the birds, the farflung journeys of migration now are a thing of the past. For us, the winter—all but a few short weeks—has run its course.

MARCH 1. How high the sun! How bright the sky! But how chill the wind! Under this sun and this sky and in this wind we enter the month of March, the month when another spring begins.

Behind us now recedes the dead of winter. Ahead of us swells and surges closer the season of change. We seem standing with one foot in retreating winter, the other in advancing spring. Wherever we walk today, we confront effects of the sun's higher path across the sky. Already we sense the hidden rivers of sap flowing around us in the woods. Already we notice a difference in the calling of the birds.

Standing beside the pond, where the wind has swept away the powdery snow that drifted down in the night, we see the molds of dead leaves, some descending to a depth of two inches into the ice. Through the greater amount of heat absorbed by their darker forms, they have melted their way downward. Once we observe the perfect mold of an oak leaf, another time the outline of a maple leaf. For a minute or two we are mystified by several perfectly round holes, looking as though they had been bored into the ice with a brace and bit. We peer down into them. At the bottom of each lies the dark, round pellet of a rabbit's dropping.

Now, so near to winter's end, our acres are inhabited by *experienced* rabbits just as the birds that will come streaming back in the weeks ahead—returning from other scenes, even from other continents—will be *experienced* birds. We can remember the rabbits when they were baby rabbits with wide and innocent eyes; we can recall the far-ranging migrants when they were fledglings, breaking home ties and trusting themselves to their untried wings.

It is nearing noon when we trace the last of our digressions in the snow and turn toward home. I call attention to the fact that the fine snow looks like confectioner's sugar. Then I observe how a gleaming, snow-clad mound ahead looks just like an angel food cake covered with white icing. A little later I note that the ringing "okaleee" of a redwing sounds like "broccoli" to me. Nellie decides I must be starving and we hasten up the hill. Hurrying to get lunch, she observes that as a cook she was born just in the nick of time—when frozen food came on the market.

MARCH 2. "The trouble with looking at birds," an all-around naturalist friend of mine once observed, "is that you miss so much else while you are looking at birds."

The same might be said for looking at ferns or rocks or wild flowers or deer. Robert Louis Stevenson's famous two-line poem in A Child's Garden of Verses runs: "The world is so full of a number of things, / I'm sure we should all be as happy as kings." For the field naturalist, however, this happiness is diluted from time to time by reflections on the "number of things" slipping by while he is looking elsewhere. When Nellie and I are circling the pond, we are missing what is going on along the Old Woods Road; when we explore the area of Ground Pine Crossing, we have no idea what is occurring in Firefly Meadow.

I have just been watching the pell-mell flight of a pursued bird hurling itself into a tangle of bushes. In shifts of direction—made in mere splinters of time—it avoids the many branches and selects the one branch in the maze on which it alights. Both it and I, both bird and man, lead lives eternally filled with choices, with decisions. Even in so little a thing as my use of the present moment, if I watch the bird I cannot watch the squirrel; if I concentrate on this particular individual I cannot observe the activity of the many other birds. In every choice we lose something and we gain something.

In the manner in which our minds dart in swift transitions, I find myself recalling the quandary of a country dog. I heard the story years ago, when I was visiting Henry Beston at his farm on the coast of Maine.

When this dog was still young, his wife, Elizabeth Coatesworth, recalled, the family that owned it moved away to another farm, a considerable distance from the place it had known from puppyhood. There it grew so homesick for the original farm it took to the road and found its way back to the old familiar scenes. The new owners recognized it and treated it kindly. But this was not enough. For here it missed the family as much as there it had missed its old surroundings. It pined and grew sick and in the end was returned by train to the new farm and the old family. Even in the relatively simple existence of a dog, the same law applies: If you do this, you can't do that. If you choose one thing, you cannot choose another. Loss and gain are part of every choice.

M A R C H 3 .   Where a dense stand of staghorn sumac once grew at the edge of Wild Apple Glade, only the silvered trunks and branches now remain. They lie fallen, jumbled together like discarded and weathered antlers.

As I am standing looking down at their interlacing mass rising above the snow, a low-pitched booming sound reverberates through the air. It comes from the direction of the pond. Hollow, rolling, it reaches me like the noise of a muffled concussion. We have heard it several times of late. Now I hear it again. It suggests some submerged timber being thrust violently upward against the underside of the ice.

Last night, the temperature dropped to zero. Already, on this morning of sunshine, it has risen to twenty-two degrees. I assume this sudden change, accentuating the tensions inherent in the ice, is responsible for the dull cannonading that I hear.

Beside that larger body of water, Walden Pond, Henry Thoreau listened to the same sound. "The pond began to boom about an hour after sunrise, when it felt the influence of the sun's rays slanted upon it from over the hills . . . ," he writes in the "Spring" chapter of *Walden*. Local anglers fishing through the ice believed the "thundering of the pond" scared the fish and prevented them from biting. This booming of the ice we hear but rarely at Trail Wood.

Another sound, more appropriate to the size of our pond, is the "chirping" of the ice when it is thickening late in autumn. It occurs only occasionally, under certain conditions of temperature change and certain thicknesses of the ice cover. I remember the first time I heard it one late-November day. At the time the pond was closed in with a solid lid of ice, a covering only an inch or so thick. Daybreak had come that morning with a temperature of twenty degrees. But when I descended the path to the pondside at about nine thirty, the mercury was rising rapidly.

Halfway down the slope I stopped and looked overhead. I was arrested by several bright, birdlike chirps. I saw no birds. Then, as I stood by the pond, the same small, clear sounds came from across the ice—here, there, singly, and several together. It was a long time before I could convince myself I was not hearing small birds flying overhead. Then the sound came from almost at my feet. Apparently this chirping—which continues for some time— has an origin similar to that which brings the booming sound to the pond. Both are the product of altering stresses within the layer of the ice.

MARCH 4.   A month from now, a year from now, this is the squirrel I will remember. It is one among many that are feeding in the yard. Yet it stands out from among all the others. When it advances along the wall top under the hickory trees, I see it fall on its side and recover itself. Its right leg hangs almost useless. Perhaps a ligament is torn, an injury slow to heal. Terribly handicapped in these harder days of winter, the little animal appears healthy, bright of eye, alert, quick of movement, filled with energy. But what will be its fate? Will it survive to enjoy the easier days of spring? No doubt its chances are increased by the ample cracked corn and birdseed on the wall. It may well be that its courageous battle against adversity will reach a happy ending. For a happy ending is no more false than an unhappy ending. Both happen sometimes—neither always.

MARCH 5.   Four black wings flail the snow. A clamor of raucous cawing fills the air. Near the bridge, at one side of the lane, two crows are locked in battle. One lies struggling on its back while the other hammers it over and over with its bill. A third crow alights beside the combatants, moving about like a referee at a prizefight. A fourth remains in the treetops—spectator and lookout. The two birds break away, flap to the top of a large ash tree; there the quarrel flares up anew, and again they tumble in a flutter of wings to the snow below.

Watching from a distance, Nellie and I can make no conjecture as to what the contention is about. As we have learned long ago, the world of the crow is filled with question marks. We recall an autumn day in North Dakota when we observed a flock of 300 of these birds beginning their southward movement of fall. They milled about. The flock broke up, then came together again. Some started south, some north. The cawing was incessant. In the end all the birds streamed away southward. We could easily have convinced ourselves that we had witnessed a kind of election day in an avian democracy, a time when a decision affecting the whole flock was being voted on and decided. But in dealing with crows we always hold a tight rein on our imaginations. I have read unbridled accounts of observers who have reported witnessing crows conducting town meetings and holding trials complete with judge and jury.

In fact, in every puzzling thing we see in nature, our interpretation, colored by the bias of our human viewpoint, needs to be subjected to careful scrutiny. Otherwise we may go as wildly astray as the violinist who excitedly wrote to a musical magazine about a miracle he had witnessed. On a bitter cold night in winter, he said, he had been practicing on his violin until a late hour. The next morning he found the windows of his room thickly covered with frost and among the patterns on the panes, he reported, he distinctly saw the notes of the music he had played the night before!

I once listened to a woman professor, distinguished in a field unrelated to natural history, being interviewed on the air. In the course of her comments, she recalled an amazing experience she had had during a vacation in the north woods of Maine. Evening after evening, she said, she heard the same magic song, the voice of some unseen and unknown bird coming from the darkening forest. One day, after sunset, when the song began again, she crept toward its source. When she glimpsed the singer, she recognized it as some kind of thrush, apparently a hermit thrush. It was resting on the ground as it sang. And as she watched, she said, it lifted one foot and began keeping time with the music. The woman obviously saw something, but just as obviously she had not seen what she thought she saw.

One more instance of somewhat less-than-accurate deduction from an observation. In an early book of travels through Italy that I once read, the author tells of looking down from the window of his room one day and seeing a tomcat squeeze through a small hole in a fence and invade the territory of another tomcat. The two animals, falling on each other, bit and clawed and rolled until the resident tomcat fled and the invader paced about victorious. However, so aroused was it, the author assures us, that it had to wait for several minutes before its tail became small enough for it to return through the hole in the fence.

Not only seeing what we look at—accuracy of observation—but truth in conclusion is a first obligation of a naturalist. To see clearly where others observe inattentively; to see familiar things sharply in all their details where others see only generalities or indistinct, mentally out-of-focus objects; to note correctly what is taking place; and then to interpret accurately all that is seen—this has seemed the goal, in a way the lifework, of certain writers in the field of nature, such as Gilbert White and Henry Thoreau. Thoreau went about noting *just* how the trees look when the wind ruffles their leaves, *exactly* how the hawk mounts in the air, *precisely* how the spring flowers spread their petals.

This exactness of personal observation in a world of hurried, unseeing glances is a thread that runs through the work of the best of the succeeding generations of nature writers. When John Muir started his thousand-mile walk to the Gulf in the wake of the Civil War, his botany professor at the University of Wisconsin offered him a copy of Virgil's poems to take with him. Muir declined. He said he wanted to see nature freshly and exactly and through his *own* eyes.

M A R C H 6 .   The trail I follow to the brook, the trail the foxes follow in the night, is the same meadow trail that was laid out in days when cattle fed in our pasture fields. Their heavy tread, when they moved in single file, tramped and packed down the earth. I remember as I walk in their footsteps this afternoon how from time to time these great farm animals—whose ancestors for centuries had lived lives tamed and regimented by man—still show flashes of wild wisdom inherited from those other, long, long ago ancestors from which they originally sprang.

During one historic blizzard in this region, several cows on a farm a mile from here were stranded in a distant field, unable to break their way through the snow and return to the barn. For two days they remained without food and without shelter. Each night the temperature dropped into the low teens or single numbers. When at last the owner, using a bulldozer, cut his way through the drifts, he found all the animals standing knee-deep in a brook where swifter current kept it open. The water was warmer than the surrounding air. Through some intuitive sagacity they had chosen this spot where conditions were slightly in their favor to aid them in their ordeal.

M A R C H 7 .   Black and white intermingle over the yard. A flock of more than 200 grackles and redwings, intermixed with cowbirds, comes streaming down through a flurry of snow.

We see them alighting, spreading out in loose carpets over the white floor of the snow-clad yard. All feed hungrily. The scattered seed disappears under their attack. We stand watching the dense concentrations of these birds that not so long ago were finding their food in the sunshine and warmth of the South. In time the black thins out. More and more, white predominates as the red-wings, with their chucking calls, and the grackles, with a harsh and ragged sound, begin taking to the air and making off in the direction of the North Woods.

The last bird disappears and we come in. The sight of the long-tailed, dark-plumaged grackles has recalled a mystery that persisted for weeks one year at the edge of a town a dozen miles from here. In the latter part of May, a woman telephoned to ask if I could explain a strange occurrence in her back yard. She had discovered four house sparrows, each with its severed head lying beside it. What had decapitated them? I could not give the answer nor could two professional ornithologists I consulted.

A few weeks later, I received another call. Now the total of beheaded sparrows had risen to nine. Still others had appeared in a neighbor's yard. On the nineteenth of July, that year, a third call reported the mystery had been solved. The neighbor had witnessed the act of beheading. The executioner was a grackle.

None of the dead birds had been eaten. Why was the victim always the same kind of bird? Why had the grackles—or it may have been a single grackle—attacked the sparrows? Why had the birds, or bird, kept pecking at their necks until the heads were severed? All these are riddles that remain.

M A R C H 8 . Our lane forms the great seasonal sundial of our year. Running east and west, it marks each day the position of the rising sun. All winter, we have seen it appearing well to the right of the lane, toward the south. Now it lifts above the treetops straight down the lane. So far has the sun come north.

On this day we hear the cooing of a mourning dove, the singing of a song sparrow. The bell music of the juncoes is increasing. The bills of the evening grosbeaks are becoming tinged with apple green. Junco bells, grosbeak bills, the soft voice of the mourning dove—all these are signs of the great change now imminent. The current is surging toward spring. We feel the tug within ourselves.

Once, in a Maine camp, I met a man who added a new word to my vocabulary. The word was tellurian. He said I was a tellurian and Nellie was a tellurian and he was a tellurian. The word means "a dweller on the earth." Many a person lives a life span on this globe without ever really dwelling on the earth. He may come to the end of his days without ever having appreciated or understood or loved or found affinity with this green and beautiful world even in the wonder of its springtime. And he may leave this unique planet unaware of all he has missed. Anyone, it seems to me, who loves anything in nature simply and sincerely will find a measure of joy in life. And those who are the tellurians, who dwell on the earth and feel a oneness with it throughout all their lives, who know a deep emotional attachment to it—these are bound together in a special way. Neither death nor time nor distance separate completely those who have experienced in common this close relationship with the earth on which we live.

M A R C H 9 . When we first came to Trail Wood, nearly twenty years ago, we noticed a gray thicket of low, scraggly trees down the slope to the west where the trail we have named Veery Lane has its beginning. On this March day of moderating weather, I pass through this thicket and pause to look about me. All the crooked trunks, all the twisted, interlacing branches are clothed in thin, smooth, silvery bark. Projecting in pairs flanking the leaf scars along the twigs are spines as sharp as the thorns of roses. These are the characteristics of the prickly ash, the northern toothache tree, *Xanthoxylum americanum*.

Its spines contribute the "prickly" to its common name, the shape of its leaves—resembling those of the ash trees—the "ash." The "toothache tree" is derived from the fact that the bark and berries have a biting and anesthetic effect when chewed and were long employed as a home remedy for toothache. Still another colloquial name is a tribute to the effectiveness of its spines: wait-a-bit.

In the course of the seasons each year, we have observed the same sequence take place in this thicket of sprawling trees. First of all, even before the leaves, come the tiny, inconspicuous whitish or greenish flowers. From them, in time, develop the red fruit, about one sixth of an inch long and containing one or two seeds. Becoming brown, then, dry, the little pods burst and hurl out these black, shining, oblong seeds. I examine the wood where a branch is broken. It is soft and pale brown in color.

We assumed, during our first years here, that such thickets were common in the region. The range of the tree extends from Quebec south to Virginia and west to Kansas. But its distribution is spotty and it is rather rare in Connecticut. During the intervening years, we have looked for other stands without success. This gray Trail Wood tangle—on this day still in its stark winter bareness—instead of being one of innumerable other clusters of the kind, has proved to be almost a rarity in the region.

M A R C H 1 0 .   Five crows are ranged around the feeding gray squirrel. They all are concentrated near the Wild Plum Tangle where Nellie has scattered cracked corn for the mourning doves—food that is shared by the jays, the crows, and the squirrels.

Squatting in the middle of the corn, monopolizing the banquet, is the squirrel. On all sides, keeping their distance, are the crows. They strive to get the feeding animal to move. They caw at it. They sidle toward it, taking one sidewise step and then another. They walk around it, keeping a yard or so away. I wonder why they do not make a concerted rush, why they all do not take wing and dive-bomb the animal. Either tactic, in all probability, would dislodge it and frighten it away. But extreme caution is the habit of the crow. The birds do not underestimate the swiftness of movement or the sharpness of tooth of the gray squirrel. They hold back. The animal, its eyes watchful, its jaws working at top speed, continues feeding.

For five minutes I stand spectator to the cautious maneuvering of the crows. Then the gray squirrel leaves. It leaves of its own volition. In three unhurried jumps it disappears into the plum tangle. The crows, in airborne leaps, come rushing in. For several minutes they gobble down corn. Then the squirrel comes charging back.

Anyone who has watched gray squirrels and small birds feeding in the same area has noted how the little animals delight in making sudden dashes through a flock, sending a cloud of fluttering birds above them. Now I see the heavy crows similarly flap into the air. They mill around with a raucous chorus of cawing. Then they drop down, once more stationing themselves in a circle around the feeding gray squirrel. And the little scene begins again. A crow takes few chances. Before it leaps, it looks—and looks several times. Wariness is a dominant trait in its nature. But it is as persistent as it is wary.

MARCH 11. On this day in March almost ninety years ago, there began in New England the storm of winter storms, the blizzard of blizzards. Today is the anniversary of the start of the historic blizzard of 1888 in which 400 persons lost their lives.

The dawn breaks slowly. Dark and heavy, the lid of the clouds trails ragged vapor low above us. The smell of snow is strong in the air. History appears about to repeat itself. But the morning passes; the storm threatens but fails to materialize. Our wild birds, as always in the face of a change for the worse in the weather, gorge themselves on seed and suet. Gray squirrels come pouring from the woods. They, too, represent living barometers, anticipating the coming of a storm. All morning, it seems to Nellie and me, we are carrying out food and more food for our wild poultry and our bushy-tailed livestock.

We have commenced our lunch when we notice the first lazy flakes drifting down. Steadily their numbers multiply. By the time the meal is over, the woods to the west and the trees along the brook to the east have changed from black to gray. We watch them grow more vague and wavering behind the increasing curtain of the snow. But this is a storm without wind, an almost gentle storm, with only the unceasing fall, hour after hour, of large, soft, drifting flakes, clusters of feathery ice. All sounds—the screaming of a jay, the barking of a dog—are muffled, the direction of the source in doubt, amid the gray-white veils and clouds of the descending snow.

Early in the afternoon, the gray squirrels begin drifting away. But the birds feed on. Snug in hollow trees or curled up in burrows in the ground, the larger animals can sleep out the storm and emerge in clearer weather. But the small songbirds, with their higher metabolism, their higher temperatures, their faster consumption of fuel, must continue to supply their bodies with food. As long as they are well fed, they can endure the cold, can survive our northern winters. So we see them scratching in the deepening snow, hunting every available seed. But on this darker day, before this earlier dusk, their numbers quickly diminish. One by one and in small flocks they disappear, retiring to secret hiding places well in advance of the night.

Before we go to bed, late this evening, I switch on the floodlight illuminating the terrace. Into its glare, drifting down, widely spaced, we see the last flakes of the lessened storm. Snowfall is almost over. The gentle storm, on this anniversary of the greater, more violent, and historic storm, has reached its quiet end.

MARCH 12. On this morning in the wake of the quiet storm, Nellie and I wade on a wandering course through the woods, along the brook, and across fields clad in immaculate white. We take our time. The sun shines. The day grows warmer. The calling of titmouse and redwing charges the air with the emotion of spring.

Twice on the way home we stop beside the stems of last year's goldenrods. Each stem exhibits a round, balloonlike swelling. And each time the swelling has been punctured by the chisel bill of a downy woodpecker. On some cold day in winter, the bird had excavated the round, beveled hole and had extracted the pupa from within the sphere of the goldenrod gall.

This handiwork of the little woodpecker revives in my memory an experience that I relate as we cross the final field toward home. In a French restaurant just off Times Square, in New York, an elderly lady at an adjoining table had talked learnedly to a companion about woodpeckers, then about other phases of natural history. She ended by saying: "You know, if anyone does as I have done and reads the encyclopedia a little each day, it is possible to learn a great many interesting things." Apparently, she had skimmed through the volumes at least beyond "W"—unlike the relative of a friend of mine, who as a young man started to read from end to end the multivolumed *Encyclopaedia Britannica.* He got only as far as "D" before he gave up. But for the rest of his life he was a brilliant conversationalist on any subject so long as it began with the four first letters of the alphabet.

MARCH 13. The feature of our walk in the March sunshine today is creatures doing something different. In the Wild Plum Tangle we see a cottontail rabbit carefully going over the surface of the snow. It is gleaning fragments of cracked corn scattered by the wind. Out in the open, downwind from the catalpa tree, as we come home later on, we discover jays, grosbeaks, and smaller birds picking at the snow. We turn aside to investigate. Several of the dark, slender, cigar-shaped seedpods dangling on the catalpa have split open. The birds are consuming the wafery seeds that speckle the snow. A red squirrel that has found a snug winter home where I have stacked gunny sacks under the roof of the garage provides the third of these instances of creatures engaged in unusual activity. I see where it has torn apart a mud dauber's nest and added meat to its diet by consuming the wasp pupae within.

On a summer day, I remember stopping on the path coming up the slope from Veery Lane to watch a bluejay in the midst of a performance that was new to me. I saw the bird alight heavily on a raspberry cane, then drop to the ground. This sequence it repeated half a dozen times. Mystified by what was happening, I worked closer and saw that each time the jay struck a cane with force, Japanese beetles, feeding on the leaves, let go their hold and dropped to the ground. The bird followed them and snapped them up in quick succession.

One last instance of the unusual in the actions of wild creatures. In a letter that arrived a few days ago, a woman on Long Island related what she thinks must be a million-to-one chance in watching a white-footed mouse wash its face. At the end of cleaning up, the little animal washed its face much as a cat does, but in this instance, instead of using its forepaw, it licked a hind foot and vigorously scrubbed its face with it. "I wonder," my correspondent concluded, "if anyone else ever saw a mouse wash its face with its hind foot?"

MARCH 14. Hour after hour today a chill rain has pelted down, eating away the snow. Near sunset, the clouds peel back from the western horizon and in their wake expands a widening band of clear blue sky. At about the same time I drive up the lane.

For most of this dark, depressing day I have been packed into a crowded room at the state Capitol. With others I have been testifying at a legislative hearing on a conservation measure. It is something worth doing, something I wanted to do, something that has resulted in success. But walking here in the sunset after the rain, I feel myself expanding. I am returned to my proper habitat. Worlds away seem the cramped quarters of that narrow room as I strike out across our open land; worlds away the dead rebreathed air as I take in deep lungful after lungful of the moist, fresh air of the country. All this, that means so much to me, can be affected by what happens in the rooms where legislation is decided. There even a soft voice among loud and dominating ones, when it enunciates facts that are solidly based, will carry weight.

But even in the midst of a successful struggle of this kind there is always a sense of impermanence. Here are victories but no victory—no stable, enduring decision. Each triumph is a transitory point in space and time. Other antagonists will arise. Lewis Mumford was once asked if he were still hopeful about the profound change in attitude necessary if man is to save himself. He replied: "I'm always hopeful because there are many possibilities still open to us. But I'm a pessimist about probability; because there are many forces which will resist any sufficient change in our habits to the bitter end."

MARCH 15. Black and white, with short legs and small and dainty feet, a symbol of change has emerged from the dusk into the bright illumination of our terrace floodlight. A skunk, coming from the woods, has picked its way across the darkened fields now almost bare of snow. Its increased activity as March progresses heralds the retreat of winter and the advance of spring.

We watch it nosing about under the apple tree hunting for fragments of food left from the scraps we have thrown out for the fox. I decide to add something new to its menu. I toss a doughnut in a high arc toward it. It strikes a limb, ricochets to the trunk of the tree, bounces off, and rolls in a wandering course down the slope and finally comes to rest against the nose of the skunk. In a thousand tries, I could not repeat that performance. As I watch it consuming this new food, it occurs to me that I may be watching the only skunk in the world eating a doughnut at this particular moment in time.

At a farm that lies on the other side of the village, there is a skunk that comes every evening to the kitchen door and lifts up its nose to be fed a spoonful of catfood. If I were a skunk, I have decided, I would want to live in Hampton.

MARCH 16. We change and as we change our viewpoint alters. Nothing is ever the same again—not even the past. We see all in a different light.

I have been remembering, as I have plodded along trails slushy with melted snow under this somber sky, and what I have remembered has had a somber cast. I remember the bluejay dying its lingering death from tongue worm, a time of dying that may drag month after month over a period of years. I recall the starling, the flesh torn from its back and neck by the sparrow hawk, pushing itself into the weeds to await its end. I see again the wasted body of the walking grackle as it spent its final hours in ceaseless activity. And the question of a lifetime returns with renewed force: Why should there be so much suffering in the world?

On this March day, I come to this old question from a different viewpoint, the viewpoint of nature. So far as we can see, nature's goal for each individual and each species is for it to survive as long as possible. The bird or animal fights on until it is overwhelmed and destroyed. No matter how severe its injury, there is the chance it may escape and recover and perhaps aid in continuing the species. If it expired as soon as pain and suffering began—if it never had to endure suffering that continued on—this aid to species continuity would be lost.

The keenness of the nervous systems of living creatures, which saves their lives so often, is not something that can be turned off when sensations become unpleasant. It is unpleasant sensations that warn them of danger. Conscious sensitivity forms the foundation on which life exists. If creatures were mere machines incapable of feeling, if—as an earlier school of scientists believed—the dog that howls in pain really is feeling no pain but is merely giving a mechanical reaction to a stimulus, this intense personal awareness that brings a conscious feeling of both pleasure and pain in life would be lost. Viewing in this light the old dilemma of why so much suffering is in the world, we find a measure of understanding. The suffering of the individual, which so often stirs our pity, has its practical value, plays its essential part in the functioning of nature.

MARCH 17. Nine fox sparrows, toward sunset, drop down on the carriage stone to feed among the juncoes. The longer days, the increasing hours of sunshine have oriented them toward the north. These largest of our sparrows—often mistaken for thrushes—commence their spring migration as early as late February. In this seasonal advance up the map it has been noted that they tend to arrive in northern states in small groups numbering from eight to twelve birds.

Fox sparrows are late feeders. With the fading of the light, we watch their striped bodies and fox-red plumages transform into indistinct shapes and indistinguishable hues. In the silence, amid this swift descent of the night, we catch, at intervals, the fragments of a song—sweet, flowing, full toned—a song we hear but rarely. Some feature of the fragmentary melody, some clear flute note, some quality in the fox sparrow's voice reminds us of the song of that most beautiful avian vocalist among the thrushes of Britain, the English blackbird.

Portions of the fox sparrow's song have been likened to "the soft tinkling of silver bells." In 1883, the American ornithologist William Brewster wrote of this northern sparrow: "What the mockingbird is to the South, the meadowlark to the plains of the

West, the robin and song sparrow to Massachusetts, and the white-throated sparrow to northern New England, the fox sparrow is to the bleak regions bordering the Gulf of St. Lawrence. At all hours of the day, in every kind of weather late into the brief summer, its voice rises among the evergreen woods filling the air with quivering, delicious melody, which at length dies softly, mingling with the soughing of the wind in the spruces, or drowned by the muffled roar of the surf beating against the neighboring cliffs."

Nellie and I have never heard this bird singing through the days on its northern nesting ground. We envy those who have.

MARCH 18. Yesterday the fox sparrows. Today the journeying geese. Tomorrow and tomorrow, in the weeks when spring arrives and spring advances, we will see the vast movement of the homecoming birds—with ripples mounting into waves—flowing toward us and sweeping around us.

Like an entering wedge, this early V of northward-cleaving waterfowl moves in the vanguard, reversing the direction of the "geese-going days" of October. We listen to their calling grow fainter. We strain to see their diminishing forms as long as possible. They pass on, a moving sign in the sky, certain assurance of the nearness of spring. A hundred geese were in the flock. How rapidly the news spreads this morning through the village! A little later we hear a killdeer in the sky.

Hearing and seeing these advance migrants revives memories of the great flocks of varied species we have seen assembling in the south, tens of thousands of birds, excited, active, stimulated by the approach of their long journey to their breeding grounds in the north. I remember a day when, as far as Nellie and I could see across the saw grass of the Everglades, the air was filled with clouds of swirling tree swallows, shooting down, twisting, climbing, their white breasts catching the sun, birds innumerable, swallows beyond counting. We wondered where they all would end their northward travels. They seemed like the milling throng in some Grand Central Station, each traveler with a ticket for a different destination.

How the migrants cross the sky, by day or night, how they guide themselves along the way to their separate destinations is a riddle that has been attacked from many sides. One of the latest approaches has been through the field of "infrasound"—sound waves too low for human ears to hear. In tests in an insulated chamber, pigeons showed they could detect sound frequencies lower than one cycle per second. In contrast, the lower limit of human hearing is about twenty cycles per second. Low-frequency sounds made by thunderstorms, ocean waves, winds, and other phenomena—all inaudible to humans—can be heard distinctly by birds. Moreover such sounds, unlike those we can detect, may travel thousands of miles. Perhaps, it is suggested, the way birds seem to anticipate changes in the weather may be explained by their ability to hear storms coming. To the research workers who have been experimenting in the field, this keenness of hearing at the lower end of the sound frequencies appears to provide a new and promising approach in the long quest for the answer to the still baffling mystery of how migrating birds orient themselves in the sky.

MARCH 19. As the hours pass on this next to the last day of our walk through the year, Nellie and I range widely. We follow the old trails, visit old friends among the trees, note the effects of the winter and its storms. The day is filled with sunshine; the mercury is rising.

We find the redwings scattered over the lowland woods. We listen to the scream of the red-shouldered hawk above the trees. We see an emerging woodchuck nibbling on tender leaves of the new green grass. We catch the richly nostalgic scent of the warming soil, the reviving earth. We stand beside the widening band of open water along the edges of the pond, noticing how transparent it is after the months of calm beneath the ice.

For the fun of it, we visit all our different kinds of birches— the lone white birch on the slope of the seven springs in the South Woods, the leaning gray birches at the top of Juniper Hill, the yellow birch that seems perched on stilts where in the swampy woods near Hyla Rill a stump has rotted away beneath it, the large black birch on the slope below the North Boulderfield beyond Old Cabin Hill.

We inspect the three skunk cabbages lined up in the little pool among the rocks below the Brook Crossing cascade and turn aside down the slope where the lady's slippers bloom in May. Among the smooth-barked trunks of the water beeches, or horn-beam trees, we descend to the beaver pond to check on how the dam survived the winter. And when, coming back from the Far North Woods, we stop by the towering chestnut stub, I run my hand over its smooth exterior. It seems to have taken on additional polish through storms of another winter.

Polished, too, appear the few remaining dry shells of the milk-weed pods, all their seeds now long scattered by the wind. Beside Fern Brook Trail, where in bright October weather we spent so many happy hours breaking up dry branches, gathering fireplace kindling, winter buds are swelling. Everywhere along the trails the varied mosses are their brightest green. Wild Apple Glade, Whippoorwill Spring, Lichen Ridge, where we saw the burrows of the *Andrena* bees in April, Ground Pine Crossing, where I was accompanied by drifting yellow clouds of lycopodium powder in November—we see them all.

Wherever we go, all our walks today are memory walks. When we descend the slope of Firefly Meadow, I come to the spot where I stood in the snow photographing the stars at mid-night and to the area we traversed in wandering in the midst of the fireflies in June. When we look back, we see the hickory trees towering above us, the trees where the katydids wrangled end-lessly during warm and moonlit nights. On other paths, we come again to the place where the flicker and I took sunbaths, where Nellie stepped aside to watch the skunk parade go by, where along the Shagbark Hickory Trail we went nutting in the fall.

On a slow circle of the pond, we walk with other memories. It was here that the chipmunk set out to swim across the water; here that I found the tiny warbler nest built among feathery ferns; here that Nellie waded into the pond trying to rescue the floating grackle; and here, near the rustic bridge, that she encountered a black and white warbler in the midst of its broken-wing perfor-mance, and, another time, two woodcock bathing in the shallow flow of Stepping Stone Brook.

After we have ascended Juniper Hill and stopped among the sprawling masses of the clumps to examine a haircap moss that so resembles tiny junipers that the German botanist Carl Ludwig Wil-lenow long ago gave it the scientific name of *juniperinum,* we sit on a fallen log in the woods beyond. Here we examine a cluster of Indian pipes, now dead and dry. The seed vessels are upthrust, each suggesting a small sculptured urn balanced at the top of its stem. I break open one of these dry containers. Snuffbrown dust, the fine powder of the seeds, streams into my palm. What will be the fate of such particles, seemingly without number? How far will they be carried by the wind and water in the storms of spring?

So we walk through this March day, as we have walked through days in the other months of the year, wondering about what we encounter. How many times have we beguiled ourselves along these trails by speculating about such things as what it would be like to be a cicada maturing slowly underground or a spiderling ballooning through the sky on a thread of gossamer or a flying squirrel gliding from tree to tree in the twilight. How alien to us have seemed all those people who are interested in no other species except their *own* species. As for us, we end as we began. For us it is always life—all life—large and tiny life, dull and brilliant-hued life, life as rooted as the lichen gardens, as intensely active as the hummingbird, that holds our interest and augments our enjoyment of the out-of-doors through all our successive years on this New England farm.

MARCH 20. Daybreak and sunrise. Starlight and night. Between the two we live the final day and make the final walks while the earth is rushing through space on the final miles that will complete its ring around the sun.

When evening comes—Spring's Eve—we are waiting in the dusk to watch once more the spectacle that has accompanied the end of recent days. We hear the whistle of wings; we catch sight of a chunky form with long, downtilted bill lifting from the darkened meadows; we follow it up and up as in wide, speeding circles it climbs into the still glowing sky. We stand with heads thrown back watching it diminish in size. We strain our ears to catch the beginning notes of that sweet, twittering warble—the flight song of a woodcock in the days that mark the transition of the seasons into spring. All during the wild, tumbling ecstasy of its plunge to earth, while it whirls like a gust-blown leaf, tilting and veering with such sudden changes we have difficulty keeping it in our glasses, the air is filled with the beauty of its song. A time of silence follows. Then the buzzing, nasal call of its parading on the ground in preparation for another flight begins. We watch the strange little crepuscular singer as long as the light lasts. Its tumbling, liquid notes are symbolic of the flaming up of the fires of spring.

Waiting on, we see the stars peep out. In the sky to the southwest, Orion, the constellation of winter nights, glitters with blue-white Rigel and reddish Betelgeuse. To us, each brilliant star seems shedding its light only in our direction. It is hard to realize that its beams shoot out in all directions from the ball of its incandescence, that the starlight we see is such a minuscule proportion of the whole. Similarly in our wanderings through this year on this land set among the hills, all the things—great and small, old and new, animate and inanimate—we have seen represent a tiny fraction of what has passed unobserved.

The singer leaves the sky. The stars of Orion wheel toward the horizon. The night moves on. And as the darkened hours pass the span of our walk through the year draws closer and closer toward its end.

AFTER
THE
END

BOUNDED by the years that have gone before and the years that will come after, these twelve months on an old farm have run their course.

Nearly two decades have passed since we first came to Trail Wood. Aside from two long trips away from home—one getting material for *Wandering Through Winter* and the other for *Springtime in Britain*—we have followed our trails through all the seasons with enjoyment unabated. We have found what Emerson found on the outskirts of Concord. "When I bought my farm," he observes in his journal, "I did not know what a bargain I had in the bluebirds, bobolinks and thrushes; as little did I know what sublime mornings and sunsets I was buying."

As I have said, this sequence of our Trail Wood days was set down at first primarily to please myself—to preserve from the sliding flow of time such things as I would find pleasure in recalling, to catch something of the joie de vivre of these outdoor days as we would wish we could know them always. Perhaps my aim has already been set down by Joseph Wood Krutch: "To achieve for myself and perhaps pass on, delight and joy."

John Constable, the English landscape painter, once described his work as giving "to one brief moment, caught from fleeting time, a lasting . . . existence." My aim as a writer has been mainly much the same. If you will feel, when you close this book, that you have lived with us through a year of sunshine and rain, spring flowers and autumn leaves, snowstorm and gale, sunrises and sunsets—if they all come alive for you, with some of the thoughts they have evoked—that end will be attained.

As I come to these final sentences, I sit here wondering if the time will ever come when such a book as this will seem like a letter from another world. Will the richness of the natural world be overrun and more and more replaced with a plastic artificial, substitute? The threat is real. And the outcome seems to depend on the wisdom and courage and endurance of those who are on the side of life—the original, natural life, the life of the fragile, yet strong, out-of-doors.

Looking back, Nellie and I will always remember that final evening—the fading sunset, the woodcock's song, the emerging stars—as we have had it so short a time ago. We will always experience it again as though in the present tense. Watching it recede into the past, we view it with mixed emotions. The walks of our year are over. But merely a dawn away they will begin again. For sometime during the hours of darkness the unending sequence of the seasons will bring another spring. Always the other spring.

INDEX

Adirondacks blackfly, 32–33
Albinos, 277–278
Alcott, Louisa May, 197
Alder, 142, 172, 222
Alderflies, 96, 118–119
Aleppo gall (or gallnut), 78
Aleutian Islands, 292
Alfoxden Journal, The (Wordsworth), 243
Algae, 30, 145, 150
American Museum of Natural History, 161, 175, 191, 264
American Ornithologists' Union, 86, 252, 274
American Wildlife and Plants (Martin, Zim, and Nelson), 166, 357
Amiel, Henri, 240
Andersen, Hans Christian, 225
Andrenid bees, 41–42, 71, 80, 396
Anemones, 48–49
Antaeus, fable of, 177
Ants, 63, 64, 73, 90, 92, 139, 215, 228, 236, 342
Apple trees and apples, 64–65, 66, 67, 76, 90–91, 92, 133, 134, 140, 151, 200, 258, 302, 323, 328, 358, 362
Aquatic larva, 32–33
Assassin bugs, 299–300
Asters, 78, 222, 225

Aucassin and Nicolette, 13–14
Auk, The (journal), 274
Auks, 363–364
Azalea Shore, 14–15, 30, 43, 52, 55, 64, 72, 87, 105, 112, 126, 152, 157, 173, 203, 239, 261, 328
Azure butterfly, 21–22, 73

Bailey, Liberty Hyde, 221
Baker Farm, 134
Ball's Hill, 136
Baltimore oriole, 159, 166
Baneberry, 188, 204
Bannertail squirrel, 293
Barberry bush, 251, 253
Bashō (poet), 191
Bass (fish), 157–158
Bats, 154, 290–292
Bayberry, 320, 334
Bear's-head, 192
Beaufort, Rear Admiral Sir Francis, 241
Beaufort scale, 241–242
Beaver, 105, 106, 308
Bedlam Road, 12
Bee weed, 225
Beebe, William, 201
Bees, 41–42, 57, 65, 71, 125, 126, 132, 225, 339, 396
Beetles, 99, 102–103, 153–154, 178–179, 210, 235–236, 341
Bent, Arthur Cleveland, 357

Beston, Henry, 380
Betulinus fungus, 29
Beyond the Aspen Grove (Zwinger), 169, 241
Big Grapevine Trail, 28, 29, 30, 68, 143, 221, 235
Birch trees, 29, 30, 337, 342
Black carpenter ant, 64
Black warbler, 53–55
Black-and-white warbler, 199
Blackberry canes, 362
Blackbirds, 49, 75, 136, 185, 269–270, 272, 343–344, 378–379
Blackcaps, 143
Black-eyed Susans, 126, 138, 188
Blackflies, 32–33, 34, 38
Blacksnake, 89, 158
Black-throated green warbler, 199
Blue flag iris, 93, 188
Bluebirds, 57, 107, 128–129, 166, 218–219, 314, 348–349
Bluegill sunfish, 145
Bluejays, 61, 95, 138, 216, 238–239, 270, 271, 294–295, 302, 306, 313, 316, 319, 326, 333, 338, 340, 342, 354–355, 358, 359, 366–367, 369
Bluets, 57, 62

Blue-winged warblers, 53–55

Bobolinks, 106–107, 184–186

Bobwhite quail, 58, 179, 360

Book of Common Prayer, The, 314

Bootlegger's Notch, 308

Boott's fern, 197, 198

Boswell, James, 180

Boulderfield, 67, 153

Bourbon rose, 115

Brauer (scientist), 146, 147

Brewster, William, 393–394

British soldiers, 155, 361

Broad Beech Crossing, 169, 230, 239, 280

Broadwings, 58–59, 60–61

Brook Crossing, 65, 162, 189–190, 249, 262, 276, 350, 374, 395

Brook trout, 39

Brooks Range, 327

Broom moss, 60

Brower, Lincoln P., 191

Brown thrashers, 107, 124, 166

Buck, Dr. John, 141–142

Buller, Dr. A. H. R., 207

Bullfrogs, 66, 87, 98–99, 111–112, 135, 141, 168, 201, 202

Bumblebees, 42, 57, 62, 125, 126, 132, 225

Burroughs, John, 61, 286, 373

Burs, 208–209

Buttercup, 48, 188

Butterflies, 21–22, 25, 68, 73, 76, 93, 101–102, 114, 136, 140, 157, 172, 188, 190–191, 196, 292

Caddis fly, 173

Caged animals, 100–101

California Institute of Technology, 256

Calopogon, 79

Canada geese, 136, 230–231

Canterbury Tales, The (Chaucer), 47

Cardinal flower, 172, 196, 221

Cardinals, 57, 166, 293, 294, 302, 313, 342

Carolina locust, 180, 211

Carolina wren, 370–371

Carpenter ants, 30, 33

Carson, Rachel, 241

Cartier, Jacques, 363

Catbirds, 106, 141, 142–143, 199–200, 341

Caterpillar, 95, 170, 299–300

Catnip, 138, 298

Cats, 131–132, 142, 152, 290–291

Cattail Corner, 75, 98

Cattail-pollen pancakes, 98–99

Cattails, 98, 136, 208

Cedar Swamp Road, 167–168

Cedar trees, 317, 332

Cedar waxwings, 166, 340

Chaucer, Geoffrey, 47

Check-List of North American Birds (American Ornithologists' Union), 252

Cherry tree, 46, 76, 166–167, 237

Chestnut-sided warblers, 53–55, 199

Chickadees, 12, 65–66, 94, 159, 306–307, 323, 326, 328, 242

Child's Garden of Verses, A, 380

Chimney swift, 53, 139

China (ancient), 78

Chipmunks, 59, 69, 70, 82–83, 85, 94–95, 166, 174–175, 218–219, 225, 233, 238, 248, 275

Christmas fern, 49, 196, 317

Christmas green, 266–267

Chrysolina beetles, 178–179

Churchill, Winston, 314

Cicadas, 138, 151, 162, 201, 202

Cinquefoil, 155, 298

Cladonia lichen, 71, 361

Cleaves, Howard H., 45

Clethra, 172

Climbing fern, 196

Clinton's fern, 196

Cloak butterfly, 25

Club moss, 317

Coatesworth, Elizabeth, 380

Cock grouse, 44–45

Coleridge, Samuel Taylor, 243

Coles, William, 264

Colorado River, 327

Colostoma cinnabarinum (fungi), 175–176

Concord River, 136, 219, 266

Constable, John, 401–402

Cooper's hawk, 319, 344

Cottam, Clarence, 293

Cottontail rabbits, 20, 61, 73–74, 80–81, 89–90, 104, 123–124, 137–138, 175, 320, 334, 355–356, 360, 371

Country of the Pointed Firs, The (Jewett), 229

Cowbirds, 35, 95, 124–125, 326, 385

Coyotes, 100, 308

Crab spider, 157

Craven, Margaret, 107

Creeping Jennie, 266–267

Crickets, 94, 146, 155–156, 173–174, 193–194, 201, 202, 218, 245, 251

Crocker Lake, 246, 255, 307

Crowfoot, 266–267

Crows, 34, 36, 166, 223–224, 287–288, 313, 315, 340, 345, 360–361, 364, 382–383, 387–388

Cruickshank, Allan D., 291

Cuckoo wasps, 164–165

Curculio beetles, 236

Cynipid wasp, 83–84

Daisies, 78, 112

Damask rose, 114

Dandelion, 208, 281

Davis, William T., 97

Day lilies, 156

Days Afield on Staten Island (Davis), 97

DDT, 186, 370

Death (or destroying) angel, 192

Deer, 15, 29, 50, 121–122, 162, 271, 308, 316, 320, 343, 349

Description of New England, A (Smith), 149

Devil's-pitchfork, 208

Dogs, 31, 151, 167–168, 338

Donati (naturalist), 290

Downy woodpecker, 72

Dragonflies, 23, 81, 102, 126, 143, 186–188, 203, 257–258

Driftwood Cove, 61–62, 64, 87, 101

Ducks, 15, 34, 103, 135, 136, 274–275

Dufresne, Frank, 292

Durand, Charles, 296

Dutch elm disease, 324

Dytiscus diving beetle, 99

Earthworms, 35, 44, 120, 215

Elm gall, 77
Elm tree, 324
Emerson, Ralph Waldo, 4, 253–254, 401
Estabrooks, Evelyn, 167

Fabre, J. H., 81, 122, 188
Fairhaven Bay, 134, 146
False mantis (or false rear-horse), 146–147,
Far Away and Long Ago (Hudson), 54, 184
Far North Woods, 65, 73, 90, 106, 131, 143, 144, 155, 162, 169, 192, 221, 276, 350, 374
Fellowship of the Ring, The (Tolkien), 54
Fern Brook, 26, 31, 58, 176, 281, 352
Fern Brook Trail, 11–12, 208, 306, 350, 396
Ferns, 101, 117, 121, 162, 197, 203, 249, 250, 317
Ferns of the Northeastern United States (Wiley), 196
Festoon pine, 266–267
Field Guide to the Birds, A (Peterson), 173
Field sparrow, 52, 96–98, 150, 168
Finches, 45–46, 58, 68–69, 124, 306–307, 376–377
Fireflies, 74, 104, 111, 112–113, 210
Firefly Meadow, 11, 19, 24, 26, 34, 44, 52, 63, 68, 79, 111, 127, 165, 190, 205, 209, 218, 233, 279, 287, 343, 345, 354, 396
FitzGerald, Edward, 226
Flaw-flower, 48
Fleabane, 78, 188
Flickers, 57, 63, 166, 168, 215, 333, 342
Floral spikes, 144–145, 159
Flycatchers, 124, 126–127
Fox, Charlie, 167, 264–265
Fox sparrows, 393–394
Foxes, 24–25, 103, 132, 133–134, 147–148, 152, 163, 166, 204, 223–224, 293, 315, 334, 371–372, 373
Fragrant liverwort, 115–117
Francis of Assisi, St., 374
Frazer, Sir James G., 339
Frisch, Dr. Karl von, 202
Froghopper foam, 78

Frogs, 15–17, 20–21, 43, 67, 87, 98–99, 104–105, 111–112, 113, 135, 141, 157, 158, 168, 201, 202, 224–225
Fuller, Eunice, 307
Fungus, 29, 30, 117
Furnessville peony, 76

Galls, 76–77, 84
Geese, 18, 195–196, 230–231, 394–395
General Jacqueminot rose, 115
George, Henry, 181
Gillard, Ellen, 120
Glazier, Susan G., 191
Glowworms, 209–210
Gold Bug, The (Poe), 225
Golden Bough, The (Frazer), 339
Golden garden spider, 296
Goldenrod, 126, 146, 169, 196, 202, 222, 295, 296, 389
Goldfinches, 298, 307, 313–314, 323, 342, 364
Goshawk, 333–334, 335–336, 355, 358–359
Grackles, 75–76, 118, 165–166, 269–270, 385
Grand Lake, 346
Grandma Way's rose, 115
Grasshoppers, 146, 211–212
Great Meadows, 136–137
Great North Woods, 307
Greece (ancient), 78
Green heron, 168, 173
Griffin Road, 73, 350
Grosbeaks, 46, 68, 141, 166, 258, 316, 326, 342, 375–377, 386
Ground pine, 266–267, 317
Ground Pine Crossing, 12, 72, 83, 114, 130, 281, 305, 380, 396
Ground Pine Crossing Trail, 155, 176, 263, 266–267, 320, 374
Groundnut vine, 231–232
Grouse, 44–45, 63, 158, 204, 216, 249, 320, 331, 333
Grout, A. J., 116

Hampton Bird Club, 297
Hampton Brook, 3, 4, 13, 31, 33, 35, 39, 40, 53, 56, 58, 62, 63, 65, 79, 80, 90, 94, 104, 106, 118, 131, 162, 171, 172, 179, 198, 204, 221, 263,

275, 276, 321, 325, 334, 350, 351, 370, 375
Hampton Reservoir, 269–270
Hampton Ridge, 151
Hardhack, 345
Harding, Walter, 134, 135, 136, 172
Hardy, Thomas, 142, 202
Harison's yellow rose, 114
Harrison, Vernon, 210
Hawks, 27, 58, 59–61, 84, 100, 131, 158, 180–181, 196, 270–271, 319, 328, 333–334, 335–336, 344, 351, 366–367
Hawthorne, Nathaniel, 219
Heath aster, 222
Hellebore, 50–51, 70–71, 90
Hellebore Crossover, 50
Helleborine, 79
Hemlock Glen, 276
Hermit Lake, 255
Hermit thrush, 37, 256
Heron, 173
Hesiod, 259
Hickory nuts, 239
Hickory trees, 220, 234, 237, 295, 304, 314, 323, 334, 343, 359
Hilltop Farm, 228
Hogbed, 266–267
Holway, Charles, 345
Homer, 336
Honeybees, 42, 65
Hudson, W. H., 54, 184
Hummingbird, 171
Hyla Pond, 15, 22, 49, 161
Hyla Rill, 49, 395
Hylas, 224

I Heard the Owl Call My Name (Craven), 107
Ichneumon fly, 189
Indian potato, 231–232
Indians, 217, 259, 262, 283, 285, 303
Inkberries, 215–216
Insect Garden, 64–65, 67, 76, 80–82, 87, 96, 102, 115, 140, 165, 171, 304, 328, 372
International Meteorological Committee, 241
Iris, 93

Jack-in-the-pulpit, 95–96, 169, 188, 204
Jacques Cartier rose, 115
Jaeger, B., 172
James L. Goodwin State Forest, 106, 349

Jefferies, Richard, 53, 125
Jewett, Sarah Orne, 229
John the Divine, St., 178
Johnson, Samuel, 180
Journal Intime (Amiel), 240
"July Grass, The," 125
Juncoes, 221, 222, 277,
    278, 286, 298, 299, 302,
    303, 319, 342
Juniper, 53, 293, 317, 334,
    337, 345
Juniper Hill, 71, 117, 118–
    120, 142, 151, 152–153,
    165, 210, 222, 239, 259,
    293, 320, 332, 342, 351,
    361, 395, 396

Katydids, 181–182, 193,
    200–201, 205, 209, 212,
    244–245, 250–251
Kazanlik rose, 114
Keats, John, 4, 313
Kenyon Road, 17, 26, 95,
    107, 307, 344, 345, 358,
    367
Kestrel, 347–348
Khayyám, Omar, 301
Killdeer, 25, 86
Kingbirds, 72, 75, 101,
    151, 205, 223
Kingfisher, 40, 137, 173
Kirtland's warbler, 86, 124
Klamath weed, 178–179
Krutch, Joseph Wood, 401

Ladies tresses, 79, 144, 188
Ladybird beetle, 153–154
Ladybugs, 153–154
Lady's slipper, 79, 80, 396
Larvae, 32–33
Lawrence, Charles W., 256
Lawrence, Helen Mary, 256
Leafhoppers, 173
Leakey, Louis S. B., 137
Lee's Bridge, 135
Leopold, Aldo, 73, 177,
    219
Lichen Ridge, 41, 71, 72,
    222, 261, 396
Lichens, 23, 41, 84, 117,
    140, 361
Life of Birds, The (Welty),
    325–326
Life of the Fly, The
    (Fabre), 81, 122
Life of North American In-
    sects, The (Jaeger), 172
Lilac bushes, 117, 120
Linnaeus, Carl, 21, 48, 159,
    178, 290
Little River, 39, 151, 180,
    276, 303, 332
Liverwort, 115–117

Lives of the Game Animals
    (Seton), 262
Lohengrin, 162–163
Lone Oak Farm, 76
Long-legged water striders,
    34–35
Lost Spring, 3, 281
Lost Spring Swamp, 176
Lost Woods, The (Teale),
    153
Luna moth, 169–170
Lutz, Frank E., 42

Mallards, 15, 34, 103, 136
Mantis, 146, 236
Maple trees, 48, 220, 237–
    238, 324, 332
Marcus, Margaret, 234
Massachusetts Audubon So-
    ciety, 270
Mathews, Helen, 324–325
Mayfly, 82, 96
Meadow mouse, 150, 357–
    358
Meadow wink, 185
Merritt Lake, 309
Milkweed, 132, 208, 252–
    253
Minnows, 30, 123, 157–
    158
Moccasin flower, 79
Mockingbirds, 166, 293,
    302
Moles, 35, 279–280
Monarch butterfly, 136,
    190–191
Monument Pasture, 58,
    143, 180, 196, 205, 321,
    360, 372
Moose River, 307
Mooseback Mountains, 246,
    308
Moss, 60, 71, 83, 117, 317
Moss rose, 114, 115
Mosses with a Hand Lens
    (Grout), 116
Moths, 63–64, 132–133,
    169–170, 186
Mount Washington, 255
Mountain mint, 113–114
Mourning doves, 168, 302,
    323, 333, 338, 355, 359,
    386
Muir, John, 52, 76, 219,
    253–254, 384
Mulberry Meadow, 169,
    205, 221, 357, 373
Mullein, 159, 160, 342
Mumford, Lewis, 391
Mushroom fritter, 206
Mushrooms, 151, 191–193,
    204

Musk rose, 114
Muskrats, 136, 261–262,
    317

Nanabojou (god), 262
Narrow Road to the Deep
    North, The (Bashō), 191
Natchaug River, 36
Natchaug State Forest, 276
National Park Service, 282
Naturalist Buys an Old
    Farm, A (Teale), 4
Near Horizons (magazine),
    67
New York Botanical
    Garden, 175
Newfoundland robins, 251–
    252, 278
Nickajack Cave, 163–165
Nighthawk, 168
Nighthawk Hill, 11, 107,
    253, 345–346, 352
Nightingales, 351
Nipmuck Trail, 303
North Boulderfield, 395
North Woods, 12, 18, 21,
    31, 33, 50, 51–52, 295,
    306, 335, 339, 341, 354,
    365
Nuthatches, 61, 94, 217,
    323, 342, 350

Oak trees, 84, 324, 332
Old Cabin Hill, 12, 15, 22,
    36, 46, 67, 79, 85, 153,
    175, 176, 192, 216, 281,
    326, 395
Old Colonial Road, 12, 15,
    21, 62, 73, 79, 161, 176,
    202, 281
Old Woods Road, 3, 21,
    22, 79–80, 161, 176,
    245, 294, 350, 380
Opossums, 166, 280, 362
Orchids, 79
Orioles, 46, 52
Our Planet, Its Past and
    Future (Denton), 23
Ovenbirds, 85–87
Owls, 36, 42, 152

Palmyra mourning dove,
    275
Parasol mushroom, 192
Parks, Edmund and
    Dorothy, 301
Parmelia lichen, 84, 375
Pearl crescent butterfly,
    172
Pectinatella organisms,
    182–184
Peddler beetle, 225–226

Pee-a-wee bird, 51–52
Peters, George, 17
Peterson, Roger Tory, 173, 217
Peterson Field Guide Series, 173
Pigeon berries, 215–216
Pigeons, 168
Pill bug, 59
Pill clams, 150
Pine Acres Tree Farm, 245
Pine siskins, 365–366
Pinecone mushroom, 192
Pixie cups, 361
Poe, Edgar Allan, 225
Poison ivy, 238
Poison mushroom, 91
Poison sumac, 238
Pokeberry, 215–216
Polistes wasp, 19, 20, 146, 243, 271, 341
Portland rose, 115
Powell, Major John Wesley, 327
Prairie warblers, 53–55, 199
Prickly ash, 386–387
Puffballs, 206–207
Purple finches, 45–46, 58, 68–69, 306
Purple fringed orchis, 79
Purslane (or pussley), 148–149
Pussy Willow Corner, 95, 169, 360, 367, 373
Pussy willows, 25, 35, 367–368, 377

Quail, 35, 179, 322, 323, 333, 360
Queen Anne's lace, 138, 188

Rabbits, 20, 26, 61, 73–74, 80–81, 89–90, 91, 104, 106, 123–124, 137–138, 158, 175, 198–199, 320, 333, 334, 355–356, 360, 371, 379
Raccoons, 103, 106, 166, 204, 218, 293, 308
Ragged fringed orchis, 79
Rand, Dorothy Freeman, 241
Rattailed maggot, 66–67
Rattlesnake plantain, 79, 144
Réaumur, René de, 331
Red-eyed verios, 166, 223
Red-legged locusts, 196–197, 211–212
Redpolls, 365–366

Red-winged blackbirds, 25, 35, 49–50, 75, 136, 137, 185, 269–270, 272, 351, 352, 378–379, 385, 389
Robins, 37, 41, 51, 57, 73, 106, 107, 117, 118, 120, 121, 138, 166, 168, 200, 215, 216, 223, 251–252, 278, 302, 333, 394
Rocky Mountains, 42
Rogerson, Dr. Clark T., 175
Rome (ancient), 78
Rose-breasted grosbeak, 46, 68
Roses, 114–115, 233
Royal Photographic Society of Great Britain, 210
Rubáiyát of Omar Khayyám, The, 226
Rue anemone, 48–49
Ruffed grouse, 63, 204
Rusty blackbirds, 343–344

Saint-John's-wort, 78, 88, 177–179, 298, 356
Salamanders, 37, 38, 39
Salet rose, 115
Salt Lake City zoo, 100, 101
Sandy Stream, 307
Sarsaparilla, 204
Sawflies, 189–190
Scarlet tanagers, 62, 141, 166
Schnierla, Dr. T. C., 161
Science (magazine), 191
Sea Around Us, The (Carson), 241
Seton, Ernest Thompson, 262
Seven Springs Slope, 58, 68
Seven Springs Swamp, 28, 334
Shadbush, 57
Shagbark Hickory Trail, 71, 239, 292, 396
Shakespeare, William, 224, 267
Shasta Valley, 274
Short-horned grasshopper, 211
Shrews, 227–228
Shrikes, 302–303
Sierra Club, 241
Silky swallowwort, 252–253
Skipper butterflies, 157, 196
Skunk cabbage, 46, 47, 50–51, 70–71, 90, 254, 368, 374, 377
Skunks, 114, 204, 270, 293, 369, 391–392

Slidedown Mountain, 307, 308
Smith, Captain John, 149, 232
Snails, 173, 259
Snakes, 55–56, 62
Snout beetles, 102, 235–236
Snow fleas, 63
Snow River, 283
Soft Whisper flower, 156–157
Song sparrows, 35, 107
Sopp, Harry and Ann, 307
South Woods, 28, 29, 30, 58, 221, 235, 282, 334, 374, 395
Souvenirs Entomologiques (Fabre), 81
Sparrow hawks, 347–348, 392
Sparrows, 17, 52, 57–58, 96–98, 124, 125, 150, 168, 180–181, 194, 221, 256, 274, 277, 298, 299, 302, 319, 322, 326, 328, 340, 342, 347, 353, 356, 357, 366, 385, 386
Sphinx moth, 186
Spicebush, 34, 47, 142, 222–223
Spicebush swallowtail butterfly, 157
Spider silk, 207–208
Spiders, 67, 77, 87–89, 94, 112, 133, 150, 164, 296, 298, 341
Spittlebugs, 78
Spotted coralroot, 79
Spotted wintergreen, 275
Spring peeper frogs, 20–21, 43–44
Springtails, 63, 341
Springtime in Britain, 301
Squirrels, 14, 33, 38, 39, 70, 95, 131, 139, 163, 166, 235, 238, 248, 254–255, 268, 280–281, 293, 295, 305–306, 326, 333, 334–335, 341, 350, 351, 359–360, 368–361, 382, 387–388, 389
Starfield, 11, 22, 24, 26, 30, 36, 111, 113–114, 133, 138, 165, 204, 205, 215, 252, 253, 270, 317, 350, 351, 352, 354, 365
Steeplebushes, 345
Stepping Stone Brook, 18–19, 43, 62, 71, 101, 106, 119, 231, 261, 396

Stevenson, Robert Louis, 380
Stocking, Bill, 69, 229
Stocking, Edson, 12
Stocking, Vinnie, 69
Sudbury River, 134
Sumac, 238, 332, 334, 381
Summerhouse Rock, 15, 55, 62, 75, 103, 168, 257–258, 285, 317
Sunfish, 145
Suwannee River, 100
Swallows, 103–104, 120, 131–132, 169, 353
Sweet-pepper bush, 172
Syrphid fly, 66–67

Tackhead minnows, 30
Tadpoles, 61–62, 87, 99, 123, 158, 168
Tagore, Rabindranath, 151
T'ao Ch'ien, 236–237
Terres, John K., 27–28
Thistles, 78, 208
Thoreau, Henry, 4, 52, 53, 86, 106, 135, 136, 137, 149, 155, 172, 177, 231, 373–374, 381, 384
Thoreau Society, 134
Thrashers, 107, 124
Thrushes, 37, 84, 141, 223, 256, 341, 351
Timber Slough, 282–283
Time of Little Brooks, 22
Tinbergen, Niko, 195
Titmice, 293, 342, 389
Toads, 61–62, 64, 101
Tolkien, J. R. R., 54
Tolstoy, Leo, 39, 40
Toothache tree, 386–387
Tortoise beetle, 225
Towhees, 58, 71, 125, 166
Tree frogs, 67, 104–105, 113
Tremex larvae, 30
Trevelyan, G. M., 300–301
Tupelo tree, 220, 237
Turkey vulture, 201, 202
Turtle Rock, 30
Turtles, 19–20, 156, 194–195, 205, 225, 234
Twayblade, 79
Twig Hill, 262–263
Twin Rocks, 105

U.S. Fish and Wildlife Service, 136
U.S. Geological Survey, 153

Veeries, 84, 223, 256
Veery Lane, 24, 72, 102, 386, 390
Verbascum thapsus (Linnaeus), 159
Violet cortinarius, 192
Violets, 51, 192
Virgil, 384
Virginia beard grass, 259
Virginia silk, 252–253

Warblers, 53–55, 85–87, 124, 138, 199, 250
Ward's Natural History Establishment (Rochester, N.Y.), 220–221
Wasps, 53, 83–84, 88, 114, 146, 147, 161–162, 163–165, 236, 243, 271, 288–289, 341
Water snakes, 55–56, 62, 145
Water striders, 34–35, 39, 90, 105, 228
Watson, Geraldine, 282–284
Weevils, 102–103
Welder (wildlife refuge), 293
Welty, Joel Carl, 325–326
West Woods, 217, 374
Wet Weather Brook, 95, 373
Wheel bug, 299–300
Wheel mushroom, 192
Whippoorwill Brook, 152, 257
Whippoorwill Cove, 25–26, 43, 82, 103, 141, 152, 173, 195, 258
Whippoorwill Spring, 25–26, 46, 71, 83, 84, 107, 192, 396
Whippoorwills, 74, 220
White, Gilbert, 384
White baneberry, 204
White warblers, 53–55
White-breasted nuthatches, 61, 217
White-crowned sparrows, 256
White-throated sparrows, 194, 221, 256, 274, 342, 394
Wild allspice, 222–223
Wild Apple Glade, 142, 217, 230, 280–281, 381, 396
Wild cherry trees, 76, 166–167

Wild cotton, 252–253
Wild grapevine, 28, 127–128, 142
Wild pasture rose, 115
Wild Plum Tangle, 14, 328, 387–388, 390
Wild Realm, The: Animals in East Africa (Leakey), 137
Wild strawberries, 84
Wiley, Farida A., 175, 191, 193, 194, 196, 197, 198
Willenow, Carl Ludwig, 396
William of Orange, 190
Wilson, Alexander, 200
Windflower, 48, 188
Winship, Doris, 314
Wintergreen, 144–145
Witch Hazel Hill, 22, 49, 50, 161, 221, 294, 350, 354
Wood anemones, 48–49
Wood pewee, 51–52
Wood thrushes, 84, 166, 256, 341
Woodchucks, 56–57, 65, 70, 123, 138, 201, 202, 204, 216, 248, 259–260, 279, 352, 353, 356
Woodcock, 22, 26–27, 45, 74, 260–261
Woodcock Brook, 93
Woodcock Pasture, 24, 26–27, 79, 111, 127
Woodland broadwings, 59–60
Woodpeckers, 22, 30, 33, 45, 72–73, 86, 139, 159–160, 215, 265, 298, 302, 318, 323, 342, 351, 389–390
Wordsworth, Dorothy, 243
Wordsworth, William, 243
Wrens, 138, 340–341, 370–371

Yellow lady's slipper, 80
Yellow-rumped myrtle warblers, 199
Yellow-shafted flickers, 63
Yellowthroat warblers, 53–55, 199

Zahnizer, Howard, 369–370
Zepherine Drouhin rose, 115
Zwinger, Ann, 169, 241